LONDON MATHEMATICAL SOCIETY LECTURE NOTE SERIES

1. General cohomology theory and K-theory, P.HILTON
4. Algebraic topology: a student's guide, J.F.ADAMS
5. Commutative algebra, J.T.KNIGHT
8. Integration and harmonic analysis on compact groups, R.E.EDWARDS
9. Elliptic functions and elliptic curves, P.DU VAL
10. Numerical ranges II, F.F.BONSALL and J.DUNCAN
11. New developments in topology, G.SEGAL (ed.)
12. Symposium on complex analysis, Canterbury, 1973, J.CLUNIE and W.K.HAYMAN (eds.)
13. Combinatorics: Proceedings of the British combinatorial conference 1973, T.P.McDONOUGH and V.C.MAVRON (eds.)
14. Analytic theory of abelian varieties, H.P.F.SWINNERTON-DYER
15. An introduction to topological groups, P.J.HIGGINS
16. Topics in finite groups, T.M.GAGEN
17. Differentiable germs and catastrophes, Th.BROCKER and L.LANDER
18. A geometric approach to homology theory, S.BUONCRISTIANO, C.P.ROURKE and B.J.SANDERSON
20. Sheaf theory, B.R.TENNISON
21. Automatic continuity of linear operators, A.M.SINCLAIR
23. Parallelisms of complete designs, P.J.CAMERON
24. The topology of Stiefel manifolds, I.M.JAMES
25. Lie groups and compact groups, J.F.PRICE
26. Transformation groups: Proceedings of the conference in the University of Newcastle upon Tyne, August 1976, C.KOSNIOWSKI
27. Skew field constructions, P.M.COHN
28. Brownian motion, Hardy spaces and bounded mean oscillation, K.E.PETERSEN
29. Pontryagin duality and the structure of locally compact abelian groups, S.A.MORRIS
30. Interaction models, N.L.BIGGS
31. Continuous crossed products and type III von Neumann algebras, A.VAN DAELE
32. Uniform algebras and Jensen measures, T.W.GAMELIN
33. Permutation groups and combinatorial structures, N.L.BIGGS and A.T.WHITE
34. Representation theory of Lie groups, M.F.ATIYAH et al.
35. Trace ideals and their applications, B.SIMON
36. Homological group theory, C.T.C.WALL (ed.)
37. Partially ordered rings and semi-algebraic geometry, G.W.BRUMFIEL
38. Surveys in combinatorics, B.BOLLOBAS (ed.)
39. Affine sets and affine groups, D.G.NORTHCOTT
40. Introduction to H_p spaces, P.J.KOOSIS
41. Theory and application of Hopf bifurcation, B.D.HASSARD, N.D.KAZARINOFF and Y-H.WAN
42. Topics in the theory of group presentations, D.L.JOHNSON
43. Graphs, codes and designs, P.J.CAMERON and J.H.VAN LINT
44. Z/2-homotopy theory, M.C.CRABB
45. Recursion theory: its generalisations and applications, F.R.DRAKE and S.S.WAINER (eds.)
46. p-adic analysis: a short course on recent work, N.KOBLITZ

London Mathematical Lecture Note Series. 45

Recursion Theory: its Generalisations and Applications

Proceedings of Logic Colloquium '79, Leeds, August 1979

Edited by

F.R.DRAKE and S.S.WAINER

Department of Pure Mathematics, Leeds

CAMBRIDGE UNIVERSITY PRESS

CAMBRIDGE

LONDON NEW YORK NEW ROCHELLE

MELBOURNE SYDNEY

CAMBRIDGE UNIVERSITY PRESS
Cambridge, New York, Melbourne, Madrid, Cape Town, Singapore, São Paulo

Cambridge University Press
The Edinburgh Building, Cambridge CB2 8RU, UK

Published in the United States of America by Cambridge University Press, New York

www.cambridge.org
Information on this title: www.cambridge.org/9780521235433

First published 1980
Re-issued in this digitally printed version 2008

A catalogue record for this publication is available from the British Library

ISBN 978-0-521-23543-3 paperback

CONTENTS

PREFACE

Logic Colloquium '79 was held at the University of Leeds, England, from 5-14 August, 1979. It was organised by the British Logic Colloquium and recognised as the European Summer Meeting of the Association for Symbolic Logic, and was co-sponsored by the School of Mathematics of Leeds University.

Financial assistance was received from the British Academy, the British Council, the International Union for the History and Philosophy of Science, the London Mathematical Society, and the Royal Society. We wish to record our thanks to these bodies.

Short courses of lectures and one-hour lectures were given as follows:

R.I. Soare (4 lectures) *Recursively enumerable sets and degrees.*
S.G. Simpson (2 lectures) *Admissible recursion theory.*
R.A. Shore (2 lectures) *Admissible recursion theory.*
D. Normann (2 lectures) *Recursion on the countable functionals.*
G. Sundholm (2 lectures) ω-*Arithmetic and recursive progressions.*
R. Ladner (4 lectures) *Complexity theory and the complexity of logical theories.*
D.B. Posner, *The non-r.e. degrees* $< 0'$.
C.G. Jockusch, *Degrees of generic sets.*
M. Lerman, *Recent results on the degrees of unsolvability.*
G.E. Sacks, *Three theorems on recursive enumerability in a normal object of finite type.*
J.V. Tucker, *The computable functions of abstract algebra.*
C.H. Smith, *Applications of recursion theory to computer science.*
D. Alton, *Natural programming languages and complexity measures for subrecursive programming languages.*
W. Paul, *Time and space bounded computations.*
L. Valiant, *Non-computational reducibilities.*
K. McAloon, *On the structure of models of arithmetic.*

Papers based on 11 of the above talks are published in this volume. In addition, there were 19 contributed papers, abstracts of which are published in the Journal of Symbolic Logic, vol. 45.

The organisation of the conference was under a programme committee consisting of: Professor D.S. Scott and Dr. R.O. Gandy (Oxford); Professor J.C. Shepherdson (Bristol); and Professor C.E.M. Yates (Manchester); together with the logicians at Leeds, who also formed the local arrangements committee: S.B. Cooper, J. Derrick, A.B. Slomson, L.M. Smith, and the undersigned.

Our special thanks are due to Mrs. Audrey Landford for assistance with all aspects of the conference and to Mr. D. Tranah of Cambridge University Press, for his help in the preparation of this volume.

F.R. DRAKE S.S. WAINER

LIST OF PARTICIPANTS

ADAMSON, A.
ALTON, D.A.
BOOS, W.
BRACKIN, S.H.
CHUDACEK, J.
CICHON, E.A.
COOPER, S.B.
DALHAUS, E.
DANICIC, I.
DAWES. A.M.
DERRICK, J.
DRABBE, J.
DRAKE, F.R.
FARRINGTON, P.
FLOTHOW, C.
FOURMAN, M.
FRENCH, G.
GANDY, R.O.
GOLD, A.
GRIFFOR, E.R.
GUMB, R.D.
HEIDEMA, J.
HINDLEY, J.R.
HINMAN, P.
HODES, H.
HYLAND, J.M.E.
JOCKUSCH, C.G.
JOHNSON, N.
KIROUSIS, L.M.
KRAJEWSKI, S.
KRASNER, M.
KUCERA, A.
LADNER, R.E.
LERMAN, M.
LOWESMITH, B.
McALOON, K.
McBETH, R.
McCOLL, W.F.
MANDERS, K.L.
MIKULSKA, M.
MITCHELL, R.
NORMANN, D.
NORMANN, S.
OIKKONEN, J.
O'LEARY, D.J.
ONYSKIEWICZ, J.
PAUL, W.
PORTE, J.
POSNER, D.B.
PREST, M.
PRYS-WILLIAMS, A.G.
QUINSEY, J.
ROLLETSCHEK, H.
ROSE, H.E.
ROTHACKER, E.
SACKS, G.E.
SELDIN, J.P.
SHORE, R.A.
SIMMONS, H.
SIMPSON, S.G.
SLOMSON, A.B.
SMART, J.J.C.
SMITH, C.H.
SMITH, L.M.
SMORYNSKI, C.
SMYTH, M.B.
SOARE, R.I.
STIEBING, H.N.
STOLTENBERG-HANSEN, V.
SUNDHOLM, G.
TAYLOR, J.M-
TERLOUW, J.
THOMPSON, S.
TUCKER, J.V.
TURNER, R.
WAINER, S.S.
WEGLORZ, B.
WETTE, E.W.
WILKIE, A.J.
WILLIAMSON, J.
VALIANT, L.
VETULANI, S.
VAANANEN, J.
VELDMAN, W.
YATES, C.E.M.

FUNDAMENTAL METHODS FOR CONSTRUCTING RECURSIVELY ENUMERABLE DEGREES

Robert I. Soare
University of Chicago

Contents

INTRODUCTION

In these lectures we survey some of the most important results and the fundamental methods concerning degrees of recursively enumerable (r. e.) sets. This material is similar to our C. I. M. E. lectures (Soare, 1980) except that there § 7 was on the Renaissance in classical recursion theory, the Sacks density theorem and beyond, while here § 7 deals with noncappable degrees and § 8 with nonbranching degrees.

We begin § 1 with Post's simple sets and a recent elegant generalization of the recursion theorem. In § 2 we give the finite injury priority method, the solution of Post's problem, and the Sacks splitting theorem. In § 3 the infinite injury method is introduced and applied to prove the thickness lemma and the Sacks density theorem. In §4 and §5 we develop the minimal pair method for embedding distributive lattices in the r. e. degrees by maps preserving infimums as well as supremums. In §6 we present the non-diamond theorem which asserts that such embeddings cannot always preserve greatest and least elements. For background reading we suggest Rogers (1967), Shoenfield (1971), and Soare (1982).

Our notation is standard as in Rogers (1967), with a few additions. For sets $A, B \subseteq \omega$ we say B is <u>recursive</u> in A ($B \leq_T A$) if there is a Turing reduction Φ such that $\Phi(A) = B$. Let $A \equiv_T B$ denote that $A \leq_T B$ and $B \leq_T A$. The <u>degree</u> of A, $dg(A) = \{B: B \equiv_T A\}$. Lower case bold-face letters $\mathbf{a}, \mathbf{b}, \ldots$ denote degrees. Let $\{\Phi_e\}_{e \in \omega}$ be an acceptable numbering of all Turing reductions. Let $W_e^A = \{x : \Phi_e(A; x)$ converges$\}$, and $W_e = W_e^{\emptyset}$, so $\{W_e\}_{e \in \omega}$ is an acceptable numbering of all r. e.

sets. The jump of A is $A' = \{e : e \in W_e^A\}$. The jump is well-defined on degrees. An r.e. set A is complete if $W_e \leq_T A$ for all e. For example, $K = \emptyset' = \{e : e \in W_e\}$ is complete. We write $dg(A) \leq dg(B)$ if $A \leq_T B$. Let $dg(A) \cup dg(B) = dg(A \oplus B)$ where $A \oplus B = \{2x : x \in A\} \cup \{2x+1 : x \in B\}$. A degree is r.e. if it contains an r.e. set. The r.e. degrees $(\underset{\sim}{R}, \leq, \cup)$ form an upper semi-lattice with least element $\underset{\sim}{0} = dg(\emptyset)$ and greatest element $\underset{\sim}{0}' = dg(\emptyset')$. Let $\Phi_{e,s}(A; x) = y$ if $e, x, y < s$ and Φ_e with oracle A and input x yields output y in $\leq s$ steps. We also use $\{e\}_s^{A_s}(x) = y$ and $\{e\}^A(x) = y$ for $\Phi_{e,s}^{A_s}(x) = y$ and $\Phi_e(A; x) = y$, respectively. If ψ is a partial function, let $\psi(x)\downarrow$ $(\psi(x)\uparrow)$ denote $\psi(x)$ converges (diverges). Fix a 1:1 recursive function from $\omega \times \omega$ onto ω and let $\langle x, y\rangle$ denote the image of the ordered pair (x, y) under this map. We identify a set A with its characteristic function, and let A[x] denote the restriction of this function to arguments $\leq x$.

§1. SIMPLE SETS, FIXED POINT THEOREMS, AND A COMPLETENESS CRITERION FOR R.E. SETS

The study of degrees of r.e. sets began with the famous paper of Post (1944). Post's problem was to find an r.e. set A of degree different from $\underset{\sim}{0}$ and $\underset{\sim}{0}'$. Post attempted to construct such an incomplete r.e. set A by considering structural properties on A such as simplicity which he hoped would guarantee incompleteness. Although unsuccessful, this approach led ultimately to a very pleasing fixed-point theorem which generalizes Kleene's recursion theorem and which yields a necessary and sufficient condition for an r.e. set to be complete.

An r.e. set C is creative if there is a recursive function f(x) such that if $W_x \subseteq \overline{C}$ then $f(x) \in \overline{C} - W_x$. For example, K is creative via the identity function. For every creative set C, $\overline{C}$ contains an infinite r.e. set $B = \{b_1, b_2, \ldots\}$ where $W_{b_0} = \emptyset$, and $W_{b_{n+1}} = W_{b_n} \cup \{f(b_n)\}$. Post defined a coinfinite r.e. set A to be simple if $\overline{A}$ contains no infinite r.e. set. He hoped that if $\overline{A}$ were sufficiently "thin" with respect to containment of infinite r.e. sets then A would be incomplete. Post constructed simple sets and proved them incomplete with respect to a stronger reducibility called m-reducibility. A set A is m-reducible to B ($A \leq_m B$) if there is a recursive function f such that, for all x, $x \in A$ iff $f(x) \in B$. It is easy to see that if $C \leq_m A$ and C is creative, then A is creative (and thus not simple). Hence no simple set A can be m-complete.

Theorem 1.1 (Post): There exists a simple set S.

Proof. Fix a recursive enumeration $\{a_s\}_{s \in \omega}$ of the r.e. set

$A = \{\langle e,x\rangle : x \in W_e \;\&\; x > 2e\}$. Define an r.e. set $B \subseteq A$ by enumerating $a_s = \langle e,x\rangle$ into B if there is no $a_t = \langle e,y\rangle$ for $t < s$ and $y \neq x$. Since B is single-valued, it represents the graph of some partial recursive (p.r.) function ψ; i.e., $\psi(x) = y$ iff $\langle x,y\rangle \in B$. Let $S = \text{range } \psi$. (Intuitively, enumerate W_e until the first element $\psi(e) > 2e$ appears in W_e and then put $\psi(e)$ into S.) The following facts give the simplicity of S.

(1) S is r.e. (S is the range of a p.r. function.)

(2) $\overline{S}$ is infinite. To see this, note that S contains at most e elements out of $\{0,1,2,\ldots,2e\}$, namely, $\psi(0), \psi(1),\ldots,\psi(e-1)$. Hence $|\overline{S} \cap [0, 2e]| > e$, so $\overline{S}$ is infinite.

(3) If W_e is infinite, then $W_e \cap S \neq \emptyset$, because $\langle e,x\rangle \in A$ for some $x > 2e$ so $\psi(e)$ is defined and $\psi(e) \in S \cap W_e$.

A simple set A is effectively simple if there is a recursive function f such that

$$(\forall e)[W_e \subseteq \overline{A} \implies |W_e| \leq f(e)],$$

where $|W_e|$ is the cardinality of W_e. Note that Post's simple set is effectively simple via $f(e) = 2e$.

Post realized that simple sets could be complete (indeed S is complete by Corollary 1.5), and so he defined coinfinite r.e. sets with still thinner complements called hypersimple and hyper-hypersimple. A coinfinite r.e. set A is hypersimple (h-simple) if there is no recursive function f which majorizes $\overline{A}$ in the sense that for all n, $f(n) \geq p_{\overline{A}}(n)$, where $p_{\overline{A}}$, the principal function of $\overline{A}$, is defined by $p_{\overline{A}}(n) = a_n$, where $\overline{A} = a_0 < a_1 < \ldots$. (This is not Post's original definition but can easily be shown equivalent to it (Rogers, 1967, p. 139).) Clearly, h-simple sets are simple. The converse is false since for Post's simple set S, the function $f(x) = 2x$ majorizes $\overline{S}$. Furthermore, h-simple sets are not necessarily incomplete. Indeed, Dekker (1954) proved that every nonzero r.e. degree contains an h-simple (and therefore simple) set.

Theorem 1.2 (Dekker): For every nonrecursive r.e. set $\overline{A}$ there is an h-simple set $B \equiv_T A$.

Proof. Let $A = rng(f)$, f a 1:1 recursive function, and let $a_s = f(s)$, and $A_s = \{f(0), \ldots, f(s)\}$. Define $B = \{s: \exists t > s)[a_t < a_s]\}$, the deficiency set of A for the enumeration f. Clearly, B is Σ_1 and hence r.e., and $\overline{B}$ is infinite. Next, B is h-simple, for if $g(x) \geq p_{\overline{B}}(x)$ for some recursive function g, then $x \in A$ iff $x \in \{a_0, a_1, \ldots, a_{g(x)}\}$, which would imply A recursive. Similarly, $A \leq_T B$ because $x \in A$ iff $x \in \{a_0, a_1, \ldots, a_{p_{\overline{B}}(x)}\}$. Finally, $B \leq_T A$ since to test whether $s \in B$, we A-recursively compute t such that $A_t[a_s] = A[a_s]$. Now $s \in B$ iff $a_u < a_s$ for some u such that $s < u \leq t$. □

The crucial point about the deficiency set B is that any non-deficiency stage $s \in \overline{B}$ is a "true" stage in the enumeration $\{A_s\}_{s \in \omega}$ of A in that $A_s[f(s)] = A[f(s)]$. These stages will prove very useful in our study of the infinite injury priority method in §3.

Martin (1966) found a fairly general sufficient condition for an r.e. set to be complete. His condition applied to many "effectively" nonrecursive r.e. sets including creative sets and effectively simple sets. Lachlan (1968) modified Martin's condition so that it became both necessary and sufficient. Arslanov (1977) then converted it to the following form which can be viewed as a generalization of Kleene's recursion theorem. It asserts that not only recursive functions but all those functions of r.e. degree less than $\underset{\sim}{0}'$ have a fixed point. (Recall that the recursion theorem (Rogers, 1967, p. 180) asserts that every recursive function f has a "fixed point" n such that $W_n = W_{f(n)}$.

<u>Theorem 1.3</u> (Arslanov): An r.e. set A is complete iff there is a function $f \leq_T A$ such that $W_{f(x)} \neq W_x$ for all x.

<u>Proof.</u> ($\Longrightarrow$) Trivial since $\{x: W_x = \emptyset\} \equiv_T \emptyset'$.

($\Longleftarrow$) Assume $(\forall x)[W_{f(x)} \neq W_x]$ where $f(x) = \{e\}^A(x)$ and g(x) is the greatest element used in the latter computation. Let $\{A_s\}_{s\in\omega}$, $\{K_s\}_{s\in\omega}$ be recursive enumerations of A and K. Let m be the partial recursive function such that $m(x) = \mu s[x \in K_s]$ if $x \in K$, and $m(x)\uparrow$ otherwise. Define the recursive function $\hat{f}(s,x) = \{e\}_t^{A_t}(x)$ where $t \geq s$ is minimal such that $\{e\}_t^{A_t}(x)$ is defined. By the recursion theorem with parameters, define the recursive function h by

$$W_{h(x)} = \begin{cases} W_{\hat{f}(m(x), h(x))} & \text{if } x \in K \\ \emptyset & \text{otherwise} . \end{cases}$$

Since $g \leq_T A$ we can define the function $r \leq_T A$ by

$$r(x) = (\mu s)[A_s[gh(x)] = A[gh(x)]] .$$

Now if $x \in K$ and $r(x) \leq m(x)$ then $A_{m(x)}[gh(x)] = A[gh(x)]$, so $\hat{f}(m(x), h(x)) = f(h(x))$ and $W_{f(h(x))} = W_{h(x)}$ contrary to the hypothesis on f. Hence, for all $x \in K$, $m(x) < r(x)$. Thus for all x,

$$x \in K \Longleftrightarrow x \in K_{r(x)}$$

so $K \leq_T A$. $\square$

Corollary 1.4. Given an r.e. degree $\underset{\sim}{a}$, $\underset{\sim}{a} < \underset{\sim}{0}'$ iff every function $f \in \underset{\sim}{a}$ has a fixed point.

Corollary 1.5. An r.e. set A is complete if A is either: (a) creative; or (b) effectively simple.

Proof. (a) Let A be creative via $g(x)$. Choose $a \in \overline{A}$ and define $f \leq_T A$ by

$$W_{f(x)} = \begin{cases} \{a\} & \text{if } g(x) \in A \\ \{g(x)\} & \text{if } g(x) \notin A \end{cases}$$

(b) Let A be effectively simple via $g(x)$. Define $f \leq_T A$ by

$$W_{f(x)} = \{a_0, a_1, \ldots, a_{g(x)}\},$$

where $\overline{A} = a_0 < a_1 < \ldots$. In each case, it is clear that $W_{f(x)} \neq W_x$ for all x so A is complete by Theorem 1.3. □

After Post's problem was solved by an entirely different method, some structural properties guaranteeing incompleteness were discovered (Marchenkov, 1976), (Soare, 1977, p. 550).

§2. THE FINITE INJURY PRIORITY METHOD

A positive solution to Post's problem was finally achieved by Friedberg (1957) and independently by Muchnik (1956). In their method, known as the priority method, the desired r.e. set is constructed by stages to meet a certain sequence of conditions $\{R_n\}_{n\in\omega}$ called requirements. If $n < m$, requirement R_n is given given priority over R_m and action taken for R_m at some stage s may at a later stage $t > s$ be undone for the sake of R_n, thereby injuring R_m at stage t. The original priority method of Friedberg and Muchnik has the property that each requirement is injured at most finitely often.

We illustrate the finite injury method by proving the Friedberg-Muchnik theorem using a variation of Sacks (1963a) which is more powerful than the standard method and which will be used in §3. In our constructions, the requirements $\{R_e\}_{e\in\omega}$ will be divided into the negative requirements $N_e = R_{2e}$ which attempt to keep elements out of the r.e. set A being constructed, and positive requirements $P_e = R_{2e+1}$ which attempt to put elements into A. The negative requirements will be of the form $C \neq \Phi_e(A)$, where C is a fixed nonrecursive r.e. set, so that the negative requirements together assert that $C \not\leq_T A$. Sacks observed that the requirement N_e can be met by attempting to preserve agreement between $C_s(x)$ and $\Phi_{e,s}(A_s;x)$ rather than disagreement as one might suppose. (The point is that if we preserve this agreement sufficiently often and if $C = \Phi_e(A)$ then C will be recursive contrary to hypothesis.) The positive requirements will ensure as in Theorem 1.1 that A is simple and hence nonrecursive.

Theorem 2.1 (Friedberg-Muchnik): For every nonrecursive r.e. set C there is a simple set A such that $C \not\leq_T A$ (and hence $\emptyset <_T A <_T \emptyset'$).

Proof. It clearly suffices to construct A to be coinfinite and to satisfy, for all e, the requirements:

$$N_e : C \neq \Phi_e(A) ,$$

$$P_e : W_e \text{ infinite} \Longrightarrow W_e \cap A \neq \emptyset .$$

Let $\{C_s\}_{s \in \omega}$ be a recursive enumeration of C. Define $A_0 = \emptyset$. Given A_s define the following three recursive functions whose roles are obvious from their names:

$$\text{(use function)} \quad u(e,x,s) = \begin{cases} \text{maximum element } z \text{ used in the computation } \Phi_{e,s}(A_s;x) & \text{if } \Phi_{e,s}(A_s;x)\downarrow \\ 0 & \text{otherwise} \end{cases}$$

$$\text{(length function)} \quad \ell(e,s) = \max\{x : (\forall y < x)[C_s(y) = \Phi_{e,s}(A_s; y)]\}.$$

$$\text{(restriaint function)} \quad r(e,s) = \max\{u(e,x,s) : x \leq \ell(e,s)\} .$$

(Note that if $\Phi_{e,s}(A_s;x) = y$ then $e, x, y, u(e,x,s) < s$.) For each $e \leq s$, if $W_{e,s} \cap A_s = \emptyset$ and

(2.1) $\quad (\exists x)[x \in W_{e,s} \ \& \ x > 2e \ \& \ (\forall i \leq e)[x > r(i,s)]]$,

then enumerate the least such x in A_{s+1}. Define $A = \bigcup_s A_s$.

(Intuitively, u(e, x, s) is the maximum element <u>used</u> in the above computation, and the elements $x \leq r(e,s)$ are <u>restrained from</u> A_{s+1} by requirement N_e in order to preserve the length of agreement measured by $\ell(e,s)$.) The negative requirement N_e is <u>injured at stage</u> s+1 <u>by element</u> x if $x \leq r(e,s)$ and

$x \in A_{s+1} - A_s$. These elements form an r.e. set:

$$\text{(injury set)} \quad I_e = \{x : (\exists s)[x \in A_{s+1} - A_s \ \& \ x \leq r(e,s)]\}.$$

Note that each I_e is finite because N_e is injured at most once for each P_i, $i < e$, whereupon P_i is satisfied thereafter. (Positive requirements, of course, are never injured.)

<u>Lemma 2.2.</u> $(\forall e)[C \neq \Phi_e(A)]$.

<u>Proof.</u> Assume for a contradiction that $C = \Phi_e(A)$. Then $\lim_s \ell(e,s) = \infty$. Choose s' such that N_e is never injured after stage s'. We shall recursively compute $C(x)$ contrary to hypothesis. To compute $C(p)$ for $p \in \omega$ find some $s > s'$ such that $\ell(e,s) > p$. It follows by induction on $t \geq s$ that

$$(2.2) \qquad (\forall t \geq s)[\ell(e,t) > p \ \& \ r(e,t) \geq \max\{u(e,x,s) : x \leq p\}],$$

and hence that $\Phi_{e,s}(A_s; p) = \Phi_e(A_s; p) = \Phi_e(A; p) = C(p)$. Since $s > s'$, (2.2) clearly holds unless $C_t(x) \neq C_s(x)$ for some $t \geq s$ and $x \leq p$; but if x and t are minimal then our use of "$\leq \ell(e,t)$" rather than "$< \ell(e,t)$" in the definition of $r(e,t)$ insures that the <u>disagreement</u> $C_t(x) \neq \Phi_{e,t}(A_t; x)$ is preserved forever, contrary to the hypothesis that $C = \Phi_e(A)$. Note that even though the Sacks strategy is always described as one which preserves agreements, it is crucial that we preserve at least one <u>disagreement</u> as well whenever possible.

<u>Lemma 2.3.</u> $(\forall e)[\lim_s r(e,s)$ exists and is finite$]$.

<u>Proof.</u> By Lemma 1.1 choose $p = \mu x[C(x) \neq \Phi_e(A; x)]$. Choose Choose s' sufficiently large such that, for all $s \geq s'$,

$$(\forall x < p)[\Phi_{e,s}(A_s; x) = \Phi_e(A;x)], \ (\forall x \leq p)[C_s(x) = C(x)], \text{ and}$$

$$N_e \text{ is not injured at stage } s.$$

Case 1. $(\forall s \geq s')[\Phi_{e,s}(A_s; p)$ undefined]. Then $r(e, s) = r(e, s')$ for all $s \geq s'$.

Case 2. $\Phi_{e,t}(A_t; p)$ is defined for some $t \geq s'$. Then $\Phi_{e,s}(A_s; p) = \Phi_{e,t}(A_t; p)$ for all $s \geq t$ because $\ell(e,s) \geq p$, and so, by the definition of $r(e, s)$, the computation $\Phi_{e,t}(A_t; p)$ is preserved and N_e is not injured after stage s'. Thus $\Phi_e(A; p) = \Phi_{e,s}(A_s; p)$. But $C(p) \neq \Phi_e(A; p)$. Thus

$$(\forall s \geq t)[C_s(p) \neq \Phi_{e,s}(A_s; p) \;\&\; \ell(e,s) = p \;\&\; r(e,s) = r(e,t)].$$

Hence, $r(e, t) = \lim_s r(e,s)$.

Lemma 2.4. $(\forall e)[W_e$ infinite $\Longrightarrow W_e \cap A \neq \emptyset]$.

Proof. By Lemma 1.2, let $r(e) = \lim_s r(e, s)$ and $R(e) = \max\{r(i): i \leq e\}$. Now if $(\exists x)[x \in W_e \;\&\; x > R(e) \;\&\; x > 2e]$ then $W_e \cap A \neq \emptyset$.

Note that $\overline{A}$ is infinite by the clause "$x > 2e$" in (2.1), and hence A is simple. □

Sacks invented the above preservation method (which plays a crucial role in the later infinite injury argument) to prove the following theorem.

Theorem 2.5 (Sacks Splitting Theorem (Sacks, 1963a)): Let B and C be r.e. sets such that C is nonrecursive. Then there exist r.e. sets A_0 and A_1 such that

(a) $A_0 \cup A_1 = B$ and $A_0 \cap A_1 = \emptyset$, and

(b) $C \not\leq_T A_i$, for $i = 0, 1$.

Proof. Let $\{B_s\}_{s \in \omega}$ and $\{C_s\}_{s \in \omega}$ be recursive enumerations of B and C such that $B_0 = \emptyset$ and $|B_{s+1} - B_s| = 1$ for all s. It suffices to give recursive enumerations $\{A_{i,s}\}_{s \in \omega}$, $i = 0, 1$, satisfying the single positive requirement

$$P: x \in B_{s+1} - B_s \Rightarrow [x \in A_{0,s+1} \text{ or } x \in A_{1,s+1}],$$

and the negative requirements for $i = 0, 1$ and all e,

$$N_e^i : C \neq \Phi_e(A_i).$$

Define $A_{i,0} = \emptyset$. Given $A_{i,s}$ define the recursive functions $\ell^i(e,s)$ and $r^i(e,s)$ as above but with $A_{i,s}$ in place of A_s. Let $x \in B_{s+1} - B_s$. Choose $\langle e', i' \rangle$ to be the least $\langle e, i \rangle$ such that $x \leq r^i(e, s)$ and enumerate $x \in A_{1-i', s+1}$. If $\langle e', i' \rangle$ fails to exist, enumerate $x \in A_{0, s+1}$. This defines A_i, $i = 0, 1$.

To see that the construction succeeds, define the injury set I_e^i as above but with A_i in place of A. It follows by simultaneous induction on $\langle e, i \rangle$ that, for $i = 0, 1$ and all e,

(1) I_e^i is finite,

(2) $C \neq \Phi_e(A_i)$, and

(3) $\lim_s r^i(e, s)$ exists and is finite. □

It can be shown (Soare, 1976, p. 525) that the r.e. sets A_i are automatically <u>low</u>, namely $A_i' \equiv_T \emptyset'$, where the <u>jump of</u> A was defined earlier to be $A' = \{e : e \in W_e^A\}$. By setting C = B in Theorem 2.5, it follows that any nonrecursive r.e. set B can be split as the disjoint union of low r.e. sets A_0 and A_1 which are Turing incomparable and such that $dg(B) = dg(A_0) \cup dg(A_1)$. Thus, there is no minimal r.e. degree.

Finite injury arguments are characterized by the fact that the injury set I_e is finite for each e. In §3 we will consider cases where I_e is infinite although usually recursive. Note that Lemma 2.2 holds by virtually the same proof as above if we assume "I_e recursive" in place of "I_e finite". This is what allows the infinite injury method to succeed.

§3. THE INFINITE INJURY PRIORITY METHOD

Shoenfield (1961) and, independently, Sacks (1963b), (1964), (1966) discovered a technique for handling a requirement which may be injured infinitely often (namely, the injury set I_e may be infinite). Sacks considerably developed this technique into what he called the "infinite injury priority method" and he used it to prove many important results on r.e. degrees, the most striking of which is the density theorem which asserts that for any r.e. degrees $\underset{\sim}{d} < \underset{\sim}{c}$ there is an r.e. degree $\underset{\sim}{a}$ such that $\underset{\sim}{d} < \underset{\sim}{a} < \underset{\sim}{c}$. We now give a brief sketch of the method. More details and applications can be found in (Soare, 1976).

We wish to construct an r.e. set A where the negative requirements N_e are as in §2 and the positive requirements are of the form

$$P_e: W_{p(e)} \subseteq^* A$$

(where $X \subseteq^* Y$ denotes that $X - Y$ is finite) so that a single positive requirement may contribute infinitely many elements to A. In the simplest cases, the r.e. sets $\{W_{p(e)}\}_{e \in \omega}$ will be recursive. For each N_e we would like a restraint function $\hat{r}(e, s)$ so that exactly as in §2 we can enumerate x in A_{s+1} for the sake of P_e just if $x \in W_{p(e),s+1}$ and $x > \hat{r}(i, s)$, for all $i \leq e$. The negative requirement N_e can now be injured infinitely often by those P_i, $i < e$, but the recursiveness of $W_{p(i)}$, $i < e$, will enable us to meet N_e as in Lemma 2.2. The main difficulty will be that some P_e remains unsatisfied because of the restraint functions $\hat{r}(i, s)$, $i \leq e$, which may now be unbounded in s (i.e., $\limsup_s \hat{r}(i, s) = \infty$). To satisfy P_e it clearly suffices to define $\hat{r}(e, s)$ such that

$$\text{(3.1)} \qquad \liminf_s \hat{R}(e, s) < \infty ,$$

where $\hat{R}(e, s) = \max\{\hat{r}(i, s): i \leq e\}$, because then P_e has a "window" through the negative restraints at least infinitely often.

The first obstacle to achieving (3.1) is that if we let $\hat{r}(e,s)$ be $r(e, s)$ as defined in § 2, then we may have $\lim_s r(e, s) = \infty$ for some e. (For example, suppose $\Phi_{1,s}(X; 0) = C(0)$ just if $n \notin X$ for some even $n < s$, but P_0 eventually forces every even number into A so that $\Phi_1(A; 0)$ is undefined. Then N_1 is satisfied by divergence but $C(0) = \Phi_{1,s}(A_s; 0)$ for almost every s, so $\lim_s u(1, 0,s) = \infty$, $\lim_s r(1,s) = \infty$ and P_1 is not satisfied.) This difficulty arises only if there are infinitely many stages s such that $A_s[u] \neq A_{s+1}[u]$ where $u = u(1, 0, s)$. Thus, we can easily remove the first obstacle by replacing $\Phi_{e,s}$ everywhere by $\hat{\Phi}_{e,s}$ defined below, and letting $\hat{r}(e, s)$ be the resulting restraint function. If $C \neq \Phi_e(A)$ we then have $\liminf_s \hat{r}(e,s) < \infty$.

The second obstacle to (3.1) is that $\lim_s \hat{R}(e, s) = \infty$ even though $\liminf_s \hat{r}(i, s) < \infty$ for each $i \leq e$. (For example, N_1 and N_2 may together permanently restrain all elements because their restraint functions do not drop back simultaneously.) Surprisingly, the $\hat{\Phi}_{e,s}$ solution to the first obstacle automatically removes the second, as Lachlan (1973) first observed.

Suppose we wish to give a recursive enumeration $\{A_s\}_{s \in \omega}$ of an r.e. set A. Given $\{A_t : t \leq s\}$, define

$$a_s = \begin{cases} \mu x[x \in A_s - A_{s \dot{-} 1}] & \text{if } A_s - A_{s \dot{-} 1} \neq \emptyset \\ \max(A_s \cup \{s\}) & \text{otherwise;} \end{cases}$$

$$\hat{\Phi}_{e,s}(A_s; x) = \begin{cases} \Phi_{e,s}(A_s; x) & \text{if defined and } u(e, x, s) < a_s, \\ \text{undefined} & \text{otherwise ;} \end{cases}$$

$$\hat{u}(e,x,s) = \begin{cases} u(e,x,s) & \text{if } \hat{\Phi}_{e,s}(A_s;x) \text{ is defined,} \\ 0 & \text{otherwise;} \end{cases}$$

and

$$T = \{s : A_s[a_s] = A[a_s]\} .$$

If $\{A_s\}_{s \in \omega}$ is any recursive enumeration of an r.e. set A we refer to T as the set of true (nondeficiency) stages of this enumeration. Note that T is infinite and $T \equiv_T A$ uniformly in A. If $\Phi_e(A;x) = y$ then clearly $\lim_s \hat{\Phi}_{e,s}(A_s;x) = y$ as before. The crucial point about $\hat{\Phi}_{e,s}$ is that for any true stage t any apparent computation $\hat{\Phi}_{e,t}(A_t;x) = y$ is a true computation $\Phi_e(A;x) = y$. Namely, using the fact that $u(e,x,t) \leq t$, we have

(3.2) $$(\forall t \in T)[\hat{\Phi}_{e,t}(A_t;x) = y \Longrightarrow$$
$$(\forall s \geq t)[\hat{\Phi}_{e,s}(A_s;x) = \Phi_e(A;x) = y \ \&\ \hat{u}(e,x,s) = u(e,x,t)]],$$

because if $\Phi_{e,t}(A_t;x)$ is defined then $u(e,x,t) < a_t$ and $A_t[a_t] = A[a_t]$.

The simplest application of this method is the thickness lemma. For any set A and $x \in \omega$, define the "column" $A^{(x)} = \{\langle y,z\rangle : \langle y,z\rangle \in A \ \&\ y = x\}$, and $A^{(<x)} = \bigcup\{A^{(z)} : z < x\}$. A subset $A \subseteq B$ is a thick subset of B if $A^{(x)} =^* B^{(x)}$ for all x, and B is piecewise recursive if $B^{(x)}$ is recursive for all x. (We write $X =^* Y$ if the symmetric differerence $(X - Y) \cup (Y - X)$ is finite.)

Theorem 3.1 (Thickness Lemma - Shoenfield (1961)): Given a nonrecursive r.e. set C and a piecewise recursive r.e. set B there is a thick subset A of B such that $C \not\leq_T A$.

Proof. Fix recursive enumerations $\{B_s\}_{s \in \omega}$, $\{C_s\}_{s \in \omega}$ of B and C. Let $A_0 = \emptyset$. Given $\{A_t : t \leq s\}$ define $\hat{\Phi}_{e,s}(A_s)$ as above.

Define the remaining functions as in § 2 with $\hat{\Phi}_{e,s}$ in place of $\Phi_{e,s}$ namely,

(length function) $\hat{\ell}(e,s) = \max\{x: (\forall y < x)[C_s(y) = \hat{\Phi}_{e,s}(A_s;y)]\}$,

(restraint function) $\hat{r}(e,s) = \max\{\hat{u}(e,x,s): x \leq \hat{\ell}(e,s)\}$,

(injury set) $\hat{I}_e = \bigcup_s \hat{I}_{e,s}$, where

$$\hat{I}_{e,s} = \{x: (\exists v \leq s)[x \leq \hat{r}(e,v) \ \&\ x \quad A_{v+1} - A_v]\}.$$

To meet the requirements

$$P_e: B^{(e)} =^* A^{(e)} \quad \text{and} \quad N_e: C \neq \Phi_e(A),$$

we enumerate x in $A^{(e)}_{s+1}$ just if $x \in B^{(e)}_{s+1}$ and $x > \hat{r}(i, s)$ for all $i \leq e$. Let $A = \bigcup_s A_s$.

Note that $\hat{I}_e \subseteq A^{(<e)}$ because N_e is injured by P_i only if $i < e$. Thus we have $\hat{I}_e \leq_T A^{(<e)}$ because if $x \in A^{(<e)}$, say $x \in A_s^{(<e)}$, then $x \in \hat{I}_e$ just if $x \in \hat{I}_{e,s}$.

Fix e and assume by induction that $C \neq \Phi_i(A)$ and $A^{(i)} =^* B^{(i)}$ for all $i < e$. Then $A^{(<e)} =^* B^{(<e)}$ is recursive and hence $\hat{I}_e$ is recursive.

<u>Lemma 3.2</u> (Injury Lemma): $C \neq \Phi_e(A)$.

<u>Proof.</u> Assume for a contradiction that $C = \Phi_e(A)$. Then $\lim_s \hat{\ell}(e,s) = \infty$. Fixing $\hat{I}_e$ as an oracle we compute C, so C is recursive contrary to hypothesis. To compute C(p) for $p \in \omega$ find some s such that $\hat{\ell}(e,s) > p$ and

$$(\forall x \leq p)(\forall z)[z \leq u(e, x, s) \Rightarrow [z \notin \hat{I}_e \text{ or } z \in A_s]].$$

Such s exists since $C = \Phi_e(A)$. By the same remarks as in Lemma 2.2, it follows by induction on $t \geq s$ that

$$(\forall t \geq s)[\hat{\ell}(e,t) > p \ \&\ \hat{r}(e,t) \geq \max\{u(e,x,s): x \leq p\}],$$

and hence that

$$\Phi_{e,s}(A_s; p) = \Phi_e(A_s; p) = \Phi_e(A; p) = C(p).$$

Lemma 3.3 (Window Lemma): Let T be the set of true stages in the enumeration $\{A_s\}_{s \in \omega}$ of A. If $C \neq \Phi_i(A)$ then $\lim_{t \in T} \hat{r}(i,t) < \infty$. (Hence, if $C \neq \Phi_i(A)$, for all $i \leq e$, then $\lim_{t \in T} \hat{R}(e,t) < \infty$, where $\hat{R}(e,s) = \max\{\hat{r}(i,s) : i \leq e\}$ thereby satisfying (3.1).)

Proof. We know $C \neq \Phi_i(A)$ for all $i \leq e$. Fix $i \leq e$. Define $p = \mu x[C(x) \neq \Phi_i(A; x)]$. Choose s' sufficiently large such that, for all $s \geq s'$,

$$(\forall x < p)[\hat{\Phi}_{i,s}(A_s; x) = \Phi_i(A; x)] \text{ and } (\forall x \leq p)[C_s(x) = C(x)].$$

Case 1. $(\forall t \geq s')[t \in T \Rightarrow \hat{\Phi}_{i,t}(A_t; p)$ undefined]. Then for any $t \geq s'$, such that $t \in T$, we have $\hat{\ell}(i,t) = p$ and $\hat{r}(i,t) = \max\{u(i,x,s') : x < p\}$.

Case 2. $\hat{\Phi}_{i,t}(A_t; p)$ is defined for some $t \in T$, $t \geq s'$. Then $\hat{\Phi}_i(A; p) = \Phi_{i,s}(A_s; p)$ for all $s \geq t$ by (3.2). But $C(p) \neq \Phi_i(A; p)$. Hence, we have

$$(\forall s \geq t)[\hat{\ell}(i,s) = p \ \& \ \hat{r}(i,s) = \hat{r}(i,t)]. \qquad \square$$

Theorem 3.4 (Thickness Lemma - Strong Form): Given a nonrecursive r. e. set C and an r. e. set B there is an r. e. set $A \subseteq B$ such that $A \leq_T B$ and

(a) $(\forall e)[C \not\leq_T B^{(<e)}] \Rightarrow [C \not\leq_T A \ \& \ A$ is a thick subset of B],

(b) $(\forall e)[C \not\leq_T B^{(<e)} \Rightarrow (\forall i \leq e)[C \neq \Phi_i(A) \ \& \ A^{(i)} =^* B^{(i)}]]$.

Furthermore, an index for A can be computed uniformly in induces for B and C.

This strong form (Soare, 1976, p. 520) follows by carefully examining the above proof and by replacing $\hat{\ell}(e,s)$ in the definition of $\hat{r}(e, s)$ by

(modified length function) $\hat{m}(e,s) = \max\{x: \exists v \leq s)[x \leq \hat{\ell}(e,v)$

$\& \ (\forall y \leq x)[A_s[\hat{u}(e,y,v)] = A_v(\hat{u}(e,y,v)]]\}.$

Corollary 3.5 (Sacks, 1966): Let $\underset{\sim}{d}_0 < \underset{\sim}{d}_1 < \underset{\sim}{d}_2 < \ldots$ be an infinite sequence of simultaneously r. e. degrees. Then there exists an r. e. upper bound $\underset{\sim}{a}$ such that $\underset{\sim}{d}_0 < \underset{\sim}{d}_1 < \ldots < \underset{\sim}{a} < \underset{\sim}{0}'$. (Hence, $\underset{\sim}{0}'$ is not a minimal upper bound for the sequence.)

Proof. Fix a recursive function h such that $dg(W_{h(x)}) = \underset{\sim}{d}_x$ for all x. Define the r. e. set B by $B^{(x)} = \{\langle x, y\rangle : y \in W_{h(x)}\}$. Let C = K and apply Theorem 3.4 to obtain a thick r. e. subset $A \subseteq B$ such that $K \not\leq_T A$. By thickness, $W_{h(x)} \equiv_T B^{(x)} =^* A^{(x)}$ so that $\underset{\sim}{d}_i < dg(A)$ for all i. □

The Sacks density theorem can be derived from the thickness lemma using the following results of Yates on index sets (1966, pp 312, 314) which can be proved without priority methods. If V is an r.e. set, let $S \in \Sigma_3^V$ denote that there is a predicate R^V recursive in V such that $x \in S$ iff $(\exists y)(\exists u)(\exists w)\ R^V(x,y,u,w)$.

Lemma A (Yates): $\{x: W_x \equiv_T V\} \in \Sigma_3^V$.

Lemma B (Yates): For any set $S \in \Sigma_3^V$ there is a recursive function h(x) such that, for all x, $W_{h(x)} \leq_T V$ and

(a) $x \in S \Rightarrow (\exists e)[W_{h(x)}^{(e)} \equiv_T V \ \& \ (\forall i < e)[W_{h(x)}^{(i)}$ is recursive]],

(b) $x \notin S \Rightarrow (\forall e)[W_{h(x)}^{(e)}$ is recursive].

Theorem 3.6) (Index Set Theorem - Yates (1966a): Given r. e. sets C and D such that $D <_T C$ and $S \in \Sigma_3^C$ there is a recursive function g(x) such that, for all x,

(a) $D \leq_T W_{g(x)} \leq_T C$, and

(b) $x \in S \iff W_{g(x)} \equiv_T C$.

Corollary 3.7 (Density Theorem - Sacks (1964)): If D and C are r.e. and $D <_T C$ then there exists an r.e. set A such that $D <_T A <_T C$.

Proof (Corollary 3.7): Let $S = \{x : W_x \equiv_T D\}$. Then, by Lemma A, $S \in \Sigma_3^D$ and hence $s \in \Sigma_3^C$. Apply Theorem 3.6 to find $g(x)$ such that $D \leq_T W_{g(x)} \leq_T C$ and $W_x \equiv_T D$ just if $W_{g(x)} \equiv_T C$. By the recursion theorem choose x_0 such that $W_{x_0} = W_{g(x_0)}$. Then $D <_T W_{g(x_0)} <_T C$. □

Proof (Theorem 3.6): Fix $V = C$, $S \in \Sigma_3^C$ and $h(x)$ the recursive function for S according to Lemma B. For each x define the r.e. set B_x by $B_x^{(0)} = \{\langle 0, y\rangle : y \in D\}$ and $B_x^{(e+1)} = \{\langle e+1, y\rangle : y \in W_{h(x)}^{(e)}\}$. (Note that $B_x \leq_T C$ for all x because $W_{h(x)} \leq_T C$ and $D \leq_T C$.) For each x apply Theorem 3.4(b) to B_x and C to find $A_x \subseteq B_x$, so that $A_x \leq_T B_x \leq_T C$. Moreover, for each x, $D \leq_T A_x$ because $A_x^{(0)} =^* B_x^{(0)} \equiv_T D$ by Theorem 3.4(b) with $e = 0$. By the uniformity of Theorem 3.4, there is a recursive function $g(x)$ such that $W_{g(x)} = A_x$. Now if $x \notin S$ then $B_x^{(e)}$ is recursive for all $e > 0$ by Lemma B(b), whence $C \not\leq_T A_x$ by Theorem 3.4(b). If $x \in S$ then, by Lemma B(a) and the definition of B, choose e such that

$$B_x^{(e)} \equiv_T C \ \& \ (\forall i)[0 < i < e \Rightarrow B_x^{(i)} \text{ is recursive}].$$

Hence, $B_x^{(<e)} \equiv_T D <_T C$. Therefore, by Theorem 3.4(b), $A_x^{(e)} =^* B_x^{(e)} \equiv_T C$, so that $C \leq_T A_x$. Thus, $A_x \equiv_T C$ because $A_x \leq_T B_x \leq_T C$. □

The density theorem can also be proved directly without index sets (Soare, 1976, p. 525) by combining the infinite injury method with a clever coding method of Sacks.

§4. THE MINIMAL PAIR METHOD

If P is any countable partially ordered set (poset), then the finite injury priority method enables us to embed P (by an order preserving map) into the r.e. degrees $\underset{\sim}{R}$ (Sacks, 1966), and the infinite injury priority method allows P to be even embedded in any interval $[\underset{\sim}{a}, \underset{\sim}{b}]$, for $\underset{\sim}{a}, \underset{\sim}{b} \in \underset{\sim}{R}$ and $\underset{\sim}{a} < \underset{\sim}{b}$ (Robinson, 1971b). If P happens to be also a lattice then these embeddings naturally preserve supremums (sups) but not necessarily infimums (infs).

We now introduce a new method for embedding certain lattices into $\underset{\sim}{R}$ preserving both infs and sups. Consider the four element Boolean algebra $\diamond$ which we call the <u>Diamond.</u> In §4 we prove that the Diamond can be embedded into $\underset{\sim}{R}$ preserving not only sups and infs but the least element as well. In §5 we replace the Diamond by an arbitrary countable distributive lattice. On the other hand, in §6 we prove the surprising fact that not even the Diamond can be embedded as a lattice in $\underset{\sim}{R}$ by a map preserving both least and greatest elements. A corollary is that the r.e. degrees are not closed under infs and thus fail to form a lattice.

The embedding technique of §4 (called the <u>minimal pair</u> method because of Definition 4.1) is also one where some requirements are infinitary, but it is quite different from the infinite injury method of §3. The version we present here (derived from Lachlan (1973)) uses special stages analogous to the non-deficiency stages of §3 so that the negative restraints drop back simultaneously, and the proof closely resembles a finite injury argument. For notational convenience we let $\{e\}_s^{A_s}(x)$ denote $\Phi_{e,s}(A_s; x)$ from now on.

Definition 4.1. Nonzero r.e. degrees $\underset{\sim}{a}$ and $\underset{\sim}{b}$ form a minimal pair if

$$(\forall \text{r.e. } \underset{\sim}{c})[[\underset{\sim}{c} \leq \underset{\sim}{a} \text{ and } \underset{\sim}{c} \leq \underset{\sim}{b}] \Rightarrow \underset{\sim}{c} = \underset{\sim}{0}].$$

Theorem 4.2 (Lachlan (1966) - Yates (1966a): There exists a minimal pair of r.e. degrees $\underset{\sim}{a}$ and $\underset{\sim}{b}$.

Corollary 4.3. The Diamond lattice can be embedded in $\underset{\sim}{R}$ preserving sups, infs, and least element.

Proof. Let $\underset{\sim}{c} = \underset{\sim}{a} \cup \underset{\sim}{b}$. Then $\{\underset{\sim}{0}, \underset{\sim}{a}, \underset{\sim}{b}, \underset{\sim}{c}\}$ embeds the Diamond lattice. □

Proof of Theorem 4.2. It suffices to construct r.e. sets A and B satisfying, for all e, i, j, the requirements

$$P_{2e}: A \neq \overline{W}_e,$$

$$P_{2e+1}: B \neq \overline{W}_e, \text{ and}$$

$$N_{\langle i,j\rangle}: \{i\}^A = \{j\}^B = f \text{ total} \Rightarrow f \text{ is recursive.}$$

The following remark allows us to simplify the form of the negative requirements.

Remark 4.4 (Posner): To satisfy all $N_{\langle i,j\rangle}$, $i, j \in \omega$, it suffices to satisfy for all e the requirement

$$N'_e: \{e\}^A = \{e\}^B = f \text{ total} \Rightarrow f \text{ is recursive.}$$

Proof. We may assume without loss of generality that we can arrange that $A \neq B$, say $n_0 \in A - B$. For each i and j there is an index e such that

$$\{e\}^X(x) = \begin{cases} \{i\}^X(x) & \text{if } n_0 \in X \\ \{j\}^X(x) & \text{if } n_0 \notin X. \end{cases}$$

The remark follows immediately. □

From now on we will replace all occurrences of negative requirements similar to $N_{\langle i,j\rangle}$ by equivalent requirements N'_e, and we will write the latter as N_e.

Given $\{A_t : t \leq s\}$ and $\{B_t : t \leq s\}$ we define as usual the functions

(length function) $\ell(e,s) = \max\{x : (\forall y < x)[\{e\}_s^{A_s}(y) = \{e\}_s^{B_s}(y)]\}$,

(maximum length function) $m(e,s) = \max\{\ell(e,t) : t \leq s\}$.

A stage s is called <u>0-maximal</u> if $\ell(0,s) > m(0, s \dot{-} 1)$. Define the <u>restraint function</u>

$$r(0,s) = \begin{cases} 0 & \text{if } s \text{ is 0-maximal,} \\ \text{the greatest 0-maximal stage } t < s & \text{otherwise.} \end{cases}$$

(Notice that we can define the restraint function in terms of a stage s rather than an element z used in a computation at stage s since we may assume $z \leq u(A_s, e, x, s) \leq s$, where $u(A_s,e,x,s)$ is $u(e,x,s)$ where A_s is the oracle used in the computation.

The strategy σ_0 for meeting a <u>single</u> negative requirement N_0 is to allow x to enter $A \cup B$ at stage s+1 only if s is 0-maximal, and at most <u>one</u> of the sets A, B receives an element x at such a stage. Thus, if x destroys one of the computations $\{0\}_s^{A_s}(p) = q$ or $\{0\}_s^{B_s}(p) = q$ for some $p < \ell(0,s)$, say $\{0\}_s^{A_s}(p)$, then the other computation $\{0\}_s^{B_s}(p) = q$ will be preserved until the A-computation is restored, and outputs q again. In this way, if $\{0\}^A = \{0\}^B = f$ is a total function then f is recursive. (To compute f(p) we find the least s such that $p < \ell(0,s)$ and we set $f(p) = \{0\}_s^{A_s}(p)$.) Furthermore, $\liminf_s r(0,s) < \infty$, since $\liminf_s r(0,s) = 0$ unless there is a largest 0-maximal stage t, in which case $r(0,s) = t$ for all $s \geq t$.

This fundamental strategy of having one side or the other hold the computation at all times is applied to the other negative requirements N_e, $e > 0$, but with some crucial modifications to force the negative restraints to drop back simultaneously, thus creating "windows" through the restraints as in § 3.

For example, to drop back simultaneously with N_0, N_1 must guess the value of $k = \liminf_s r(0, s)$. Thus, N_1 must simultaneously play infinitely many strategies σ_1^k, $k \in \omega$, one for each possible value of k. Each strategy σ_1^k is played like σ_0 but with $S^k = \{s : r(0, s) = k\}$ in place of ω as the set of stages during which it is active, and on which its length functions ℓ and m are defined. This allows σ_1^k to open its window more often since its length functions ignore the stages in $\omega - S^k$. Strategy σ_1^k still succeeds if any restraint it imposes is maintained during intermediate stages $s \notin S^k$ while σ_1^k is dormant. Thus, at stage s if $k = r(0, s)$, we play σ_1^k, maintain the restraints previously imposed by the dormant σ_1^i, $i < k$, and discard restraints imposed by σ_1^j, $j > k$. Thus if $k = \liminf_s r(0, s)$, then: (1) strategy σ_1^k succeeds in meeting N_1; (2) the strategies σ_1^i, $i < k$, impose finitely much restraint over the whole construction; and (3) the strategies σ_1^j, $j > k$, drop all restraint at each stage $s \in S^k$. Thus, the entire restraint $r(1, s)$ imposed by N_0 and N_1 together has $\liminf_s r(1, s) < \infty$.

Construction of A and B.

Stage $s = 0$. Do nothing.

Stage $s + 1$. Given A_s and B_s, define the restraint function $r(e,s)$ for N_e by induction on e as follows. Define $r(0, s)$ as above. A stage s is (e+1)-maximal if

$$(\forall t < s)[r(e, t) = r(e, s) \Rightarrow \ell(e+1, t) < \ell(e+1, s)].$$

Let $r(e+1, s)$ be the maximum of

(i) $r(e, s)$,

(ii) those $t < s$ such that $r(e, t) < r(e, s)$, and

(iii) those $t < s$ such that $r(e,t) = r(e,s)$ and t is $(e+1)$-maximal, if s is not $(e+1)$-maximal.

Requirement P_{2e} <u>requires</u> <u>attention</u> if

(4.1) $$W_{e,s} \cap A_s = \emptyset \text{, and}$$

(4.2) $$(\exists x)[x \in W_{e,s} \;\&\; 2e < x \;\&\; r(e, s) < x]$$

and likewise for P_{2e+1} with B in place of A. Choose the highest priority requirement P_e which requires attention and the least x corresponding to that e. Enumerate x in A if e is even (in B if e is odd).

<u>Lemma 1.</u> $(\forall e)[\liminf_s r(e, s) < \infty]$.

<u>Proof.</u> We first prove the case $e = 0$. If there are finitely many 0-maximal stages then $\liminf_s r(0, s) = 0$. Otherwise, $\lim_s r(0, s)$ is the largest 0-maximal stage. For the inductive step, fix e and assume $k = \liminf_s r(e, s)$. Then there are only finitely many stages s such that $r(e, s) < k$. Let t be the largest such. Let $S = \{s: r(e, s) = k\}$. Either there are infinitely many $(e+1)$-maximal stages in S, in which case $\liminf_s r(e+1, s) = \max\{t, k\}$, or else there is a largest $(e+1)$-maximal stage $v \in S$, in which case $\liminf_s r(e+1, s) = \max\{t, k, v\}$.

<u>Lemma 2.</u> Every positive requirement is satisfied and acts at most once.

<u>Proof.</u> Consider requirement P_{2e} (since P_{2e+1} is similar). First $\overline{A}$ is infinite as usual by the second clause of (4.2). Now if W_e is infinite then W_e contains some $x > \liminf_s r(e, s)$, and some such x is eventually enumerated in A satisfying P_{2e}.

Lemma 2. Every positive requirement is satisfied and acts at most once.

Proof. Consider requirement P_{2e} (since P_{2e+1} is similar). First $\overline{A}$ is infinite as usual by the second clause of (4.2). Now if W_e is infinite then W_e contains some $x > \liminf_s r(e,s)$, and some such x is eventually enumerated in A satisfying P_{2e}.

Lemma 3. $(\forall e)$[requirement N_e is met].

Proof. Fix e and let $k = \liminf_s r(e-1, s)$, and $S = \{s : r(e-1, s) = k\}$. (If $e = 0$ let $S = \omega$ and $k = 0$.) Choose s' such that no P_i, $i < e$, acts after stage s' and $r(e-1, s) \geq k$ for all $s \geq s'$. Now assume that $\{e\}^A = \{e\}^B = f$ is a total function. To recursively compute $f(p)$, $p \in \omega$, find an e-maximal stage $s'' \in S$, $s'' > s'$, such that $\ell(e, s'') > p$. Let $q = \{e\}_{s''}^{A_{s''}}(p) = \{e\}_{s''}^{B_{s''}}(p)$. We will prove by induction on t that for all $t \geq s''$ either

$$\text{(i)} \quad \{e\}_t^{A_t}(p) = q \text{, or}$$

$$\text{(ii)} \quad \{e\}_t^{B_t}(p) = q \text{,}$$

and hence that $f(p) = q$. Suppose that x destroys the last of the computations (i) or (ii). Now if x enters $A \cup B$ at any stage s+1 such that $s \in S$ then s must have been e-maximal, so both (i) and (ii) hold for $t = s$. But x can destroy at most one of the computations, so the other holds at $t = s+1$. Furthermore, x cannot enter $A \cup B$ at stage s+1 for $s \notin S$, $s > s''$, since $r(e, s) \geq x$ by clause (ii) in the definition of $r(e, s)$. $\square$

This construction can be modified in a number of ways. First, one can construct an r.e. sequence of r.e. degrees $\{\underset{\sim}{a}_i : i \in \omega\}$ such that $\underset{\sim}{a}_i, \underset{\sim}{a}_j$ is a minimal pair for each $i \neq j$. Next, by allowing infinitary positive requirements as in §3, one can construct a minimal pair $\underset{\sim}{a}, \underset{\sim}{b}$ of r.e. degrees which are high (i.e., $\underset{\sim}{a}' = \underset{\sim}{b}' = \underset{\sim}{0}''$) (see Lachlan (1966). By an easy modification of the method (Yates, 1966a) one can construct r.e. degrees degrees $\underset{\sim}{b}, \underset{\sim}{c}$, and an increasing sequence of r.e. degrees $\underset{\sim}{a}_0 < \underset{\sim}{a}_1 < \ldots$ such that $(\underset{\sim}{b}, \underset{\sim}{c})$ is an <u>exact pair</u> for the sequence $\{\underset{\sim}{a}_n\}$ in the sense that

$$(\forall \text{r.e. } \underset{\sim}{d})[[\underset{\sim}{d} < \underset{\sim}{b} \ \& \ \underset{\sim}{d} < \underset{\sim}{c}] \Rightarrow (\exists n)[\underset{\sim}{d} < \underset{\sim}{a}_n]].$$

It follows that $\underset{\sim}{b}$ and $\underset{\sim}{c}$ have no infimum in $\underset{\sim}{R}$ and hence $\underset{\sim}{R}$ is not a lattice.

§5. EMBEDDING DISTRIBUTIVE LATTICES IN THE R.E. DEGREES

Using a fairly easy modification of the preceding method we will now replace the Diamond lattice of Theorem 4.2 by <u>any</u> countable distributive lattice. Since any countable distributive lattice can be embedded in the countable atomless Boolean algebra it suffices to prove the following.

<u>Theorem 5.1</u> (Lerman-Lachlan-Thomason (Thomason, 1971)): There is an embedding of a countable atomless Boolean algebra $\mathfrak{B}$ into the r.e. degrees $\underset{\sim}{R}$ which preserves sups, infs, and least element.

<u>Proof.</u> Let $\{\alpha_i : i \in \omega\}$ be any uniformly recursive sequence of recursive sets (i.e., $\{\langle x, i\rangle : x \in \alpha_i\}$ is a recursive relation) which forms an atomless Boolean algebra $\mathfrak{B}$ under $\cup$, $\cap$, and complementation, contains ω and has $\emptyset$ as its only finite member. We will construct r.e. sets A_i, $i \in \omega$, and define $A_\alpha = \{\langle i, x\rangle : x \in A_i \;\&\; i \in \alpha\}$ for $\alpha \in \mathfrak{B}$. Notice that we immediately have

(5.1) $$dg(A_{\alpha \cup \beta}) = dg(A_\alpha) \cup dg(A_\beta),$$

(5.2) $$\alpha \subseteq \beta \implies dg(A_\alpha) \le dg(A_\beta), \text{ and}$$

(5.3) $$dg(A_{\alpha \cap \beta}) \le dg(A_\alpha), dg(A_\beta).$$

We will further meet for all i, j, α, β the requirements

$$P_{\langle i,j\rangle} : A_i \ne \overline{W}_j$$

$$N_{(\alpha,\beta,j)} : \{j\}^{A_\alpha} = \{j\}^{A_\beta} = f \text{ total} \implies f \le_T A_{\alpha \cap \beta}.$$

These requirements insure

(5.4) $$dg(A_{\alpha \cap \beta}) = dg(A_\alpha) \cap dg(A_\beta), \text{ and}$$

(5.5) $\qquad dg(A_\alpha) \leq dg(A_\beta) \Rightarrow \alpha \subseteq \beta.$

Note that (5.1)-(5.5) guarantee that the map $\alpha \to dg(A_\alpha)$ is the desired embedding, and (5.5) guarantees that the map is 1:1. (To see that the negative requirements insure (5.5) suppose: (1) $dg(A_\alpha) \leq dg(A_\beta)$; and (2) $\alpha \not\subseteq \beta$, say $i \in \alpha - \beta$. Then $dg(A_i) \leq dg(A_\alpha) \leq dg(A_\beta)$ by (5.2) and (1), but $dg(A_i) \leq dg(A_{\overline{\beta}})$ by (2). Hence $dg(A_i) \leq dg(A_{\beta \cap \overline{\beta}}) = dg(A_\emptyset) = \underset{\sim}{0}$ by (5.4), contradicting A_i nonrecursive.)

The strategy for meeting the negative requirements $N_{\langle\alpha,\beta,j\rangle}$ begins as before. Denoting $N_{\langle\alpha,\beta,j\rangle}$ by N_e, where $\alpha = \alpha_{i_1}$, $\beta = \alpha_{i_2}$, and $e = \langle i_1, i_2, j\rangle$, we define the restraint function by induction on e exactly as in §4. However, new difficulties in proving Lemma 3 (that N_e is satisfied) require greater care in enumerating elements for the positive requirements. To meet requirement $P_{\langle i,j\rangle}$ we will appoint <u>followers</u> $x \in \omega$ so-called because the eventual enumeration of x in A_i will satisfy $P_{\langle i,j\rangle}$ (although x may be <u>cancelled</u> before this happens). If x is a follower of P_i and y a follower of P_j then we say x has <u>higher priority</u> than y $(x \prec y)$ if $i < j$ or $i = j$ and x was appointed before y. We will arrange for all followers x and y existing at stage s that $x \prec y$ iff $x < y$.

<u>Construction of A_i, $i \in \omega$:</u>

<u>Stage s = 0.</u> Do nothing.

<u>Stage s + 1.</u> Requirement $P_{\langle i,j\rangle}$ is <u>satisfied</u> if $A_{i,s} \cap W_{j,s} \neq \emptyset$. Requirement $P_{\langle i,j\rangle}$ <u>requires attention</u> if $P_{\langle i,j\rangle}$ is not satisfied and either

(5.6) $\qquad x \in W_{j,s}$ and $x > r(\langle i,j\rangle, s)$ for some uncancelled follower x of $P_{\langle i,j\rangle}$; or

(5.7) $x \in W_{j,s}$ for every uncancelled follower x of $P_{\langle i,j\rangle}$.

Let $P_{\langle i,j\rangle}$ be the highest priority requirement which requires attention. If (5.6) holds for some x enumerate the least such x in A_i. If (5.6) fails, and (5.7) holds, then appoint x = s+1 as a follower of $P_{\langle i,j\rangle}$. In either case <u>cancel</u> all followers y of lower priority than x (i.e., $x \prec y$). (If no $P_{\langle i,j\rangle}$ requires attention, then do nothing.)

<u>Lemma 1</u>. $(\forall e)[\liminf_s r(e,s) < \infty]$.

<u>Proof.</u> Exactly as in §4, Lemma 1.

<u>Lemma 2</u>. $(\forall e)[P_e$ receives attention at most finitely often and is met].

<u>Proof.</u> Fix e and choose s_0 such that for no $e' < e$ does $P_{e'}$ receive attention after stage s_0. Let $k = \liminf_s r(e,s)$ by Lemma 1. Let $e = \langle i,j\rangle$. Now if $P_{\langle i,j\rangle}$ receives attention infinitely often, then some follower $x > k$ is appointed to follow $P_{\langle i,j\rangle}$ and x is never cancelled. Furthermore, $x \in W_j$ by (5.7). Hence, there is a stage $t+1 > s_0$ such that $r(e,t) < x$ and $x \in W_{j,t}$. Now x or some smaller follower of P_e is enumerated in A at stage t+1, P_e is met, and P_e never again requires attention. Therefore, P_e receives attention at most finitely often. Finally, $P_{\langle i,j\rangle}$ is met because otherwise $\overline{A}_i = W_j$, $x \in W_j$ for every uncancelled follower x of $P_{\langle i,j\rangle}$ and $P_{\langle i,j\rangle}$ receives attention infinitely often under (5.7).

<u>Lemma 3</u>. $(\forall \alpha)(\forall \beta)(\forall j)$[requirement $N_{(\alpha,\beta,j)}$ is met].

<u>Proof.</u> Fix $N_e = N_{(\alpha,\beta,j)}$. Choose k, S and stage s' as in Lemma 3 of §4. Assume that $\{j\}^{A_\alpha} = \{j\}^{A_\beta} = f$ is a total function. A computation $\{e\}_s^{A_{\gamma,s}}(x)$ is A_δ-<u>correct</u> if

$A_{\delta,s}[u] = A_\delta[u]$ where $u = u(A_{\delta,s'}, e,x,s)$. To $A_{\alpha\ \beta}$-recursively compute f(p), find an e-maximal stage $s \in S$, $s > s'$, such that $\ell(e,s) > p$ and both computations $\{e\}_s^{A_{\alpha,s}}(p) = q$ and $\{e\}_s^{A_{\beta,s}}(p) = q$ are $A_{\alpha\cap\beta}$-correct. We will show by induction on t that for all $t \geq s$ either

(5. 8) $$\{e\}_t^{A_{\alpha,t}}(p) = q\ , \quad \text{or}$$

(5. 9) $$\{e\}_t^{A_{\beta,t}}(p) = q\ ,$$

via an $A_{\alpha\cap\beta}$-correct computation. Now if x destroys either computation (5. 8) or (5.9) by entering $A_\alpha \cup A_\beta$ at stage t+1 then t must have been e-maximal and $t \in S$ (as in Lemma 3 of § 4) so <u>both</u> computations existed at the end of stage t. By inductive hypotheses at least one computation, say (5.8), is $A_{\alpha\cap\beta}$-correct. Suppose x is enumerated in A_α at stage t+1, destroying this computation. Then x cancels at stage t+1 all followers y such that $x < y$ (since these are exactly those followers y such that $x \prec y$). Furthermore, $z > t+1 \geq u_\beta =_{dfn} u(A_{\beta,t}, e,t,p)$ for any follower z later appointed. But $x \leq u_\alpha =_{dfn} u(A_{\alpha,t}, e,t,p)$ since the A-computation is destroyed by x. Also $A_{\alpha\cap\beta,t}[u_\alpha] = A_{\alpha\cap\beta}[u_\alpha]$ since the A_α-computation (5.8) was $A_{\alpha\cap\beta}$-correct. Hence $A_{\alpha\cap\beta,t}[t+1] = A_{\alpha\cap\beta}[t+1]$ and $u_\beta \leq t+1$ so the A_β-computation (5.9) now becomes assured of remaining $A_{\alpha\cap\beta}$-correct. □

<u>Corollary 5.2</u>. Any countable distributive lattice can be embedded into the r. e. degrees $\underset{\sim}{R}$ by a map which preserves sups, infs, and least elements.

Embedding nondistributive lattices into $\underset{\sim}{R}$ is much more difficult. Lachlan (1972) showed that the following two 5-element nondistributive lattices M_5 (a modular lattice) and N_5 (a non-modular lattice) can be embedded in $\underset{\sim}{R}$ by a map preserving sups, infs, and least element.

Figure 5.1

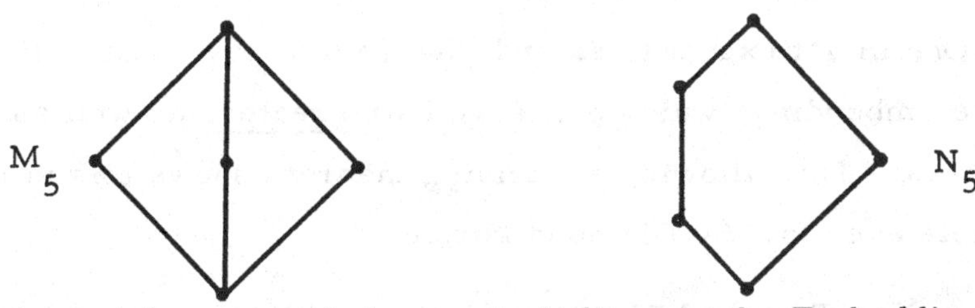

This partial success led many to believe in the <u>Embedding Conjecture</u> which asserted that every finite lattice can be embedded in $\underset{\sim}{R}$ as a finite lattice. This conjecture was recently refuted by Lachlan and Soare (1980) who showed that the following lattice S_8 cannot be embedded in $\underset{\sim}{R}$ as a lattice.

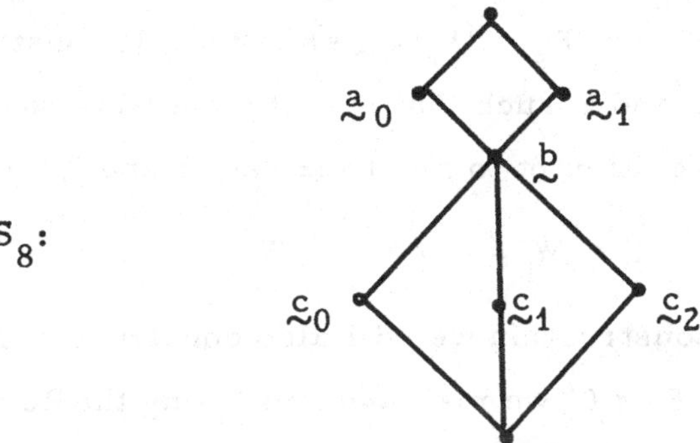

The obstacle to embedding S_8 is that for $\underset{\sim}{b}$ to be the sup of the lower M_5 lattice, an elaborate system of traces is required for enumerating elements into the set of degree $\underset{\sim}{b}$. This interferes with the delicate minimal pair machinery above which insures that $\underset{\sim}{b}$ is the inf of $\underset{\sim}{a}_0$ and $\underset{\sim}{a}_1$.

The most important open question on $\underset{\sim}{R}$ is that of the decidability of its elementary theory. Considerably more structural results (such as embedding and nonembedding theorems) will be required to meet this goal.

§6. THE NON-DIAMOND THEOREM

One might expect to extend Theorem 5.1 by constructing lattice embeddings which preserve both greatest as well as least elements. The following surprising theorem shows this is impossible even for the Diamond lattice.

6.1 Non-Diamond Theorem (Lachlan, 1966): If $\underset{\sim}{a}$ and $\underset{\sim}{b}$ are nonrecursive r.e. degrees such that $\underset{\sim}{a} \cup \underset{\sim}{b} = 0'$ then there is a nonrecursive r.e. degree $\underset{\sim}{c}$ such that $\underset{\sim}{c} < \underset{\sim}{a}$ and $\underset{\sim}{c} < \underset{\sim}{b}$.

Proof. Let A and B be r.e. sets in degrees $\underset{\sim}{a}, \underset{\sim}{b}$ respectively. and $\{a_s\}_{s \in \omega}$, $\{b_s\}_{s \in \omega}$ be recursive enumerations of A and B. Let $A_s = \{a_t : t \leq s\}$ and $B_s = \{b_t : t \leq s\}$. We will construct coinfinite r.e. sets E and F such that one of these sets has the desired degree $\underset{\sim}{c}$. We attempt to meet for each i and j, i < j, the requirement

$$P_{i,j}: E \cap W_i \neq \emptyset \text{ or } F \cap W_j \neq \emptyset.$$

As an aid in the construction we will also construct a Δ_2^0 set D. Since $dg(A \oplus B) = \underset{\sim}{0}'$ we may assume (using the Recursion Theorem) that we can fix at the beginning an index e such that $D = \{e\}^{A \oplus B}$. Let u(k, s) denote the use function $u(A_s \oplus B_s, e, k, s)$. (For X = D, E or F, let X_s denote the elements in X at the end of stage s.)

In order to insure that $E \leq_T A$ and $E \leq_T B$ we will enumerate a number x in E at stage s only if numbers $\leq x$ are enumerated in both A and B. For example, to satisfy the requirement $R_i: W_i \cap E \neq \emptyset$ suppose $x \in W_{i,s}$, $a_s \leq x$, and $u < x$ for some k such that $\{e\}_s^{A_s \oplus B_s}(k) = D_s(k)$, $u = u(k, s)$. We can now attack this requirement by inserting or extracting the attacker k from D according as $D_s(k) = 0$ or 1. Since $D = \{e\}^{A \oplus B}$ this forces a number $z \leq u$ to be enumerated in either A or B. If z is enumerated in B then the attack is successful and we enumerate x in E satisfying R_i forever.

If there are infinitely many unsuccessful attacks on R_i, we must insure that $F \cap W_j \neq \emptyset$ for all infinite W_j. Thus, we do not attack as above unless there is $t < s$ such that $y \in W_{j,t}$, $b_t < y$, and $u < y$. Now after the attack, if z is enumerated in A, then we can enumerate y in F and argue that $F \leq_T A$ and $F \leq_T B$. Thus we do do not attack R_i directly but rather requirement $P_{i,j}$ via the attacker $k = j$. An attack on $P_{i,j}$ succeeds in satisfying $P_{i,j}$ forever so $P_{i,j}$ is never attacked again. Thus, D is Δ_2^0 because each j is inserted in (or extracted from) D finitely often, namely at most once for each $P_{i,j}$, $i < j$.

<u>Construction.</u>

<u>Stage s = 0.</u> Do nothing.

<u>Stage s + 1.</u> Requirement $P_{i,j}$ requires attention if $i<j$, $P_{i,j}$ is not yet satisfied (i.e., $E_s \cap W_{i,s} = \emptyset$ and $F_s \cap W_{j,s} = \emptyset$), and there is a pair (x, y) satisfying:

(1) $\{e\}_s^{A_s \oplus B_s}(j) = D_s(j)$

(2) $x > 2i, a_s, u(j, s)$

(3) $x \in W_{i,s}$, and

(4) there is some $t < s$ such that:

(a) $y > 2j, b_t, u(j,t)$

(b) $y \in W_{j,t}$ and

(c) $(A_s \oplus B_s)[u(j,t)] = (A_t \oplus B_t)[u(j,t)]$ & $\{e\}^{A_t \oplus B_t}(j)$ converges.

(Note that these conditions imply $u(j,s) = u(j,t)$.)

Choose the least $\langle i,j \rangle$ such that $P_{i,j}$ requires attention and the least corresponding pair $\langle x, y\rangle$. Insert or extract j from D to insure $D_{s+1}(j) \neq D_s(j)$. Enumerate A and B until the first num-number $z < u(j,t)$ appears in $A \cup B$. (If no such z appears, the con construction stops. However, if e is the index obtained by the Recursion Theorem satisfying $D = \{e\}^{A\oplus B}$, then z must appear.) If z appears in A, enumerate y in F. If z appears in B, enumerate x in E.

Lemma 1. The set D is Δ_2^0 and hence $D \leq_T \emptyset'$.

Proof. An integer j is inserted or removed from D only when some $P_{i,j}$ with $i < j$ receives attention, but each $P_{i,j}$ receives attention at most once, so j is inserted at most finitely often.

Lemma 2. The sets E and F are coinfinite, $E \leq_T A$, $E \leq_T B$, and $F \leq_T A$.

Proof. The sets are coinfinite by the conditions $x > 2i$, $y > 2j$ of (2) and (4). The $\leq_T$ reductions follow by the usual permitting method. (For example, to A-recursively decide whether $x \in E$, find s such that $a_t > x$ for $t > s$. Now $x \in E$ iff $x \in E_s$.)

Lemma 3. If E is recursive, then

(i) F is nonrecursive, and

(ii) $F \leq_T B$.

Proof of (i). Fix i_0 such that $W_{i_0} = \overline{E}$. If F is recursive, choose the least $j > i_0$ such that $W_j = \overline{F}$. Choose s_0 such that for all $s \geq s_0$, $u(j,s) = u(j,s_0)$ and $(A \oplus B)[u] = (A_{s_0} \oplus B_{s_0})[u]$, where $u = u(j,s_0)$. Now since B is nonrecursive, there exist $y > u$ and $t > s_0$ such that $y \in W_{j,t_0}$ and y is permitted by B at stage t, i.e., such that (4)(a), (b) and (c) hold for y and any $s > t$. But since A is nonrecursive there exist infinite many $x \in W_{i_0}$ satisfying (1), (2) and (3) for some $s > t$. Hence requirement $P_{i_0, j}$ receives attention, and either $D \cap W_{i_0} \neq \emptyset$ or $F \cap W_j \neq \emptyset$ contrary to hypothesis.

Proof of (ii). To prove $F \leq_T B$ we may first assume that we know those finitely many y contributed to F by some $P_{i,j}$ with $j \leq i_0$. For each reamining y, to decide whether $y \in F$ we B-recursively find the largest t such that $b_t \leq y$. Consider all $j > i_0$ such that (4)(a), (4)(b) and the second clause of (4)(c) hold for y and t. For each such j find the least $s \geq t$ such that either:

(5) $$(A_s \oplus B_s)[u(j,t)] \neq (A_t \oplus B_t)[u(j,t)],$$

(in which case clause (4)(c) prevents $P_{i,j}$ for any i from putting y into F after stage s), or

(6) $$W_{j,s} \cap F_s \neq \emptyset,$$

in which case if $y \in F$ for $P_{i,j}$ then $y \in F_s$. Note that if (5) fails then (6) must hold because as in part (i) there are infinitely many $x \in W_{i_0}$ satisfying (1), (2), and (3) and thus eligible to form a pair with y for $P_{i_0, j}$, so $P_{i_0, j}$ will receive attention. ☐

Theorem 6.2 (Lachlan, 1966): If $\underset{\sim}{a}, \underset{\sim}{b}$ are r.e. degrees and $\underset{\sim}{d}$ is a degree $\leq \underset{\sim}{a}$ and $\leq \underset{\sim}{b}$ then there is an r.e. degree $\underset{\sim}{c}$ such that $\underset{\sim}{d} \leq \underset{\sim}{c}$, $\underset{\sim}{c} \leq \underset{\sim}{a}$, $\underset{\sim}{c} \leq \underset{\sim}{b}$.

Proof. Fix r.e. sets $A \in \underset{\sim}{a}$, $B \in \underset{\sim}{b}$ and indices e, i such that $\{e\}^A = \{i\}^B = D$ for some set $D \in \underset{\sim}{d}$. Let $\{A_s\}_{s \in \omega}$ and $\{B_s\}_{s \in \omega}$ be recursive enumerations of A and B. As usual define the recursive length function

$$\ell(s) = \max\{x : (\forall y < x)[\{e\}_s^{A_s}(y) = \{i\}_s^{B_s}(y)]\}.$$

For each x define an r. e. set C_x as follows. At stage s, if $\ell(s) > x$ and $\{e\}_s^{A_s}(x) \neq \{e\}_t^{B_t}(x)$ where $t = \max\{v : v < s \,\&\, \ell(v) > x\}$, then enumerate in C_x all $s' < s$. (Hence, C_x contains all those stages t such that the common value $y = \{e\}_t^{A_t}(x) = \{i\}_t^{B_t}(x)$ is later replaced by a new common value $y' = \{e\}_s^{A_s}(x) = \{i\}_s^{B_s}(x)$ at some stage $s > t$. Thus, for any $s \notin C_x$, if $\ell(s) > x$ then $\{e\}_s^{A_s}(x) = \{i\}_s^{B_s}(x) = \{e\}^A(x) = \{i\}^B(x)$.) Now set $C = \oplus \{C_x : x \in \omega\}$. Clearly C is r.e., $D \leq_T C$, and $C \leq_T A, B$. □

Corollary 6.3 (Lachlan, 1966). There are r.e. degrees $\underset{\sim}{a}, \underset{\sim}{b}$ with no infimum. Hence, the r.e. degrees do not form a lattice.

Proof. Let $\underset{\sim}{a}, \underset{\sim}{b}$ be incomparable low r.e. degrees with $\underset{\sim}{a} \cup \underset{\sim}{b} = \underset{\sim}{0}'$. (Obtain these by the Sacks splitting theorem.) Let $\underset{\sim}{d}$ be any degree below both $\underset{\sim}{a}$ and $\underset{\sim}{b}$. Now $\underset{\sim}{d}' = \underset{\sim}{0}'$ since $\underset{\sim}{a}' = \underset{\sim}{0}'$, so the relativization to $\underset{\sim}{d}$ of Theorem 6.1 produces $\underset{\sim}{c} > \underset{\sim}{d}$, such that $\underset{\sim}{c} < \underset{\sim}{a}$, $\underset{\sim}{c} < \underset{\sim}{b}$ and by Lemma 6.2 we may assume that $\underset{\sim}{c}$ is r.e. Thus $\underset{\sim}{a}$ and $\underset{\sim}{b}$ have no infimum in the upper semi-lattice of r.e. degrees or even in the upper semi-lattice of all degrees. □

Recently, Jockusch has given an easy direct proof of Corollary 6.3. Jockusch has also noted that the method of Theorem 6.1 easily yields the following generalization. Let

$$\underset{\sim}{M} = \{\underset{\sim}{a} : \underset{\sim}{a} \text{ is r.e. and one half of a minimal pair}\}.$$

No finite supremum of degrees in $\underset{\sim}{M}$ is $\underset{\sim}{0}'$. Hence, by applying the Sacks splitting theorem to K there is a low degree which is not below any finite supremum of elements of $\underset{\sim}{M}$.

Lachlan (1966) raised the question of whether incomparable r.e. degrees $\underset{\sim}{a}, \underset{\sim}{b}$ satisfying $\underset{\sim}{a} \cup \underset{\sim}{b} = \underset{\sim}{0}'$ can ever have an infimum. Shoenfield and Soare (1978) and simultaneously and independently Lachlan (1980) showed that they could. Indeed, Lachlan combined a new method of preserving infimums with the

Sacks preservation method § 2 to obtain the following pleasing generalization of the Sacks splitting theorem

Theorem 6.4 (Lachlan Splitting Theorem (Lachlan, 1980)): Let A be a nonrecursive r.e. set. There exist r.e. sets B_0, B_1, and C such that

(a) $C \leq_T A$

(b) $B_0 \cup B_1 = A$ and $B_0 \cap B_1 = \emptyset$

(c) $B_{i-1} \not\leq_T B_i \oplus C$, $i = 0, 1$

(d) $\mathrm{dg}(C) = \mathrm{dg}(B_0 \oplus C) \cap \mathrm{dg}(B_1 \oplus C)$.

Thus, for any r.e. degree $\underset{\sim}{a} > \underset{\sim}{0}$ there are incomparable r.e. degrees $\underset{\sim}{b}_0 = \mathrm{dg}(B_0 \oplus C)$ and $\underset{\sim}{b}_1 = \mathrm{dg}(B_1 \oplus C)$ which have supremum $\underset{\sim}{a}$ and infimum $\underset{\sim}{c} = \mathrm{dg}(C)$.

§7. NONCAPPABLE DEGREES

An r.e. degree $\underset{\sim}{a}$ is noncappable if $\underset{\sim}{0} < \underset{\sim}{a} < \underset{\sim}{0}'$ and $\underset{\sim}{a}$ is not half of a minimal pair (i.e., $\underset{\sim}{a}$ cannot be "capped" to $\underset{\sim}{0}$). (These degrees were first called simple in (Soare, 1978, p.1174), because they satisfy the same first order sentence as simple sets: $x > 0$ & $(\forall y > 0)[x \cap y > 0]$. However , variations of the term, such as "effectively simple", became too confusing with well-established analogous terms for simple sets.) The existence of these degrees follows from the result of Jockusch mentioned after Corollary 6.3. A straightforward direct finite injury priority construction of such a degree was first given by Yates (1966a).

Briefly, we construct an r.e. set A to meet for all e and i the following requirements:

N_e: $\Phi_e(A) \neq K$;

$P_{e,i}$: W_e nonrecursive $\Rightarrow [A^{(e)} \leq_T W_e$ & $A^{(e)} \neq \overline{W}_i]$, where
$A^{(e)} = \{\langle y,x\rangle : \langle y,x\rangle \in A$ & $x = e\}$ as in §3.

For N_e, we define the restraint function $r(e, s)$ as in §2. Let $\langle x, y, e\rangle = \langle\langle x, y\rangle, e\rangle$. For $P_{e,i}$ we enumerate some x of the form $\langle n, i, e\rangle$ in $A^{(e)}$ at stage s+1 if:

(1) $A_s^{(e)} \cap W_{i,s} = \emptyset$;

(2) $x \in W_{i,s}$;

(3) $(\exists y < x)[y \in W_{e,s+1} - W_{e,s}]$;

(4) $(\forall j \leq \langle e,i\rangle)[x < r(j,s)]$;

and x is minimal with respect to these properties. Each $P_{e,i}$ contributes at most one element to A, so N_e is injured finitely often and $\lim_s r(e,s)$ exists. Condition (3) guarantees that $A^{(e)} \leq_T W_e$, and obviously $A^{(e)} \leq_T A$. Now if W_e is nonrecursive then requirement $P_{e,i}$ is satisfied as in the usual permitting method of Yates (1965). (If $A^{(e)} = \overline{W}_i$, choose an increasing r.e. sequence of elements $x_1 < x_2 < \ldots$, in W_i of the form $\langle n, i, e \rangle$ for some n, such that $x_1 > \lim_s r(j,s)$, for all $j \leq \langle e, i \rangle$. For each k choose s_k minimal such that $x_k \in W_{i, s_k}$. Now $W_{e, s_k}[x_k] = W_e[x_k]$ so W_e is recursive.) This completes the sketch of the proof.

Since the set $\underset{\sim}{N}$ of noncappable degrees is one of the few nontrivial definable subclasses of r.e. degrees, it is interesting to study where the degrees of $\underset{\sim}{N}$ lie with respect to all the nontrivial r.e. degrees: $\underset{\sim}{R}^- =_{dfn} \underset{\sim}{R} - \{\underset{\sim}{0}, \underset{\sim}{0}'\}$. Clearly, $\underset{\sim}{N}$ is closed upward in $\underset{\sim}{R}^-$ because its complement $\underset{\sim}{M}$ in $\underset{\sim}{R}^-$ is closed downward. Two questions naturally arise:

(1) $(\forall \underset{\sim}{a} \in \underset{\sim}{R}^-)(\exists \underset{\sim}{b} \in \underset{\sim}{N})[\underset{\sim}{a} < \underset{\sim}{b}]$; and

(2) $(\forall \underset{\sim}{a} \in \underset{\sim}{N})(\exists \underset{\sim}{b} \in \underset{\sim}{N})[\underset{\sim}{b} < \underset{\sim}{a}]$?

Klaus Ambos has given a positive answer to (1) by proving it for $\underset{\sim}{a} \in M$. (If $\underset{\sim}{a} \in \underset{\sim}{N}$, then any $\underset{\sim}{b}$, $\underset{\sim}{a} < \underset{\sim}{b} < \underset{\sim}{0}'$, is in $\underset{\sim}{N}$ by upward closure of $\underset{\sim}{N}$.) Question (2) is still open along with the following questions. If $\underset{\sim}{a}, \underset{\sim}{b} \in \underset{\sim}{M}$ then is $\underset{\sim}{a} \cup \underset{\sim}{b} \in \underset{\sim}{M}$? Does there exist a strongly noncappable degree, namely an r.e. degree $\underset{\sim}{a}$, $\underset{\sim}{0} < \underset{\sim}{a} < \underset{\sim}{0}'$, such that for no r.e. degree $\underset{\sim}{b}$ with $\underset{\sim}{b} | \underset{\sim}{a}$ does the infimum of $\underset{\sim}{a}$ and $\underset{\sim}{b}$ exist?

Ambos proves (1) by combining the Yates construction above with the non-diamond method of 6. If B and C are r.e. sets whose degrees form a minimal pair, he constructs an r.e. set A with $\deg(A) \in \underset{\sim}{N}$ and $A \oplus C$ incomplete. To make A noncappable use the requirements $P_{e,i}$ and the strategy for $P_{e,i}$ exactly as above. To make $A \oplus C$ incomplete construct an r.e. set D and attempt to arrange that $D \neq \Phi_e(A \oplus C)$ by the Friedberg-Muchnik style method. Moreover, C, over which we have no control, may change too often for the usual strategy to succeed so we attempt to construct nonrecursive r.e. sets $E_e \leq_T B, C$, $e \in N$, as in the non-diamond theorem, contradicting the hypothesis that B and C form a minimal pair. The negative requirement is now

$$N_{e,i}: \quad D \neq \Phi_e(A \oplus C) \quad \text{or} \quad E_e \neq \overline{W}_i .$$

Requirement $N_{e,i}$ <u>requires attention</u> at stage $s+1$ if:

(1) there is no attack on $N_{e,i}$ in progress;

(2) for some pair (x, y) with $x = \langle n, i, e \rangle$,

$$0 = D_s(x) = \Phi_{e,s}(A_s \oplus C_s; x)$$

with use u;

(3) $y > u, 2i$;

(4) $y \in W_{i,s}$, and $W_{i,s} \cap E_{e,s} = \emptyset$;

(5) $(\exists z < y)[z \in B_{s+1} - B_s]$.

If $N_{e,i}$ is the highest priority N requirement to require attention, begin an attack on $N_{e,i}$ for putting x into D and setting the restraint $r(\langle e, i\rangle, s+1) = u$. If C later changes an argument $\leq u$, put y in E_e. One can then prove that each requirement receives attention finitely often; each requirement $P_{e,i}$ is met; and $D \neq \Phi_e(A \oplus C)$, for any e.

§8. NONBRANCHING DEGREES

An r.e. degree $\mathbf{a}$ is called <u>branching</u> if there are r.e. degrees $\mathbf{b}$ and $\mathbf{c}$ different from $\mathbf{a}$ such that $\mathbf{a}$ is the infimum of $\mathbf{b}$ and $\mathbf{c}$, and $\mathbf{a} < \mathbf{0}'$ is <u>nonbranching</u> otherwise. For example, in §4 we used the minimum pair method to prove that $\mathbf{0}$ is branching, and it follows from the lattice embedding in §5 that there are many other branching degrees. We now turn to nonbranching degrees. Such a degree was first constructed by Lachlan [1966, p. 554]. Extensions of this nonbranching method played a key role in refuting the embedding conjecture [Lachlan-Soare, 1980] by proving that any r.e. degree which is the maximum element of a lattice of the form M_5 of Figure 5.1 must be nonbranching. Recently, Peter Fejer [1980] has combined the nonbranching degree construction with the density theory to prove a density theorem for nonbranching degrees, namely for any r.e. degrees $\mathbf{c} < \mathbf{d}$ there is a nonbranching (r.e.) degree $\mathbf{a}$ such that $\mathbf{c} < \mathbf{a} < \mathbf{d}$. It follows that the nonbranching degrees generate (under $\cup$) all the nonzero r.e. degrees. This is the first nontrivial definable subset of the r.e. degrees known to be dense and hence to generate the r.e. degrees. As suggested by Lerman, we give the r.e. degrees the order topology where a typical subbasic open set has the form $[0, \mathbf{a}) = \{\mathbf{b} : \mathbf{b} < \mathbf{a}\}$ or $(\mathbf{a}, \mathbf{0}'] = \{\mathbf{b} : \mathbf{b} > \mathbf{a}\}$. The branching degrees together with $\mathbf{0}'$ are precisely the isolated points. It follows from the Fejer density theorem above for nonbranching degrees that the Cantor-Bendixion rank of the r.e. degrees with this topology is 1.

We present now the basic method for constructing nonbranching degrees. It is a finite injury argument and closely resembles the Friedberg-Muchnik method for satisfying a requirement of the form $B \neq \{e\}^A$, but it also requires the permitting method to insure that B is recursive in certain sets $\hat{W}_i$, $\hat{W}_j$.

Theorem 8.1 (Lachlan 1966). For any nonrecursive r.e. degree $\underset{\sim}{c}$ there is a nonbranching (r.e.) degree $\underset{\sim}{a} \not\geq \underset{\sim}{c}$.

Proof. We will construct an r.e. set A and define $\underset{\sim}{a} = \deg(A)$. To make $C \not\leq_T A$ we will meet for each e the negative requirement

$$N_e: \quad C \neq \{e\}^A$$

which is done exactly as in Theorem 2.1 using the restraint function $r(e, s)$ defined there. For any r.e. set W_e, define $\hat{W}_e = W_e \oplus A$. To make A nonbranching we must insure that if

$$(8.1) \qquad A <_T \hat{W}_i \quad \text{and} \quad A <_T \hat{W}_j$$

then $\deg(A)$ is not the infimum of $\deg(\hat{W}_i)$ and $\deg(\hat{W}_j)$, namely there is an r.e. set $B_{i,j}$ such that

$$(8.2) \qquad B_{i,j} \leq_T \hat{W}_i \quad \text{and} \quad B_{i,j} \leq_T \hat{W}_j \ ; \quad \text{and}$$

$$(8.3) \qquad B_{i,j} \not\leq_T A.$$

For (8.3) we must meet for each e the requirement

$$(8.4) \qquad R_{\langle e,i,j\rangle}: \quad B_{i,j} \neq \{e\}^A .$$

We sketch first the strategy for meeting a single such requirement R_n, $n = \langle e, i, j\rangle$, for fixed e, i, and j. Let $T_n = \omega^{(n)} = \{\langle x, y\rangle : x = n\}$, so $\{T_n\}_{n \in \omega}$ is a partition of ω into infinite recursive sets. For notational convenience drop the subscripts i, j on $B_{i,j}$.

We attempt to meet R_n just as in the usual Friedberg-Muchnik procedure. Namely, we:
(1) choose a fresh witness $x \in T_n$ (not restrained by any requirement of higher priority); (2) wait for a stage s such that

$\{e\}_s^{A_s}(x)\downarrow = z$; (3) define a <u>restraint function</u> $q(n,s) = u_x^s = u(A_s; e, x, s)$, and restrain with priority R_n any $z \leq q(n,s)$ from entering A; (4) enumerate x in B_{s+1} iff $z = 0$; thereby guaranteeing $B(x) \neq \{e\}^A(x)$. The problem is that to insure $B \leq_T \hat{W}_i$, and $B \leq_T \hat{W}_j$, we must not put x into B unless we simultaneously put a certain <u>trace</u> y_x into A (and therefore into $\hat{W}_i$ and $\hat{W}_j$). However, doing so would be useless in preserving (2) unless $y_x > u_x^s$, else the computation $\{e\}_s^{A_s}(x) = t$ would be destroyed. Thus we have for each x, a "movable marker" Γ_x whose position at the end of stage s, Γ_x^s, denotes our current candidate for the trace y_x. The motion of Γ_x will satisfy:

(8.5) $\qquad \Gamma_x^{s+1} \neq \Gamma_x^s \Rightarrow (\exists z \leq x)[z \in W_{i,s+1} - W_{i,s}]$; and

(8.6) $\qquad x \in B_{s+1} - B_s \Rightarrow \Gamma_x^s \in A_{s+1} - A_s$.

It follows from (8.5) that Γ_x moves only finitely often, so $\Gamma_x^\omega =_{dfn} \lim_s \Gamma_x^s < \infty$. Now by (8.5), $\lambda x[\Gamma_x^\omega]$ is a function recursive in W_i. Furthermore, $B \leq_T \hat{W}_i$ since $x \in B$ iff $x \in B_s$ where $s = \mu t[A_t[\Gamma_x^\omega] = A[\Gamma_x^\omega]]$ because of (8.6). We will have $B \leq_T W_j$ by the usual simple permitting, namely

$$x \in B_{s+1} - B_x \equiv (\exists z \leq x)[z \in W_{j,s+1} - W_{j,s}].$$

The hypothesis $A <_T \hat{W}_i$ is used to move Γ_x (according to (8.5)) to some element $y_x > u_x^s$ before performing the above Friedberg-Muchnik procedure, as we now explain.

At stage $s = 0$ we place (for each y) marker Γ_y on $\langle n, y\rangle$, the yth element of T_n, and we set $A_0 = B_0 = \emptyset$.

<u>Stage s+1.</u>

<u>Step 1.</u> For R_n find the least $x \in T_n - B_s$, $x \leq s$, (if such exists) such that:

(8.7) $\quad \{e\}_s^{A_s}(x)\downarrow$; and

(8.8) $\quad (\exists v \le x)[v \in W_{i,s+1} - W_{i,s}]$, (i.e., W_i permits on x).

Now move Γ_x to the least $y \in T_n - A_s$ such that: $y > u_x^s = u(A_s; e, x, s)$; $y \ge \Gamma_x^s$; and y is not restrained with higher priority (i.e., $y > \max\{r(m, s), q(m, s)\}$ for all $m \le n$). Also move markers Γ_z, $z > x$, in order to fresh elements of $T_n - A_s$.

Step 2. We say that x is eligible if $x \le s$; $x \in T_n - B_s$; (8.7) holds for x; and

(8.9) $\quad u(A_s; e, x, s) < \Gamma_x^{s+1}$.

Choose the least eligible x (if such exists) such that for all $m < n$,

$$\Gamma_x^s \ge \max\{r(m, s), q(m, s)\}, \quad \text{and}$$

(8.10) $\quad (\exists v \le x)[v \in W_{j,s+1} - W_{j,s}]$ (i.e., W_j permits on x).

If x exists we say that R_n requires attention. Enumerate Γ_x^{s+1} in A (and therefore in $\hat{W}_i$ and $\hat{W}_j$); define $q(n, s) = u(A_s; e, x, s)$; and enumerate x in B iff $\{e\}_s^{A_s}(x) = 0$. In this case we say requirement R_n receives attention at stage s+1.

If R_n receives attention it remains satisfied and does not receive further attention unless it is later injured by a higher priority requirement. To see that R_n will eventually receive attention assume $A <_T \hat{W}_i$ and $A <_T \hat{W}_j$ but $B = \{e\}^A$. For each x let $u_x = \lim_s (A_s; e, x, s)$, and $s_x = (\mu t)(\forall s \ge t)[u_x = u(A_s; e, s, x)\}$. Since $\lambda x[s_x]$ is an A-recursive function the following set is A-r.e.

$$U = \{x : x \in T_n \;\&\; (\exists s \ge s_x)(\exists z \le x)[z \in W_{i,s+1} - W_{i,s}]\}.$$

Now if U were finite then $W_i \le_T A$ since for almost every z, z $z \in W_i$ iff $z \in W_{i,s_x}$ where $x = (\mu y > z)[y \in T_n]$. Hence, U is infinite; step 1 is performed on each $x \in U$ at some stage $t_x \ge s_x$;

and x is eligible at every stage $s \geq t_x$. Now since $\{t_x : x \in U\}$ is A-r.e., so is the set

$$V = \{x : x \in U \ \& \ (\exists s \geq t_x)(\exists z \leq x)[z \in W_{j,s+1} - W_{j,s}]\}.$$

But V is infinite else $W_j \leq_T A$. Now for at most finitely many $x \in V$ is $\Gamma_x^{t_x}$ restrained from A with priority higher than R_n. Hence, R_n eventually receives attention via some $x \in V - B$, becomes satisfied; and remains satisfied thereafter (unless injured by a higher priority requirement R_m, $m < n$).

(Notice that we do not restrain $A[u_x]$ when step 1 is performed and Γ_x is moved for the sake of R_n. Hence, lower priority requirements may enumerate elements $z \leq u_x$ into A. Thus x may become alternately eligible and ineligible many times before step 2 is finally performed.)

This completes the description of the strategy for a single requirement. To handle all requirements simultaneously we restore i and j as subscripts to $B_{i,j}$ and superscripts to $\Gamma_x^{i,j}$. At stage s+1 perform step 1 for each requirement R_n, $n \leq s$. At step 2 choose the least n such that R_n requires attention and perform step 2 on R_n.

<u>Lemma</u>. $(\forall n = \langle e, i, j \rangle)[C \neq \{n\}^A$ and $[A <_T \hat{W}_i \ \& \ A <_T \hat{W}_j$

$\Rightarrow B_{i,j} \neq \{e\}^A]$ and R_n receives attention at

most finitely often].

<u>Proof</u>. Fix n and assume for all m < n that the above hold. Thus, the limits $r(m) = \lim_s r(m, s)$ and $q(m) = \lim_s q(m, s)$ exist for all m < n. Furthermore, R_n is injured at most finitely often and so is eventually satisfied forever . Thus, N_n is injured finitely often and is finally satisfied as in §2.

Remark. Carl Jockusch noticed that this proof actually establishes that for every nonrecursive r.e. set C there exists an r.e. set A, $C \not\leq_T A$, such that

$$[W_i \not\leq_T A \;\&\; W_j \not\leq_T A] \Rightarrow$$

$$(\exists \text{ r.e. } B_{i,j} \not\leq_T A)[B_{i,j} \leq_T A \oplus W_i \;\&\; B_{i,j} \leq_T W_j].$$

This work was partially supported by NSF Grant MCS-76-07033.

REFERENCES

Arslanov, M.R., Nadirov, R.F., & Solovev, V.D. (1977). Completeness criteria for recursively enumerable sets and some general theorems on fixed points. Matematica University News, 179, no. 4.

Dekker, J. (1954). A theorem on hypersimple sets. Proc. Amer. Math. Soc., 5, pp. 791-796.

Fejer, P.A. (1980). The structure of definable subclasses of the recursively enumerable degrees. Ph.D. Dissertation, University of Chicago.

Friedberg, R.M. (1957). Two recursively enumerable sets of incomparable degrees of unsolvability. Proc. Natl. Acad. Sciences, U.S.A., 43, pp. 236-238.

Lachlan, A.H. (1966). Lower bounds for pairs of r.e. degrees. Proc. London Math. Soc., 16, no. 3, pp. 537-569.

Lachlan, A.H. (1968). Complete recursively enumerable sets. Proc. Amer. Math. Soc., 19, pp. 99-102.

Lachlan, A.H. (1972). Embedding nondistributive lattices in the recursively enumerable degrees. Conf. Math. Logic, London London 1970. Lecture Notes in Math., no. 255, Springer-Verlag, Berlin and New York, 1972, pp. 149-177.

Lachlan, A.H. (1973). The priority method for the construction of recursively enumerable sets. Proc. Cambridge Summer School in Logic, 1971. Springer-Verlag, Berlin and New York.

Lachlan, A.H. (1975). A recursively enumerable degrees which will not split over all lesser ones. Ann. Math. Logic, 9, pp. 307-365.

Lachlan, A.H. (1980). Decomposition of recursively enumerable degrees. Proc. Amer. Math. Soc., to appear, 1980.

Lachlan, A.H. & R.I. Soare. (1980). Not every finite lattice is embeddable in the recursively enumerable degrees, to appear in Adv. in Math., 1980.

Marchenkov, S.S. (1976). A class of partial sets. Mathematicheskie Zametki, 20, no. 4, pp. 473-478.

Martin, D.A. (1966). Completeness, the recursion theorem, and effectively simple sets. Proc. Amer. Math. Soc., 17 938-842.

Muchnik, A.A. (1956). On the unsolvability of the problem of reducibility in the theory of algorithms. Doklady Akademii Nauk. (Russian), n. s., 108, pp. 194-197.

Post, E.L. (1944). Recursively enumerable sets of positive integers and their decision problems. Bull. Amer. Math. Soc., 50, pp. 284-316.

Robinson, R.W. (1971a). Interpolation and embedding in the recursively enumerable degrees. Annals Math., 93, pp. 285-314.

Robinson, R.W. (1971b). Jump restricted interpolation in the r.e. degrees. Annals Math., 93, pp. 586-596.

Rogers, H. Jr. (1967). Theory of Recursive Functions and Effective Computability. McGraw-Hill, New York.

Sacks, G.E. (1963). On the degrees less than 0'. Annals Math., 77, pp. 211-231.

Sacks, G.E. (1963). Recursive enumerability and the jump operator. Trans. Amer. Math. Soc., 108, pp. 223-239.

Sacks, G.E. (1964). The recursively enumerable degrees are dense. Annals of Math., 80, no. 2, pp. 300-312.

Sacks, G.E. (1966). Degrees of unsolvability. Annals of Math. Studies, rev. ed. No. 55, Princeton University Press, Princeton, New Jersey.

Shoenfield, J.R. (1961). Undecidable and creative theories. Fundamenta Mathematicae, 49, pp. 171-179.

Shoenfield, J.R. (1971). Degrees of Unsolvability. North-Holland, Amsterdam.

Shoenfield, J.R. & Soare, R.I. (1978). The generalized diamond theorem. (Abstract) Recursive Function Theory Newsletter, 19, no. 219.

Soare, R.I. (1976). The infinite injury priority method. J. Symbolic Logic, 41, pp. 513-530.

Soare, R.I. (1977). Computational complexity, speedable and levelable sets. J. Symbolic Logic, 42, pp.545-563.

Soare, R.I. (1978). Recursively enumerable sets and degrees. Bull. A.M.S., 84, no. 6, pp. 1149-1181.

Soare, R.I. (1980). Constructions in the recursively enumerable degrees. C.I.M.E. Conference on "Recursion Theory and computational complexity", Bressanone, Italy, June 1979, to appear in 1980.

Soare, R.I. (1982). Recursively enumerable sets and degrees. Springer-Verlag (Omega Series), Berlin and New York, to appear in 1982.

Thomason, S.K. (1971). Sublattices of the recursively enumerable degrees. Z. Math. Logik und Grundlagen der Math., 17, pp. 273-280.

Yates, C.E.M. (1965). Three theorems on the degree of recursively enumerable sets. Duke Math. J., 32, pp. 461-468.

Yates, C.E.M. (1966a). A minimal pair of r.e. degrees. J. Symbolic Logic, 31, pp. 159-168.

Yates, C.E.M. (1966b). On the degrees of index sets. Trans. Amer. Math. Soc., 121, pp. 309-328.

Yates, C.E.M. (1969). On the degrees of index sets, II. Trans. Amer. Math. Soc., 135, pp. 249-266.

A SURVEY OF NON-R.E. DEGREES ≤ 0'

David B. Posner [1]
San Jose State University

CONTENTS

[1] The author acknowledges the support of NSF grant MCS 7903379

D. Posner

Part II: METHODS

A. Oracle Constructions
   1. Additional Notation and Terminology
   2. The Sacks Minimal Degree Construction
   3. Sasso's Cornucopia of Minimal Degrees
   4. Complementary Minimal Degrees
   5. Minimal Degrees Below High Degrees
   6. Minimal Degrees are GL_2
   7. The Join Theorem

B. Full Approximation Constructions
   1. Cooper's Minimal Degree Construction
   2. Minimal Degrees and R.E. Degrees
   3. Minimal Degrees Below High Degrees

INTRODUCTION

Most of the research on the degrees below 0' has focused on the r.e. degrees. In part, this is a result of the origins of degree theory in logic where it was hoped that degrees of unsolvability would serve as a useful classification of the complexity of axiomatizable theories. Since we will be concentrating on the non-r.e.degrees below 0', it seems appropriate to begin with an example which illustrates that non-r.e. degrees below 0' also arise naturally in logic.

Complete Theories and Degrees Below 0'

Let T be some essentially undecidable axiomatizable first-order theory (e.g., first-order Peano arithmetic or Zermelo-Fraenkel set theory). Since T is essentially undecidable, no complete extension of T is r.e. How constructive can complete extensions of T be?

Consider the following construction of a complete extension of T. Fix some effective listing $u_0, u_1, \ldots$ of the sentences of the language of T and let $v_0, v_1, \ldots$ be an effective listing of T. Our construction takes place in stages. At stage s we effectively define a set of sentences C_s which is, in effect, a finite approximation to a complete extension of T. More precisely, we require that for each $i < s$ either u_i or $\sim u_i$ is in C_s and we require that no outright contradiction, i.e., $u\&\sim u$ for some sentence u, be derivable from $C_s \cup \{v_i: i \le s\}$ by a proof with Gödel number ≤ s.

Since no complete extension of T is r.e. it is not possible to define (effectively) C_s as described above in such a way that $C_s \subseteq C_{s+1}$ for all s. Thus at some stages a given sentence will be in the approximation and at other stages will be out. However, if we are careful, we can arrange the construction so that for each sentence u either $u \in C_s$ for all sufficiently large s or $u \notin C_s$ for all sufficiently large s. For example, for s > 0 define C_s so as to maximize (subject to the stated requirements) the largest value of i such that for all j < i, $u_j \in C_{s-1}$ iff $u_j \in C_s$. Our construction will thus define a complete extension of T, C, in the limit.

This construction strikes us as being very natural. We proceed effectively, at each step making the best approximation we can based on a necessarily finite amount of information. As the construction proceeds and we gain more information, we discover that some of our earlier actions were incorrect. We cannot in general be _certain_ about any given part of our approximation but we are able proceed so that everything works out right in the end.

How constructive is C from a degree theoretic standpoint? C is an example of a "limit computable" set. In general, let f be a function with a computable domain D. f is said to be _limit computable_ if there is a recursive function g with domain $\omega \times D$ such that for all x in D, $f(x) = \lim_s g(s,x)$. A subset of a computable domain is limit computable if its characteristic function is. By _Shoenfield's Limit Lemma_ (see Shoenfield(1971)) a function is limit computable iff it is recursive in 0'. Thus, C is recursive in 0'.

The fact that every consistent axiomatizable theory has a compete extension which is recursive in 0' was first pointed out by Kreisel (1950). (As an alternative to using the _full approximation construction_ given above, one can prove this using a _0'-oracle construction_, i.e., a construction which employs an oracle for 0'. In fact one can simply use the standard Lindenbaum construction.) Jockusch&Soare(1972) showed that every consistent axiomatizable theory has a complete extension whose degree is not only below 0' but has jump equal to 0'. Degrees with jump 0' are called low degrees and are obviously strictly less than 0'. Going along with this we have, by a result of Scott & Tennenbaum (1960), if the set of T-refutable and the set of T-provable sentences are effectively inseparable (e.g., if T is first-order Peano arithmetic or Zermelo-Fraenkel set theory) then <u>no</u> completion of T of degree $<$ 0' is of r.e. degree.

Motivation

One reason for studying the degrees below 0' is simply that they're there. Certainly 0' is there. 0' is the maximum r.e. degree and the degree of virtually every "natural" axiomatizable undecidable theory. The class of objects computable from 0' also seems to be there. The concept of limit computable is certainly natural and, as we have mentioned, being limit computable is equivalent to being recursive in 0'. Thus, for example, if we use any standard representation of real numbers as functions, we have that the reals which are obtained as limits (in the sense of analysis) of recursive sequences of rationals are exactly the reals recursive in 0'. By Post's theorem (see Rogers (1967), page 314) we can also characterize the functions and sets which are recursive in 0' as those which are Δ_2^0.

The study of the degrees below 0' is intimately related to the study of the jump operator. The jump has been included as a basic object of study in degree theory from the very beginning (i.e., in Kleene & Post (1954)). It provides the basic scale by which we measure the complexity of unsolvable problems. It is thus of interest to know how the jump behaves with respect to the partial ordering of the degrees. The study of the degrees below 0' has shed considerable light on the degree theoretic properties of the jump.

Research on the degrees below 0' has led to the development of a variety of new construction techniques. In constructing degrees below 0' we have more flexibility than we have in constructing r.e. degrees but not nearly the freedom we have in constructingdegrees in the global structure. In many constructions, for example, it is necessary to combine the kinds of priority methods used in the study of the r.e. degrees, with the sorts of forcing methods employed in the global theory. The difficulties encountered in attempting to "constructivize" results from the global theory to the degrees below 0' have led to a better understanding of the nature and complexity of certain kinds of constructions.

A special source of interest in the degrees below 0' lies in the nature of recursive approximations. Here we are interested in the kinds of properties of an approximation which influence the degree theoretic properties of the limit. Why, for example, do r.e. degrees seem to be so special?

Brief History

Along with the rest of degree theory, the study of the degrees below 0' began with Kleene & Post (1954). Using what are in effect forcing constructions (see the article by Jockusch in this volume), they showed that various order types, including the order type of the rationals, are embeddable in the degrees below 0'. In the second paper on degree theory, Spector (1956) obtained several results on the degrees below 0' including the fact that the degrees below 0' do not form a lattice (i.e., there exists a pair of degrees below 0' with no greatest lower bound).

Between 1956, following the solution of Post's problem, and around 1970 there were only a few results on the non-r.e. degrees below 0'. Several of these were extremely important however. Shoenfield (1959) characterized the range of the jump operator on the degrees below 0'. This paper was not only important because of this theorem but also because it included the Limit Lemma and because in involved the first application of the priority method to construct a non-r.e. degree. Two years later, Sacks (1961) used the priority method in order to construct a minimal degree below 0'. Shoenfield (1966) then recast the Sacks minimal degree construction in terms of partial trees and used this simplification to obtain a stronger result on minimal degrees below 0'.

Around 1970, primarily because of work of Yates, Cooper, Robinson, and Sasso , interest in the non-r.e. degrees below 0' increased greatly. The intensity of this research has increased steadily ever since. One thing to note from this is that the study of the non-r.e. degrees below 0' is still fairly new. Almost all of the results have appeared within the last ten years and most of these have appeared within the past five years.

Notation

The letters a,b,c,d are reseved for degrees of unsolvability. Unless otherwise specified other small Roman letters will denote natural numbers. Capital Roman letters will generally denote subsets of $\omega = \{0,1, \ldots \}$. 2^{ω} is the power set of ω. We identify a set of natural numbers A with its characteristic function

$$A(x) = \begin{cases} 1 \text{ if } x \in A \\ 0 \text{ if } x \notin a \end{cases}$$

We write $\leq_T$, $'$, $\oplus$ for Turing reducibility, jump, and join respectively on subsets of ω, and $\leq$, $'$, $\vee$ for the induced ordering and operations on the degrees. Iterates of the jump are defined by $a^{(0)} = a$, $a^{(n+1)} = (a^{(n)})'$. We write $a \wedge b = c$ if c is the greatest lower bound of a and b. (Not every pair of degrees has a greatest lower bound however.) $\mathcal{D}$ denotes the degrees of unsolvability and for any degree a, $\mathcal{D}(\leq a)$ is the set of degrees $\leq$ a, and similarly for $\mathcal{D}(\geq a)$. We will also use interval notation. Thus $\mathcal{D}[a,b]$ is the set of degrees $\geq$ a and $\leq$ b, $\mathcal{D}(a,b]$ is the set of degrees > a and $\leq$ b, etc.

The jump operator defines a natural classification of degrees below 0' into high and low classes. For each n $\geq$ 1 let H_n be the class of degrees $\leq$ 0' whose n-th jump is the n-th jump of 0' (i.e., $0^{(n+1)}$) and let let L_n be the class of degrees < 0' whose n-th jump is the n-th jump of 0. Thus, the degrees in H_n have highest possible n-th jump and the degrees in L_n have lowest possible n-th jump (for degrees $\leq$0' of course). The degrees in L_1 are called <u>low</u> degrees and the degrees in H_1 are called high degrees. Note that $H_i \subseteq H_j$ and $L_i \subseteq L_j$ whenever $i \leq j$ and that $H_i \cap L_j$ is empty for all i and j. By theorems of Sacks (See Sacks (1966)) the inclusions above are strict when $i < j$ and by a theorem of Lachlan and (independently) Martin (see Rogers(1967)) these classes do not exhaust the degrees below $0'$.

The classification described above can be extended to degrees not lying below $0'$ as follows. For each $n \geq 1$ let $GL_n = \{a: a^{(n)} = (a \vee 0')^{(n-1)}\}$ and let $GH_n = \{a: a^{(n)} = (a \vee 0')^{(n)}\}$. We again have $GL_i \subseteq GL_j$ and $GH_i \subseteq GH_j$ whenever $i \leq j$ and $GL_i \cap GH_j$ is empty for all i and j. Also, GH_i and H_i coincide on the degrees below $0'$ and similarly for L_i and GL_i. (Intuitively, these generalized high and low classes measure relative completeness. Degrees in GH_1 are close to complete and degrees in GL_1 are far from complete.)

We will occasionally make reference to 1-generic degrees and their properties. The reader should consult the article by Carl Jockusch in these proceedings for a detailed discussion. We will make use of the following facts about 1-generic degrees. There exist 1-generic degrees below $0'$. Every 1-generic degree is in GL_1. If a is 1-generic then $\mathcal{D}(\leq a)$ is not a lattice, every countable partial ordering can be embedded in $\mathcal{D}(\leq a)$, and every countable distributive lattice is lattice embeddable in $\mathcal{D}(\leq a)$. There exist non-0 degrees b and $c \leq a$ such that $b \wedge c = 0$.

Organization

The paper is divided into two parts. Part I surveys the known results (i.e., theorems) on the non-r.e. degrees below $0'$. Part II deals with the methods of proof. The author chose this taxonomy primarily because he felt that it would be useful to have a relatively concise catalogue of the known theorems on $\mathcal{D}(\leq 0')$. It should be pointed out however, that in the view of most who study degrees below $0'$ (including the author) "methods of proof" are results in their own right.

D. Posner

Part I: RESULTS

In this part of the paper we will survey the basic known results on the structure of the degrees below 0'. In §A we will look at the purely order theoretic properties of $\mathcal{D}(\leq 0')$. Where appropriate, we will contrast this partial ordering with that of the r.e. degrees. Both are upper semilattices but beyond that they have little in common.

In §B we discuss results which concern the jump operator. Finally, in §C we will look at results concerning the relationship between the properties of a Δ_2^0 set and the properties of its degree. Throughout Part I we have included problems which, at least to the author's knowledge, are open.

A. Order Theoretic Results

1 Embeddings. Since there are 1-generic degrees below 0' and every countable partial ordering can be embedded in the degrees below a 1-generic degree, every countable partial order is embeddable in $\mathcal{D}(\leq 0')$. In fact, the same is true in the r.e. degrees. (See Sacks (1966).)

By a lattice embedding we mean an order embedding of a lattice which preserves joins and meets. Again since there are 1-generic degrees below 0', we have that any countable distributive lattice is lattice embeddable in the degrees below 0'. By a theorem of Thomason and (independently) Lachlan the same is true in the r.e. degrees. It would seem to be almost certainly true that any countable lattice is lattice embeddable in $\mathcal{D}(\leq 0')$ but, as far as we know, this has not been proved. It does follow from a very difficult theorem of Lerman (1978),(TA) discussed below that every finite lattice is lattice embeddable in $\mathcal{D}(\leq 0')$ (in fact as an initial segment), but there should be an easier proof. In contrast, Lachlan & Soare (see the article by Soare in this volume) have shown that there is a finite lattice which is not lattice embeddable in the r.e. degrees.

2 Ideals. An ideal of degrees is an initial segment of degrees (i.e., a downward closed set of degrees) which is closed under joins. An ideal of degrees is principal if it is of the form $\mathcal{D}(\leq a)$ for some a. Obviously, ideals of degrees are upper semilattices. Lachlan & Lebeuf (1976) showed that every countable upper semilattice is isomorphic to an ideal of degrees. There has been considerable interest in determining which ideals are possible in the degrees below 0'.

There are certain known restrictions on ideals below 0'. Let a be a degree. An upper semilattice $(L, \leq_L, \vee_L)$ is said to be a-presentable if there exists an a-recursive binary relation R and an a-recursive binary function f such that $(L, \leq_L, \vee_L)$ is isomorphic to (ω, R, f). Yates (1970a) observed that an ideal of the form $\mathcal{D}(\leq a)$ is $a^{(3)}$-presentable as an upper semilattice and thus any pricipal ideal in $\mathcal{D}(\leq 0')$ must be $0^{(4)}$-presentable as an upper semilattice. Further, by a result of Jockusch & Posner (1978), if a is not GL_2 then a bounds a 1-generic degree and so, since the degrees below a 1-generic degree do not form a lattice, $\mathcal{D}(\leq a)$ is not a lattice. It follows that any principal ideal below 0' which is a lattice must be $0^{(3)}$-presentable.

As mentioned in the introduction, Sacks (1961) showed that there is a two element ideal below 0', i.e. a minimal degree below 0'. Several people including Yates (1970a), Epstein (TA), and Lerman (1978),(TA) have worked on obtaining stronger positive results. The strongest result so far is due to Lerman: every $0^{(2)}$-presentable upper semilattice is isomorphic to an ideal of $\mathcal{D}(\leq 0')$. Thus, for example, any recursive linear ordering is isomorphic to an ideal in the degrees below 0'. One consequence of this is that there are degrees below 0' with no minimal predecessors. It would be interesting to know if this is true of 1-generic degrees.

3 The First-Order Theory. Shore (1979), employing the result of Lerman given in 2 above, has shown that the first-order theory of $(\mathcal{D}(\leq 0'), \leq)$ is recursively isomorphic to the first-order theory of true arithmetic. (The undecidability of $(\mathcal{D}(\leq 0'), \leq)$ was first obtained

independently by Lerman (1979) and Epstein (TA).) Shore's result was first conjectured by Simpson (1977a) as the local analogue to his result (1977b) that the first-order theory of $(\mathcal{D},\leq)$ is recursively isomorphic to the second-order theory of true arithmetic.

Shore (1979) has also shown that $(\mathcal{D}(\leq 0'),\leq)$ is not elementarily equivalent to either $(\mathcal{D}(\leq 0''),\leq)$ or to $(\mathcal{D}[0',0''],\leq)$. The latter inequivalence is analogous to another result of Shore (TA): there is a degree a such that $(\mathcal{D}(\geq a),\leq)$ is not elementarily equivalent to $(\mathcal{D},\leq)$. The degrees thus fail to be homogeneous or even locally homogeneous.

4 $\mathcal{D}(\leq a)$ for a Non-recursive R.E. Many results on the degrees below 0' have been shown to hold in the degrees below an arbitrary non-recursive r.e. degree. By relativizing such results we obtain information about the structure of "cones" in $\mathcal{D}(\leq 0')$, i.e., sets of degrees of the form $\mathcal{D}[b,0']$ for $b < 0'$. For example using r.e. permitting, which we will discuss in §B2 of Part II, it is quite easy to show that every non-recursive r.e. degree bounds a 1-generic degree (see Epstein (1979), pp. 145 - 150). Thus any countable partial ordering can be embedded in the degrees below a non-recursive r.e. degree and any non-recursive r.e. degree bounds a minimal pair of degrees, i.e., a pair of degrees with meet equal to 0. Using the fact that 0' is r.e. and so r.e. *relative* to any degree, one can relativize these results to conclude that if $b < 0'$ then every countable partial ordering can be embedded between b and 0' and that b is a "branching" degree in $\mathcal{D}(\leq 0')$, i.e., $b = c \wedge d$ for some c and d strictly between b and 0'. In contrast to the latter result, Lachlan (1966b) showed that there exist r.e. degrees less than 0' which are non-branching in the r.e. degrees.

Yates (1970b) showed that every non-recursive r.e. degree bounds a minimal degree. Relativizing this result we conclude that every degree $< 0'$ has a minimal cover in the degrees below 0'. Not too much beyond Yates' result is known about ideals below arbitrary non-recursive r.e. degrees. Epstein (TA) has been able to embed certain linear orderings as initial segments below arbitrary non-recursive r.e. degrees and, using these results, has been able to show that the first-order theory of $\mathcal{D}(\le a)$ under ≤ is undecidable for a high and r.e. Epstein (1979) conjectures that if a is high r.e. then $\mathcal{D}(\le a)$ is isomorphic to $\mathcal{D}(\le 0')$. As far as the author knows it is possible (though unlikely) that $\mathcal{D}(\le a)$ is isomorphic to $\mathcal{D}(\le 0')$ if a is <u>any</u> non-recursive r.e. degree.

5 <u>Joins, Meets, and Complements</u>. a and b are said to be <u>complementary</u> in $\mathcal{D}(\le 0')$ if $a \vee b = 0'$ and $a \wedge b = 0$. Posner (TAa), extending results of Robinson (see Posner & Robinson (TA)), Epstein (1975), and Posner(TAb), has shown that $\mathcal{D}(\le 0')$ is complemented, i.e., for every degree a ≤ 0' there is a degree b such that a and b are complementary in $\mathcal{D}(\le 0')$. This is in strong contrast with the situation in the r.e. degrees where, as Lachlan (1966a) has shown, there are <u>no</u> complementary pairs other than 0 and 0'. Cooper (1972a) produced a pair of minimal degrees which are complementary in $\mathcal{D}(\le 0')$. It is not known whether every degree a strictly between 0 and 0' has a complement which is a minimal degree. This is the case if a is r.e. (Epstein (1975)) or high (Posner (TA)).

A set of degrees $\mathcal{A}$ is said to be <u>uniformly recursive</u> in a degree a if there exists a collection of sets $\{A_n: n\in\omega\}$ uniformly recursive in a such that $\mathcal{A} = \{\deg(A_n): n\in\omega\}$. Sasso (1970), extending a result of Shoenfield (1959), showed that if $\mathcal{A}$ is uniformly recursive in 0' then there is a minimal degree m ≤ 0' such that $m \wedge a = 0$ for every $a \in \mathcal{A} \sim \{0'\}$. This implies, for example, that there is a minimal degree below 0' which forms a minimal pair with every r.e. degree. In contrast, Yates (1965) showed that there is an r.e. degree a < 0' which does not form a minimal pair with <u>any</u> non-0 r.e. degree.

Posner & Robinson (TA) have proved the following "join theorem" for the degrees below 0'. Let $\mathcal{A}$ and $\mathcal{B}$ be sets of degrees uniformly recursive in 0' such that $0 \notin \mathcal{A}$. Then there is a low degree c (in particular c < 0') such that $c \vee a = 0'$ for all $a \in \mathcal{A}$ and $c \not\ge b$ for all $b \in \mathcal{B} \sim \{0\}$. It is not known whether there is a single degree which joins every non-0 r.e. degree to 0'.

6 Automorphism Bases. It is not known whether there are any non-trivial automorphisms of $(\mathcal{D}(\leq 0'),\leq)$. (This problem is also open for $\mathcal{D}$ and for the r.e. degrees.) One approach to problems of this sort, suggested in the context of degree theory by M. Lerman (1977), is to look for automorphism bases, where a set of degrees $\mathcal{A}$ is said to form an automorphism base for a set of degrees $\mathcal{B} \supseteq \mathcal{A}$ if the action of any automorphism of $\mathcal{B}$ is determined by its action on $\mathcal{A}$. Note that if $\mathcal{A}$ is an automorphism base for $\mathcal{B}$ then any automorphism of $\mathcal{B}$ which leaves the elements of $\mathcal{A}$ fixed must in fact be the identity function on $\mathcal{B}$. Thus, one could show that there are no non-trivial automorphisms of $\mathcal{B}$ if one could produce an automorphism base for $\mathcal{B}$ whose elements are necessarily fixed by automorphisms of $\mathcal{B}$. One way of showing that $\mathcal{A}$ is an automorphism base for $\mathcal{B}$ is to show that $\mathcal{A}$ generates $\mathcal{B}$ under joins and meets.

Jockusch & Posner (TA) have shown that any "sufficiently large", in a sense of category, set of degrees below 0' generates the degrees below 0'. We explain. Let 2^{ω} have its usual product topology. This induces a notion of category for sets of degrees by identifying a set of degrees $\mathcal{A}$ with $\{A: \deg(A) \in \mathcal{A}\}$. Recall that a subset of 2^{ω}, $\mathcal{A}$, is comeager iff player II has a winning strategy in the Banach-Mazur game for $2^{\omega} \sim \mathcal{A}$. (See, for example, Oxtoby (1971).) This led Yates (1976) to formulate the following constructive notion of category. Let a be a degree. $\mathcal{A}$ is said to be a-comeager if player II has an a-recursive winning strategy in the Banach-Mazur game for $2^{\omega} \sim \mathcal{A}$. A set of degrees $\mathcal{A} \subseteq \mathcal{D}(\leq 0')$ is said to be comeager in $\mathcal{D}(\leq 0')$ of $\mathcal{A} = \mathcal{B} \cap \mathcal{D}(\leq 0')$ for some 0'-comeager set of degrees $\mathcal{B}$. Jockusch & Posner (TA) showed that any set of degrees comeager in $\mathcal{D}(\leq 0')$ generates and so forms an automorphism base for the degrees below 0'. This result implies, for example, that the 1-generic degrees below 0' generate the degrees below 0'. Another consequence is that there exist degrees a,b in $\mathcal{D}(0,0')$ such that $\mathcal{D}[a,0'] \cup \mathcal{D}[b,0']$ generates the degrees below 0'. It is not known if there is a degree a > 0 such that $\mathcal{D}[a,0']$ is an automorphism base for $\mathcal{D}(\leq 0')$. (It follows from results in Jockusch & Posner (TA) and a recent result of Harrington & Shore (1980) that there is a degree a > 0 such that $\mathcal{D}(\geq a)$ is an automorphism base for $\mathcal{D}$.)

Jockusch & Posner (TA) have shown that for any $a \leq 0'$, $\{b \leq 0' : b' = a'\}$ generates $\mathcal{D}(\leq 0')$. Thus any of the H_n or L_n classes or the degrees below 0' not lying in any of these classes generates $\mathcal{D}(\leq 0')$. Posner (TAb) shows that the set of minimal degrees below 0' generates $\mathcal{D}(\leq 0')$ and, in fact, if $b < 0'$ then the set of minimal degrees below 0' forming a minimal pair with b generates the degrees below 0'. Jockusch & Posner (TA) show that if $0 < b \leq 0'$ then the set of degrees which join b to 0' generates the degrees below 0'.

B. The Jump Operator and $\mathcal{D}(\leq 0')$

1. Jump Inversion. Shoenfield (1959) showed that the degrees $\leq 0'$ take all "possible" jumps in that for all $c \geq 0'$ and r.e. in 0' there is a degree $a \leq 0'$ such that $a' = c$. In fact, by a theorem of Sacks (1963), a can be taken to be r.e. Further, by relativizing Robinson's Jump Interpolation Theorem (Robinson (1971)) we have that the degrees in any "cone" in $\mathcal{D}(\leq 0')$ take all possible jumps, i.e., if $a \leq 0'$ and $c \geq a'$ and r.e. in 0' then there exists $b \in \mathcal{D}[a,0']$ with $b' = c$.

2. Minimal Degrees and the Jump Operator. Cooper (1973) showed that no minimal degree is high. Thus, the range of the jump operator on the minimal degrees below 0' does not coincide with its range on all of $\mathcal{D}(\leq 0')$. However, in this same paper, Cooper also showed that minimal degrees in $\mathcal{D}$ do take all possible jumps, i.e., every degree $\geq 0'$ is the jump of a minimal degree. There thus seemed to be a stronger relationship between the order theoretic properties of a degree and its jump in $\mathcal{D}(\leq 0')$ than exists in $\mathcal{D}$. The resolution of this apparent disharmony was found in the generalized jump hierarchy. Jockusch & Posner (1978) showed that every minimal degree is in GL_2. Thus every minimal degree below 0' is low_2 (i.e., in L_2). Sasso (1974) showed that there are minimal degrees in $L_2 \sim L_1$ and of course it follows from the fact that every nonrecursive r.e. degree bounds a minimal degree (Yates (1970b)) that there are low minimal degrees. A complete characterization of the jumps of minimal degrees below 0' has not been given. Jockusch has conjectured (even before Jockusch & Posner (1978)) that minimal degrees $\leq 0'$ take all jumps consistent with being in L_2.

3. Non-L_2 Degrees. Jockusch & Posner (1978) have obtained the following results for any degree a in $\mathcal{D}(\le 0')\sim L_2$. a bounds a 1-generic degree. For every c ≥ a there is a 1-generic degree b < c such that a∨b = c. Any degree c ≥ 0' and r.e. in a is the jump of a degree below a. The first two results hold for any non-GL_2 degree. The last does as well if "c ≥ 0' " is replaced with "c ≥ (a∨0') ". It is not known whether every non-L_2 degree below 0' bounds a minimal degree. Jockusch has produced a non-GL_2 degree which has no minimal predecessor however.

4. High Degrees. Let a be high or, more generally, GH_1. Of course a is non-GL_2 so the results of 3 above apply. In addition a has the following properties. a bounds a minimal degree. (Cooper (1973) proved this for high degrees. Jockusch (1977)extended the result to arbitrary GH_1 degrees.) a is in fact the join of a pair of minimal degrees (Posner(TAb)). Further, the join theorem of Posner & Robinson (TA) (see §IA5) holds when 0' is replaced by a (Posner (TAc)). A consequence (pointed out by Jockusch) of the last result mentioned is that every GH_1 degree is high and r.e. relative to some predecessor (see Posner (1977)). Epstein (1979) is able to conclude from this that the first-order theory of $\mathcal{D}(\le a)$ is undecidable when a is in GH_1. It seems likely that the result of Lerman (1979) on ideals in $\mathcal{D}(\le 0')$ (see §IA2) will extend to $\mathcal{D}(\le a)$ for a in GH_1. It would then follow as in Shore (1979) that the first-order theory of $\mathcal{D}(\le a)$ for a in GH_1 is at least as complex as the first-order theory of true arithmetic.

5 The Question of the Definability of the Jump. A major question in degree theory is whether or not the jump operator is order theoretically definable. The results which have been discussed show that there are at least close ties between the ordering of the degrees and the jump. An obvious first step toward proving that the jump is definable would be to show that 0' is definable. (Given that most results in degree theory do seem to relativize it is likely that any proof of the definability of 0' would in fact give the definability of the jump.) It seems to the author that there is good reason to believe that 0' and in fact the jump are definable.

For example, by an observation of Jockusch (see Posner & Robinson (TA)), it follows from the relativization of the join theorem for $\mathcal{D}(\leq 0')$ that every incomplete degree has a GL_1 successor. Of course, every complete degree is GH_1. Thus, 0' is the least degree which has only GH_1 successors and is also the least degree with no GL_1 successor. Thus, for example, any order theoretic formula distinguishing between GL_1 degrees and complete degrees would yield a definition of 0'.

The author conjectures that 0' is the least degree with the property that every degree above it is the join of a pair of minimal degrees. More specifically, suppose a is not ≥ 0'. We may assume that a is GL_1. Let b be "very generic", for example suppose B ∈ b meets every a-arithmetic dense set of strings (see Jockusch's article in this volume). Then b∨a is a GL_1 successor of a. It seems to us highly unlikely that b∨a can be the join of a pair of minimal degrees.

C. Δ_2^0-Sets and Degrees

1 <u>Computation Functions</u>. Let A be a Δ_2^0 set and let $\{A_s\}$ be an effective approximation for A, i.e., $\{A_s\}$ is uniformly recursive and for all x, $\lim_s A_s(x) = A(x)$. The <u>computation function</u> C for $\{A_s\}$ is defined by

$$C(n) = \mu s[A_s(x) = A(x) \text{ for all } x \leq n].$$

It is important to distinguish the computation function for $\{A_s\}$ from the <u>modulus function</u> M defined by $M(n) = \mu s[\forall t{\geq}s \forall x{\leq}n(A_t(x){=}A(x))]$. The difference is that C(n) gives the <u>first</u> stage at which the approximation is correct for all arguments ≤n while M(n) gives the stage at which the approximation becomes "finally" correct on all arguments ≤n. At first glance this might appear to be a minor distinction but in fact it is not. Notice that we always have deg(C) = deg(A). However, in general, all we can say about M is that $A \leq_T M$. In fact we have the following theorem of Shoenfield (1959): A is of r.e. degree if and only if A has an effective approximation whose modulus function is recursive in A.

Computation functions are important because the

complexity of a Δ_2^0 set is closely related to the rate of growth of its computation functions. (See §IIB2 and IIB3 on r.e. and Δ_2 permitting.) A function f is said to <u>dominate</u> a function g if $f(n) > g(n)$ for all but finitely many n. Miller & Martin (1968) showed that if C is a computation function for A (i.e., the computation function for some effective approximation for A) and f dominates C then A is recursive in f. This of course implies that if A is non-recursive then no recursive function dominates any computation function for A. It follows that every non-0 degree below 0' contains an hyperimmune set (Miller & Martin (1968)).

A Δ_2^0 set A is said to be <u>high</u> if A has a computation function which dominates every recursive function. Martin (1966) showed that a degree a contains a function dominating every recursive function iff $a' \geq 0''$. In particular, any high set is of high degree. Cooper (1973), extending a result of Robinson (1968), showed that in fact a degree is high if and only if it contains a high set. A related characterization of the high degrees follows from results of Jockusch (1969) and Cooper (1972b): a degree $\leq 0'$ is high iff it contains a hyperhyperimmune set. (See Rogers (1967), Chapter 12 for a discussion of hyperhyperimmune sets.)

2 <u>The Difference Hierarchy</u>. Note that if A is r.e. and $\{A_s\}$ is the effective approximation corresponding to some recursive enumeration of A, then for any x $\{A_s(x)\}$ changes at most once, i.e., there is at most one s such that $A_s(x) \neq A_{s+1}(x)$. This leads to a natural generalization of the class of r.e. sets. A set A is said to be <u>n-r.e.</u> if A has an effective approximation $\{A_s\}$ such that A_0 is empty and for all x

$$|\{s: A_s(x) \neq A_{s+1}(x)\}| \leq n.$$

(The history of this area is unclear. Addison (1965), Putnam (1965), and Gold (1965) formulated related concepts at about the same time. See Epstein, Haas, & Kramer (TA).) Note that the 1-r.e. sets are just the r.e. sets and that $\bigcup_n\{A: A \text{ is n-r.e.}\}$ is just the Boolean algebra generated by the r.e. sets. Note also that A is n-r.e. iff there exist r.e. sets $A^1, \ldots, A^n$ such that $A = A^1 \Delta A^2 \Delta \ldots \Delta A^n$ where Δ denotes the symmetric difference operator.

A SURVEY OF THE NON-R.E. DEGREES $\leq 0'$

By an n-r.e. degree we of couse mean the degree of an n-r.e. set. From the characterization of the n-r.e. sets in terms of the difference operator it is clear that $U_n\{A: A \text{ is n-r.e.}\}$ is uniformly recursive in $0'$ and so, for example by the theorem of Sasso (1970) discussed in §IA5, there are degrees below $0'$ which are not n-r.e. for any n. In fact, every n-r.e. degree > 0 bounds a non-recursive r.e. degree (this is apparently due to Lachlan, see Epstein et al.) and so no n-r.e. degree is minimal. For each n there are (n+1)-r.e. degrees which are not n-r.e. (This result is apparently due to Cooper (1971) but proved independently by Lachlan. See Epstein (1979), Appendix 2.) Other than non-minimality, there has been little success in extending results from the r.e. degrees to even the 2-r.e. degrees. For example, it is not known whether the 2-r.e. degrees are dense.

Ershov (1968) extended the hierarchy of n-r.e. sets to transfinite levels as follows. Let $\mathcal{O}$ denote Kleene's system of constructive ordinal notations. (See, for example, Rogers (1967) §11.7) Let u be an element of $\mathcal{O}$. For each partial recursive function φ with domain $\{v: v <_{\mathcal{O}} u\}$ let $W(\varphi)$ be

$\{x: \exists v <_{\mathcal{O}} u(x \in W_{\varphi(v)})$ & the ordinal denoted by the $<_{\mathcal{O}}$-least such v is even $\}$.

(The motivation here is the characterization of the n-r.e. sets in terms of differences of r.e. sets.) Let $\Sigma(u)$ denote the class of all sets obtainable from u in this way. From the properties of $\mathcal{O}$ it follows that the elements of $\Sigma(u)$ are Δ^0_2. If $u,v \in \mathcal{O}$ and $|u|^{\mathcal{O}}=|v|^{\mathcal{O}}<\omega^2$ then $\Sigma(u) = \Sigma(v)$. However, for every Δ^0_2 set A there is a u in $\mathcal{O}$ with $|u|^{\mathcal{O}} = \omega^2$ and $A \in \Sigma(u)$. For each ordinal $\alpha \leq \omega^2$ let $\Sigma(\alpha) = U\{\Sigma(u): |u|^{\mathcal{O}} = \alpha\}$ and define the $\Sigma(\alpha)$ degrees to be the degrees of $\Sigma(\alpha)$ sets.

Epstein et al. (extending the analogous result for sets by Ershov (1968)) show that if $u <_{\mathcal{O}} v$ then there are $\Sigma(v)$-degrees which are not $\Sigma(u)$. It thus follows that $\langle\Sigma(\alpha)\text{-degrees} : \alpha \leq \omega^2\rangle$ defines a proper and exhaustive hierarchy on the degrees below $0'$. Also, every non-0 $\Sigma(\alpha+1)$ degree has a non-0 $\Sigma(\alpha)$ predecessor. Thus minimal degrees must first appear at limit levels. There are $\Sigma(\omega)$ minimal degrees (Epstein et.al.) This result follows from the proof that every r.e. degree > 0 bounds a minimal degree and the following characterization of the $\Sigma(\omega)$ sets: A is $\Sigma(\omega)$ iff there exists an effective approximation for A, $\{A_s\}$, and a recursive function f such that for all x, $|\{s : A_s(x) \neq A_{s+1}(x)\}| \leq f(x)$.

Part II: METHODS

In this Part of the paper we will describe some of the basic methods used in the construction of degrees below 0'. We will do so by sketching the proofs of some representative theorems. Our intention is primarily to give the flavor and underlying ideas of these constructions though we have tried to give enough clues to allow the really ambitious reader to put together complete proofs

The proofs sketched in §A employ what are called "oracle constructions". In such a construction the characteristic function for a set B is defined as the union of a sequence of finite strings $\{\beta_s\}$. It is required that $\beta_s \subset \beta_{s+1}$ for all s. Thus each β_s is an initial segment of the final set B and so B is recursive in $\{\beta_s\}$. Hence the constructions are in general non-effective.

In the constructions described in §B sets are constructed as the pointwise limit of recursive sequences of strings. In these constructions there is no longer the requirement that each string be contained in all succeeding strings. It is only required that the sequence eventually settle down on each initial segment of ω. Such constructions are dubbed "full approximation constructions". In general the full approximation constructions are more difficult than the oracle constructions.

A. Oracle Constructions

1 Additional Notation and Terminology. A string of length n is a function $\sigma : \{i: i < n\} \rightarrow \{0,1\}$. Small Greek letters are reserved for strings. $lh(\sigma)$ is the

length of σ. $\sigma*\tau$ denotes the concatenation of σ and τ. σ^m is defined by $\sigma^0 = \emptyset$ (the empty string) and $\sigma^{m+1} = \sigma^m*\sigma$. If σ is not the empty string then σ^- denotes the unique substring of σ of length $lh(\sigma)-1$. We assume that all finite objects such as strings, pairs of strings, computation procedures, etc. are Godel numbered in some reasonable way. We can thus extend the μ - operator to relations on strings, pairs of strings, etc. Let $<.,.>$ be a 1-1 recursive coding of $\omega\times\omega$ onto ω and let $(.)_0,(.)_1$ be the corresponding component functions.

We write $[e](\sigma;x) = y$ if $x < lh(\sigma)$ and the e-th Turing reduction procedure, given input x and oracle information σ, gives output y within $lh(\sigma)$ many steps. (Note that from e and σ we can effectively compute the finite set of x such that $[e](\sigma;x)$ is defined.) $[e](B;x) = y$ means that $[e](\beta;x) = y$ for some $\beta \subseteq B$. We use $[e](\sigma)$ to denote the partial function defined by e and σ and similarly for $[e](B)$. Following Shoenfield (1971) we use = between two expressions to mean that either both expressions are defined and equal or both expressions are undefined. For a partial function φ and integer $n \geq -1$ we use $\varphi[n]$ to denote the restriction of φ to $\{m : m \leq n\}$. (Thus, $\varphi[-1]$ denotes the empty function.)

A <u>tree</u> is a partial function T from strings to strings such that for any σ if one of $T(\sigma*0),T(\sigma*1)$ is defined then all of $T(\sigma),T(\sigma*0),T(\sigma*1)$ are defined and $T(\sigma*0),T(\sigma*1)$ are incompatible extensions of $T(\sigma)$. A string is <u>on</u> T if it lies in the range of T. A set B is <u>on</u> T if β is on T for infinitely many $\beta \subseteq B$. A tree T' is a <u>subtree</u> of T if the range of T' is contained in the range of T. The <u>identity tree</u> is the identity function on strings and is denoted by Id. If T is a tree and $\delta = T(\sigma)$ then the <u>full subtree of T above δ</u>, denoted $Fu(T,\delta)$, is defined by $Fu(T,\delta)(\gamma) = T(\sigma*\gamma)$. Thus the range of $Fu(T,\delta)$ is the set of strings on T which contain δ.

A pair of strings γ,δ is said to <u>e-split ρ</u> if γ and δ extend ρ and $[e](\gamma)$ and $[e](\delta)$ are defined and unequal for some argument. Note that by our convention concerning computations from strings we can effectively determine whether or not a given pair of strings e-splits. ρ is said to be e-splittable on a tree T if there exists a pair of strings on T which e-split ρ. T is e-splitting if for all σ, if $T(\sigma*0), T(\sigma*1)$ are defined then they e-split. See Shoenfield (1971) Chapter 11 for a proof of the following standard lemma.

Lemma 1.1. Suppose that T is a partial recursive tree and that B is on T.

(a) If T is e-splitting and [e](B) is total, then B is recursive in [e](B).

(b) If there is some $\beta \subseteq B$ such that β is not e-splittable on T then [e](B) is partial recursive.

A tree T is said to satisfy the e-th minimal degree requirement for B if B is on T and T is partial recursive either T is e-splitting or there is some $\beta \subseteq B$ which is not e-splittable on T. By a result of Posner & Epstein, if for all e there is a tree T satisfying the e-th minimal degree requirement for B then B is not recursive. Combining this with Lemma 1.1 we thus have

Lemma 1.2. Suppose that for every e there is a partial recursive tree satisfying the e-th minimal degree requirement for B. Then B is of minimal degree.

Let T be a partial recursive tree and let β be on T. A partial recursive tree T' is said to be an e-splitting subtree of T above β if T' is a subtree of T, $T'(\emptyset) = \beta$, T' is e-splitting, and whenever $T'(\sigma)$ is defined and e-splittable on T, $T'(\sigma*0)$ and $T'(\sigma*1)$ are defined. Fix some effective procedure (e.g., the one on page 52 of Shoenfield (1971)) which given a Gödel number for T and a string β on T produces a Godel number for an e-splitting subtree of T above β. In most of our constructions a partial recursive tree is "defined" by its Godel number. In such constructions (where Godel numbers are implicit) we will refer to the e-splitting subtree of T above β meaning of course the tree defined by our fixed procedure.

2 The Sacks Minimal Degree Construction. Spector (1956) was the first to construct a minimal degree in $\mathcal{D}$. His construction is essentially a forcing construction in which the forcing conditions are recursive total trees. The minimal degree construction of Sacks (1961) differs from this in two major respects. First, partial trees replace total trees and, second, the construction

itself requires the use of the finite injury priority method. (The formulation of these constructions in terms of trees is due to Shoenfield (1966).)

The construction is $0'$-effective and defines an ascending sequence of strings $\{\beta_s\}$. The object of course is to ensure that the set $B = \cup_s \beta_s$ satisfies the hypothesis of Lemma 1.2 (§IIA1), i.e., to ensure that for each e there is a tree T_e satisfying the e-th minimal degree requirement for B. For this reason, at each stage s, in addition to β_s, we define a nested sequence of partial recursive trees

$$Id = T_{-1,s} \supseteq T_{0,s} \supseteq T_{1,s} \supseteq \ldots \supseteq T_{s,s}$$

Each $T_{e,s}$ represents a "guess" at a tree T_e satisfying the e-th minimal degree requirement for B. These trees are defined so that the following conditions hold.

<u>Conditions 2.1</u>. For each $s \geq 0$ and $e \leq s$:

(a) β_s is on $T_{e,s}$;

(b) if $s > 0$ and β_{s-1} is e-splittable on $T_{e,s}$ then $T_{e,s}$ is the e-splitting subtree of $T_{e-1,s}$ above β for some $\beta \subseteq \beta_{s-1}$;

(c) if $s > 0$ and β_{s-1} is not e-splittable on $T_{e-1,s}$ then $T_{e,s} = Fu(T_{e-1,s}, \beta)$ for some $\beta \subseteq \beta_{s-1}$;

(d) if β_s has incompatible extensions on $T_{e,s}$ then $T_{e,s+1} = T_{e,s}$.

<u>Construction 2.2</u>.

<u>Stage 0</u>. Set $\beta_0 = \emptyset$, $T_{-1,0} = Id$, and $T_{0,0}$ = the 0-splitting subtree of Id above $\emptyset$.

<u>Stage $s > 0$</u>. (We assume, inductively, that the conditions of 2.1 above have held through stage s-1.) Let r(s) be the largest r, $-1 \leq r \leq s-1$, such that β_{s-1} has incompatible extensions on $T_{r,s-1}$. (Note that r(s) can be computed $0'$-effectively.)

For each e, $-1 \leq e \leq r(s)$, set $T_{e,s} = T_{e,s-1}$. Inductively define $T_{e,s}$ for $r(s) < e \leq s$ as follows.

If β_{s-1} is e-splittable on $T_{e-1,s}$ let $T_{e,s}$ be the e-splitting subtree of $T_{e-1,s}$ above β_{s-1}.

Otherwise, let $T_{e,s} = Fu(T_{e-1,s}, \beta_{s-1})$. (Note that we can $0'$-effectively determine whether β_{s-1} is e-splittable on $T_{e-1,s}$.)

Finally, set $\beta_s = T_{s,s}(0)$.

The reader can easily verify that the construction satisfies the conditions of 2.1. Further, since the construction is 0' effective and β_{s+1} properly contains β_s for each s, the union of the β_s defines a set B which is recursive in 0'. It remains to show that the hypothesis of Lemma 1.2 is satisfied.

We first argue by induction on e that for each e there is a stage s(e) such that $T_{e,s} = T_{e,s(e)}$ for all $s \geq s(e)$ and so $\{T_{e,s}\}$ eventually settles on a limit tree T_e. This is obviously true for e = -1. So suppose that this claim is true for e-1, e > 0. It follows from 2.1(a) that B is on T_{e-1}. If every string contained in B is e-splittable on T_{e-1} let s(e) = the maximum of e, s(e-1). Otherwise, let s(e) be the first stage $s \geq$ e and s(e-1) such that β_{s-1} is not e-splittable on T_{e-1}. Then, from 2.1(b) and (c) it follows that for some $\beta \subseteq \beta_{s(e)}$ either every string contained in B is e-splittable on T_{e-1} and $T_{e,s(e)}$ is an e-splitting subtree of T_{e-1} above β or $\beta_{s(e)}$ is not e-splittable on T_{e-1} and $T_{e,s(e)}$ is the full subtree of T_{e-1} above β. In either case, an easy induction using 2.1(d) shows that $T_{e,s} = T_{e,s(e)}$ for all $s \geq s(e)$.

It is easy to verify that each T_e satisfies the e-th minimal degree requirement for B. By 2.1(a), B is on T_e and from 2.1(b) and (c) it follows that either T_e is e-splitting or there is some string contained in B which is not e-splittable on T_e. Thus by Lemma 1.2, B is of minimal degree.

While the construction and proof given above are not terribly difficult it is still reasonable to ask whether they can be simplified. Two questions come immediately to mind. Is it possible to avoid the use of partial trees? Is it possible to eliminate the "guessing" (and the resulting priority argument)? The answer to both questions is, as we now show, no.

We can see that partial trees are essential as follows. Suppose that B is recursive in 0' and let $\{\gamma_s\}$ be some recursive sequence of strings converging pointwise to B. Let e be such that for all A,

$$[e](A;x) = \begin{cases} A(x), & \text{if } \exists s(lh(\gamma_s) > x \;\&\; \gamma_s \subseteq A) \\ \text{undefined}, & \text{otherwise.} \end{cases}$$

Then, as the reader can easily verify, B will be the only set on any tree satisfying the e-th minimal degree requirement for B and so no total tree could satisfy the e-th minimal degree requirement for B.

To see that the "guessing" is essential, let B be some set and let $\{z_e\}$ be a sequence of Gödel numbers corresponding to a sequence of partial recursive trees $\{T_e\}$ such that for each e, T_e satisfies the e-th minimal degree requirement for B. We will show that $\{z_e\}$ cannot be recursive in 0'. Let f be a recursive function such that for any i and x and any set A,

$$[f(i)](A;x) = \begin{cases} A(x), \text{ if } [i](A;y) \text{ is defined for all } y \le x \\ \text{undefined, otherwise.} \end{cases}$$

Then for any i, [i](B) is total if and only if $T_{f(i)}$ is f(i)-splitting. From the hypotheses concerning the T_e it follows that it can be determined $B \oplus \emptyset'$-effectively from z_k whether or not T_k is k-splitting. Thus, since $dg(B)'' = dg(\{i : [i](B) \text{ is total}\})$, we have that $dg(B)'' \le (dg(B) \vee 0' \vee dg(\{z_e\}))$. Since $dg(B)' \ge dg(B) \vee 0'$ we cannot have $dg(\{z_e\}) \le 0'$.

3 Sasso's Cornucopia of Minimal Degrees

Theorem 3.1 (Sasso (1970)). There is a total tree T recursive in 0' such that every set on T is of minimal degree.

The idea behind the proof of 3.1 is to carry out a "tree" of Sacks minimal degree constructions. More precisely, at each stage s of the construction we define for each string σ of length s a string β_σ and a nested sequence of partial recursive trees

$$Id = T_{-1,\sigma} \supseteq T_{0,\sigma} \supseteq \cdots \supseteq T_{s,\sigma} .$$

Stage 0 is the same as stage 0 of 2.2. Stage s then proceeds from stage s-1 as follows. Let σ be a string of length s. The sequence $T_{-1,\sigma}, \ldots, T_{s,\sigma}$ is obtained from β_{σ^-} and $T_{-1,\sigma^-}, \ldots, T_{s-1,\sigma^-}$ in exactly the way in which $T_{-1,s}, \ldots, T_{s,s}$ is obtained from β_{s-1} and $T_{-1,s-1}, \ldots, T_{s-1,s-1}$ in 2.2. (Note that the trees corresponding to σ^-*0 and σ^-*1 will be the same.) We then set $\beta_{\sigma^-*i} = T_{s,\sigma}(i)$. Note that β_{σ^-*0} and β_{σ^-*1} are incompatible extensions of β_{σ^-}.

We define T by $T(\sigma) = \beta_\sigma$. Since the construction is recursive in 0', T is also. Further, arguing just as in 2 one can show that for any set C, $U_s \beta_{C[s]}$ is of minimal degree. Thus T has the desired properties.

Corollary 3.2 (Sasso(1970)). For any $c \geq 0'$ there is a minimal degree b such that $c = b \vee 0'$.

Let $C \in c$ and let T be as in Theorem 3.1. We "code" C into a set B on T by taking $B = U_s T(C[s])$. Clearly $c \leq dg(B) \vee dg(T) \leq dg(B) \vee 0'$. On the other hand, since $c \geq 0'$ T is recursive in c and so $dg(B) \vee 0' \leq c$.

Corollary 3.3 (Sasso(1970)). Let $\mathcal{A}$ be a set of degrees uniformly recursive in 0'. Then there is a minimal degree $b < 0'$ such that b is incompatible with every degree in $\mathcal{A} \sim \{0, 0'\}$.

The proof we give is slightly different from Sasso's original proof and is based on the following extension of the result of Miller & Martin (1968) which was discussed in §IC1.

Lemma 3.3.1 (Posner (TAa)). Let S be an infinite set which is recursive in 0'. Then there is a function f of degree 0' such that no function of degree strictly less than 0 dominates f on S, i.e., if $dg(g) < 0'$ then $f(s) \geq g(s)$ for infinitely many $s \in S$.

The proof of this Lemma is quite easy. By the result of Miller & Martin (1968) discussed in IC1 we may choose h of degree 0' such that no function of degree < 0' dominates h on ω. We may assume that h is strictly increasing. Then define f by $f(x) = h(s)$ where s is the first element of $S > x$. The reader can easily verify that any function dominating f on S dominates h on ω.

We now prove Corollary 3.3. Let T be as in Theorem 3.1 and let $\{A_i\}$ be uniformly recursive in 0' and such that $\mathcal{A} = \{dg(A_i): i \in \omega\}$. In order to prove 3.3 it will suffice to construct a set B of minimal degree such that $dg(B) < 0'$ and for all $a \in \mathcal{A} \sim \{0'\}$, $dg(B) \not\leq a$. We will construct B as $U_s T(\sigma_s)$ where $\{\sigma_s\}$ is defined 0'-effectively and $lh(\sigma_s) = s$. For each i,j we have the following requirement

$R_{<i,j>}$: either $[i](A_j) \neq B$ or $deg(A_j) = 0'$.

For each s, let $m(s) = \mathrm{Max}(\{lh(T(\sigma)):lh(\sigma)=s\})$. Let $S = \{m(s):s\geq 0\}$. Since T is recursive in 0' (and T is total) S is recursive in 0'. Hence by Lemma 3.3.1 we may take f to be a function of degree 0' which is not dominated on S by any function of degree < 0'.

Construction 3.3.2. Let $\sigma_0 = \emptyset$. We obtain σ_{s+1} from σ_s as follows.

Requirement $R_{<i,j>}$ is said to "require attention" if $R_{<i,j>}$ did not "receive attention" at a prior stage and $[i](A_j[f(m(s+1))];x)$ is defined for all $x \leq m(s+1)$. If there is no requirement which requires attention then let $\sigma_{s+1} = \sigma_s*0$. Otherwise do the following. Let i,j be the pair with the least code such that $R_{<i,j>}$ requires attention. Then, since $T(\sigma_s*0)$ and $T(\sigma_s*1)$ are incompatible and have length $\leq m(s+1)$ at least one of them is incompatible with $[i](A_j[f(m(s+1))])$. Take σ to be the least of σ_s*0, σ_s*1 having this property. $R_{<i,j>}$ has received attention.

It is clear that if every $R_{<i,j>}$ is satisfied then B will have the desired properties. It is also clear that if $R_{<i,j>}$ ever receives attention then it is satisfied. Let i and j be given. We may assume that $[i](A_j)$ is total and that A_j does not have degree 0', as otherwise $R_{<i,j>}$ is obviously satisfied.

Let g be the function defined by

$$g(x) = \mu z(\ [i](A_j[z];y) \text{ is defined for all } y \leq x).$$

g is clearly recursive in A and so does not dominate f on S. Thus, there is an s such that $s \geq <i,j>$, $f(m(s)) \geq g(m(s))$, and no requirement $R_{<k,l>}$ with $<k,l> \; < \; <i,j>$ requires attention at s. (The last condition can be obtained because no requirement requires attention after once receiving attention and requirements receive attention according to code.) Then if $R_{<i,j>}$ has not received attention prior to stage s it will receive attention at s. Thus $R_{<i,j>}$ will be satisfied.

4 A Complementary Pair of Minimal Degrees.

Theorem 4.1 (Cooper(1972a)). There exists a pair of minimal degrees a and b such that a and b are complementary in $\mathcal{D}(\leq 0')$.

Cooper's original proof of this theorem uses a full approximation construction. The proof sketched here (see also Posner (TAb)) is based on the Sacks minimal degree construction. This proof in fact shows that <u>every</u> complete degree is the join of a pair of minimal degrees.

Note that if a and b are minimal degrees and $a \vee b = 0'$ then a and b also form a minimal pair. Thus, in order to prove 4.1, it suffices to construct A and B such that dg(A) and dg(B) are minimal and join to 0'. The basic idea is to construct A and B via Sacks minimal degree constructions but to "couple" the two constructions in such a way that we can code $\emptyset'$ into $A \oplus B$. Thus A and B will be defined as $\cup_s \alpha_s$ and $\cup_s \beta_s$ respectively where $\{\alpha_s\}$ and $\{\beta_s\}$ are ascending sequences of strings generated 0'-effectively by our construction. Further, at each stage s we will define two nested sequences of partial recursive trees:

$$\begin{array}{l} Id = T_{-1,s} \supseteq T_{1,s} \supseteq T_{3,s} \supseteq \cdots \\ \qquad\qquad\qquad\qquad\qquad\qquad\qquad T_{k(s),s}. \\ \qquad T_{0,s} \supseteq T_{2,s} \supseteq T_{4,s} \supseteq \cdots \end{array}$$

k(s) may be even or odd. The trees with odd indices are associated with the construction of A and the even indexed trees are associated with B. Thus, for each e, $\{T_{2e+1,s}\}$ and $\{T_{2e,s}\}$ will eventually settle on respective trees T_{2e+1} and T_{2e} satisfying the e-th minimal degree requirement for A and B respectively.

We will code $\emptyset'$ into $A \oplus B$ as follows. At stage s we code $\emptyset'(s)$ into either A or B, depending on whether k(s) is odd or even, by requiring the corresponding one of α_s or β_s to contain $T_{k(s),s}(\emptyset'(s))$. This will ensure that if the construction itself can be recovered $A \oplus B$-effectively then $\emptyset' \leq_T A \oplus B$, where by "recovering the construction" we mean recovering k(s), α_s, β_s, and the Gödel numbers for the $T_{i,s}$, $i \leq k(s)$.

In order to ensure that the construction can be recovered $A \oplus B$-effectively it will be necessary to code certain additional information such as Gödel numbers for trees into A and B. Here we run into a problem. Since α_s and β_s will of course be constrained to lie on the appropriate $T_{i,s}$, $i \leq k(s)$, information coded into A or B at stage s will have to be coded as a sequence of left and right branchings on some $T_{n,s}$. In order to decode such information it is necessary to already have a Gödel number for $T_{n,s}$.

The solution to this problem is quite easy. We use a kind of bootstrap coding procedure. $T_{-1,s}$ is always the identity tree, so we know how to decode information coded into A by left and right branchings on $T_{-1,s}$. One piece of information which can be so coded at some point is the Gödel number for $T_{0,s}$. This will allow us to recover information coded into B by lefts and rights on $T_{0,s}$ on which we will at some point have coded the Gödel number for $T_{1,s}$ from which we will obtain the Gödel number for $T_{2,s}$ etc.

Note that we do not need to recode the Gödel number for every $T_{e,s}$ at stage s (that would be impossible). In fact it will only be necessary to code the Gödel for $T_{k(s),s}$ at stage s because the $T_{i,s}$ with $i < k(s)$ will be the same as they were at stage s-1.

This still leaves us with a problem however because of the fact that the Sacks construction involves guessing. The difficulty is that we cannot know when coding the Gödel number for a given $T_{n,s}$ whether or not $T_{n,s}$ is in fact the true T_n. If at some later stage, t, it becomes necessary to change our guess at T_n then it will be necessary to update this information for A⊕B. The new guess can be coded on $T_{n-1,t}$ of course but we must have some way of signaling to A⊕B that a new guess at T_n is being made and that the Godel number for the new guess is being coded on $T_{n-1,t}$.

The method by which we signal that we are changing our guess at T_n at stage t is to push the appropriate one of α_t or β_t, depending on whether n-1 is odd or even, off of a tree $T^*_{n-1,t-1}$ on which it would otherwise lie. The *-subtree of a tree T, denoted T* is obtained by pruning every other branch of T (see Figure 4.2). (Without harm we can modify the Sacks construction so that each tree in the nested sequence is a subtree of the *-subtree of its predecessor. These *-subtrees were first used by Sasso (1974) in order to construct non-GL_2 minimal degrees. The updating method used here is also used in Posner (TAa) and appears in other guises elsewhere such as Epstein (1975), Chapter VI.)

Figure 4.2.

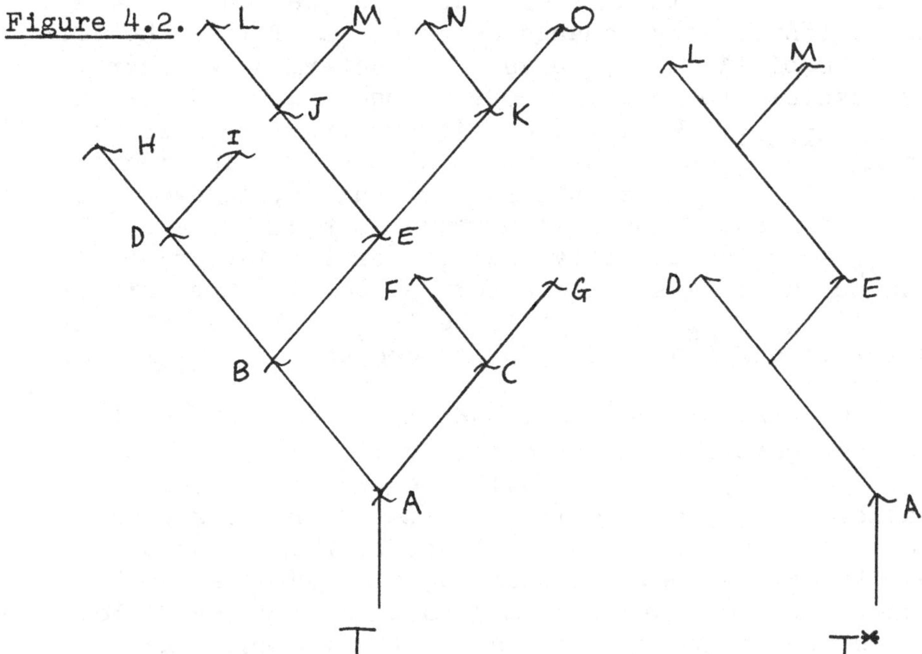

Stage s of the construction proceeds roughly as follows. We begin in essentially the same way as in the standard Sacks construction by making a new guess at some T_n (in fact n = k(s-1)+1). We then use our $\emptyset'$-oracle to determine whether it is possible to code the Gödel number for the new guess into the appropriate one of A or B by means of a sequence of branchings on $T_{n-1,s-1}$ starting immediately above the appropriate one of α_{s-1}, β_{s-1} in which the first branch in the sequence is a string which is incompatible with any string on $T^*_{n-1,s-1}$. (For example, if $T_{n-1,s-1}$ is T of Figure 4.2 and the appropriate one of α_{s-1}, β_{s-1} is the string labeled E then the first step in the coding sequence would be the string labeled K.)

The coding sequence is not possible if before being completed a string is encountered which has no proper extensions on $T_{n-1,s-1}$. If this occurs then it must be the case that there is an n' < n, of opposite parity from n, and a string γ on $T_{n',s-1}$ such that $T_{n',s-1}$ was defined to be an e-splitting subtree of $T^*_{n'-2,s-1}$, e = greatest integer $\leq n/2$, and γ is not e-splittable on $T^*_{n'-2,s-1}$. In this case we find the least such n' and, tentatively, change our guess at $T_{n'}$ to the full subtree of $T^*_{n'-2,s-1}$ above γ and determine whether it is possible to code the Godel number for this new guess on $T_{n'-1,s-1}$. We may thus discover that we need to change our guess at some $T_{n''}$, $n'' < n'$. This process may continue several times. Eventually, however, it must terminate with some n (which is k(s)) because $T_{-1,s-1}$ is the identity tree (in particular is total) and so can always be used for coding. The appropriate one of α_s or β_s is taken to be $T_{k(s),s}(\emptyset'(s))$ and the other is defined by the coding sequence on $T_{k(s)-1,s}$.

The finite injury argument of §IIA2 still applies to the construction above and so $\lim_s k(s) = \infty$. Thus for each n, $\{T_{n,s}\}$ eventually settles on a tree T_n satisfying the appropriate minimal degree requirement for the appropriate one of A or B. Thus A and B are of minimal degree. Further, we can $A\oplus B$-effectively recover the construction and so compute $\emptyset'$ as follows. To recover stage s from stage s-1, find the least k such that the appropriate one of A or B leaves $T^*_{k-1,s-1}$ immediately above the appropriate one of α_s or β_s. This is k(s). For k < k(s) we have $T_{k,s} = T_{k,s-1}$. Then, from $T_{k(s)-1,s-1}$ and A and B, we can recover the Godel number of $T_{k(s),s}$ and α_s and β_s as well.

5 <u>Minimal Degrees Below High Degrees</u>. We next describe Jockusch's construction (Jockusch (1977)) of a minimal degree below an arbitrary high degree. The original proof of this result by Cooper (1973) uses a full approximation construction and will be discussed in §B3.

Let h be high, i.e. $h \leq 0'$ and $h' = 0''$. We want to construct a set B of minimal degree as $\cup_s \beta_s$ where $\{\beta_s\}$ is generated <u>h-effectively</u> by our construction. The key to this construction is the Recursion Theorem (see, for example, Rogers (1967), Chapter 11) which allows us to make use of functions defined <u>in terms of B</u> in order to <u>construct</u> B.

Looking back at the Sacks minimal degree construction, one observes that the key idea in obtaining $T_{e+1,s}$ from $T_{e,s}$ is to ensure that for the limit trees, T_e and T_{e+1}, we have the following condition.

(*) If every string contained in B is (e+1)-splittable on T_e then T_{e+1} is an (e+1)-splitting subtree of T_e.

If it were possible to determine whether or not the antecedent of (*) is valid for a given tree--0'-effectively that is -- then it would be possible to construct a minimal degree below 0' without "guessing". Unfortunately, the problem of determining whether or not the antecedent of (*) is valid is only $(B \oplus \emptyset')'$-effective. In particular, if B is recursive in 0' then the problem is 0″ hard. In the Sacks construction we get around this (roughly) by guessing that the antecedent is valid until we discover otherwise, i.e., discover a string on the splitting tree with no proper extensions. Unless h is 0' we can't h-effectively determine whether or not a given string is e-splittable on a given tree. However, since h' = 0″ we can, by the Limit Lemma, h-effectively <u>approximate</u> the answer to **the question** of whether or not every string contained in B is (e+1)-splittable on a given tree and that, as Jockusch observed, is all we need. Of course we've been talking as if B has already been defined, but, as we said earlier, that is the point of the Recursion Theorem. A more detailed explanation follows.

Let $H \in h$. Notice that the set of triples (i,e,z) such that <u>every string β compatible with [i](H) is e-splittable on the tree with Gödel number z</u> is Π_1 in $H \oplus \emptyset'$ and so recursive in $(h \vee 0')' = h'$. Thus, by Shoenfield's Limit Lemma, there is an h-effective function S(i,e,z,s) such that $\lim_s$ S is the characteristic function for this set. We use this guessing function to ensure that (*) is satisfied as follows.

Using the Recursion Theorem we fix i such that [i](H) is the function defined by our H-effective **construction.** At stage s the trees $T_{e,s}$, $e \leq s$, are defined inductively. $T_{-1,s}$ is as usual the identity tree. We obtain $T_{e,s}$ from $T_{e-1,s}$ in roughly the following way. Tentatively set $T_{e,s}$ to be an e-splitting subtree of $T_{e-1,s}$, preferably $T_{e,s-1}$ if $T_{e-1,s}$

$= T_{e-1,s-1}$ and $T_{e,s-1}$ was defined to be an e-splitting subtree of $T_{e-1,s-1}$. We then begin looking for an e-splitting of β_{s-1} on $T_{e-1,s}$. At the same time we examine values of $S(i,e,z,t)$, for $t \geq s$ and z the Godel number of $T_{e-1,s}$. We continue until either an e-splitting is found or we find $t \geq s$ such that $S(i,e,z,t) = 0$. The definition of S and the fact that [i](H) is the function defined by our construction and so must contain β_{s-1} guarantee that this search will terminate. If the search terminates by finding an e-splitting of β_{s-1} on $T_{e-1,s}$ then we stick with our original definition of $T_{e,s}$. Otherwise we take $T_{e,s}$ to be a full subtree of $T_{e-1,s}$, preferably $T_{e,s-1}$ if $T_{e-1,s}$ has not changed and $T_{e,s-1}$ was a full subtree.

The argument that for each e the sequence $\{T_{e,s}\}$ eventually settles on a tree T_e satisfying the e-th minimal degree requirement for B is by induction. The argument for the inductive step is quite simple. Let z be the Godel number for T_{e-1} and let s be sufficiently large so that for all $t \geq s$, $T_{e-1,t}$ has Godel number z and $S(i,e,z,t) = S(i,e,z,s)$. Then at each stage $\geq s$ the guess at T_{e-1} will in fact be correct, and the information obtained from S will be correct and so we will make the correct guess at T_e.

The reader should observe, as Jockusch did, that the construction above applies to any degree h satisfying $h' = (h \vee 0')'$, i.e., to any degree in GH_1.

6 <u>Minimal Degrees are GL_2</u>. We next sketch the proof of Jockusch & Posner (1978) that every minimal degree is GL_2. This result of course implies that every minimal degree below $0'$ is in L_2. The proof in fact shows that every non-GL_2 degree in fact bounds a 1-generic degree and so, for example, bounds infinitely many degrees.

Let a be non-GL_2, i.e., $a'' > (a \vee 0')'$. We wish to construct a 1-generic set B as $\bigcup_s \beta_s$ where $\{\beta_s\}$ is generated a-effectively. Thus for each e we have the requirement

R_e: either $[e](B;e)$ is defined or there is some $\beta \subseteq B$ such that $[e](\gamma;e)$ is undefined for all $\gamma \supseteq \beta$.

We begin by developing a strategy for satisfying a single requirement R_e. Let us consider what we can do at stage s of our construction toward satisfying R_e. If $[e](\beta_{s-1};e)$ is defined then R_e is already satisfied. So suppose that $[e](\beta_{s-1};e)$ is not defined. All that we really can do is spend some amount of time searching for a string $\beta \supseteq \beta_{s-1}$ such that $[e](\beta;e)$ is defined. If such a β is found then we can satisfy R_e by pushing B along β. If we do not find such a β in the allotted time then we will just have to define β_s arbitrarily and try again at the next stage.

The crucial question of course is how much time should we allow for the search a stage s. Our construction must be a-effective. If a were 0' then we could simply determine whether or not there exists any $\beta \supseteq \beta_{s-1}$ with $[e](\beta;e)$ defined. If a were GH_1 then we could use the Recursion Theorem and employ an a-effective guessing function which in the limit correctly answers the question: is every string $\gamma \subseteq B$ extended by some string β with $[e](\beta;e)$ defined. But, all we are assuming is that a is not GL_2.

Here is the solution. First, we slow the construction down by requiring that $lh(\beta_s) = s$ for all s. Intuitively, this will give us more time in which to find a suitable β satisfying R_e. It certainly does us no harm since if at some point we do find an appropriate β then we can simply take several stages to push B along β rather than doing it all at once.

Next, we define a function f which will be used to allocate search time, i.e., at stage s we will spend f(s) amount of time searching for a suitable β. f must of course be recursive in a. In addition, we want f to grow sufficiently quickly so that

(*) if for every $\gamma \subseteq B$ there is a $\beta \supseteq \gamma$ with $[e](\beta;e)$ defined, then for infinitely many s, such a β extending β_s is found within time f(s).

It is clear that if f satisfies (*) then our strategy for satisfying R_e will succeed. (Of course our single requirement strategy all we need is a single s as in the conclusion of (*). We are requiring infinitely many s in order to successfully put our strategies together.

In defining f satisfying (*) we finally make use of the assumption that a is not GL_2. By a result of Martin (1966), the degree of any function dominating every recursive function has jump ≥ 0″. Relativizing this result to a and using the assumption that (a∨0')' < a″ we have that no function which is recursive in a∨0' dominates every function recursive in a. Let g_e be the function defined by

$$g_e(s) = \mu z\{\text{for all } \sigma,\ lh(\sigma)\le s, [(\exists\beta\supseteq\sigma([e](\beta;e)\text{defined})) \rightarrow (\exists \text{such } \beta \text{ with Godel number} < z)]\}.$$

g_e is clearly recursive in 0'. Hence we may choose f which is recursive in a and not dominated by g_e. The reader can easily verify that f must satisfy (*) and so our strategy for satisfying R_e will succeed.

Putting the single requirement strategies together requires little beyond a simple priority argument. We may choose f recursive in a such that f is not dominated by g where g is defined by g(s) = Max($\{g_e(s):e\le s\}$). Then f is not dominated by any g_e and so will satisfy (*) for every e. At stage s of the construction we simply attack the highest priority R_e(i.e., least e), e ≤ s, such that R_e is not yet satisfied and a suitable β is found within f(s) steps. (If no R_e, e ≤ s, requires attention then define β_s arbitrarily. Also, don't bother finding new β's if an old β will do.)

Note that all that was required of a in the proof above is that a contain a function which is not dominated by g. Thus we have a single function (of degree 0' even) which dominates every function of minimal degree. This is somewhat surprizing considering the fact that there are a continuum of minimal degrees.

7 <u>The Join Theorem</u>. In our last example of an oracle construction we sketch the proof of the following.

<u>Theorem</u>. For every degree a, 0 < a ≤ 0', there is a degree b < 0' such that b∨a = 0'.

This was originally proved by R.W. Robinson using a full approximation construction. The proof given here is from Posner & Robinson (TA).

Let a satisfying $0 < a < 0'$ be fixed. In order to ensure that b is strictly less than 0' we will make b the degree of a 1-generic set B. We thus have the same requirements on B as we had in the previous construction. We again use R_e to denote the requirement that either $[e](B;e)$ is defined or there exists $\beta \subseteq B$ such that for all $\gamma \supseteq \beta$, $[e](\gamma;e)$ is undefined.

The basic strategy of our construction is very direct and does not require a priority argument. At stage s we will satisfy R_s and code $\emptyset'(s)$. $\emptyset'(s)$ will be coded as the last element of β_s, i.e., $\beta_s = \beta_s^-{*}\emptyset'(s)$. Thus $\emptyset'$ will be recursive in $\{\beta_s\}$. Our problem will be to ensure that $\{\beta_s\}$ is recursive in avdg(B).

By the result of Miller & Martin (1968) discussed in §IC1, there is a function f of degree a such that f is not dominated by any recursive function. Let such an f be fixed.

We define $\{\beta_s\}$ inductively. Let $\beta_{-1} = \emptyset$. We obtain β_s from β_{s-1} as follows. For each m let $\gamma_m = 1^m{*}0$. (All that is required is that $\{\gamma_m\}$ be some recursive sequence of pairwise incompatible strings.)

Let m(s) be the least m such that either

(a) for all $\sigma \supseteq \beta_{s-1}{*}\gamma_m$, $[e](\beta;e)$ is undefined, or

(b) $\exists\beta\supseteq\beta_{s-1}{*}\gamma_m([e](\beta;e)$ is defined and the Godel number of $\beta < f(m))$.

The fact that f is not dominated by any recursive function ensures that m(s) is defined.

If condition (a) in the definition of m(s) holds then let $\beta_s = \beta_{s-1}{*}\gamma_{m(s)}{*}\emptyset'(s)$.

Otherwise, let $\beta_s = \beta{*}\emptyset'(s)$ where β is the least string $\supseteq \beta_{s-1}{*}\gamma_{m(s)}$ and satisfying $[e](\beta;e)$ is defined.

Obviously, for each s, β_s satisfies R_s and codes $\emptyset'(s)$ in the specified way. Further, with oracles for B and f we can recover β_s from β_{s-1} as follows. First, we compute m(s) as the unique m such that $\beta_{s-1}{*}\gamma_m \subseteq B$. Using f we can then determine whether condition (b) in the definition of m(s) holds. If not, then $\beta_s = B[lh(\beta_{s-1}{*}\gamma_{m(s)})]$. If (b) does hold then $\beta_s = B[lh(\beta)]$ where β is the least string satisfying (b) for m(s).

Thus $\{\beta_s\}$ is recursive in $a \vee dg(B)$ and so $0' \le a \vee dg(B)$. Since B is recursive in 0' we thus have $a \vee dg(B) = 0'$.

B. Full Approximation Constructions

1 Full Approximation Construction of a Minimal Degree. In this section we will construct a set B of minimal degree as $\lim_s \beta_s$ where $\{\beta_s\}$ is a recursive sequence of strings. The construction is essentially due to Cooper (1972a). (See Epstein (1975).)

As with the oracle construction described in §IIA2, in order to make B of minimal degree we will ensure that for each e there is a tree, which in this construction we call T_e^* (not to be confused with the *-subtrees of §IIA4 for which we will have no more use), satisfying the e-th minimal degree requirement for B. Further, at each stage s we will define a nested sequence of trees

$Id = T_{-1,s} \supseteq T_{0,s} \supseteq T_{1,s} \supseteq \ldots \supseteq T_{s,s}$

and take β_s to be $T_{s,s}(0)$. Each $T_{e,s}$ is of course associated with the e-th minimal degree requirement.

The $T_{e,s}$ in the present construction will differ from those in the oracle construction in several repects however. Each T_e,s will be defined as a total recursive tree. Also, while for each e, $\{T_{e,s}\}$ "converges" to a tree T_e, the convergence in this case is only pointwise, i.e., we only have $T_e(\sigma) = \lim_s T_{e,s}(\sigma)$. In fact, T_e will not in general even be recursive. However, T_e will contain a partial recursive subtree, T_e^*, which will satisfy the e-th minimal degree requirement for B.

Let us begin by considering how $T_{0,s}$ is defined. The idea is quite simple. We make $T_{0,s}$ as much of a 0-splitting subtree of Id as we can while limiting search time to s. More precisely, let $T_{0,s}$ be defined by $T_{0,s}(\emptyset) = \emptyset$ and for all σ,

$T_{0,s}(\sigma*0)$ and $T_{0,s}(\sigma*1)$ are defined so as to form the the least 0-splitting of $T_{0,s}(\sigma)$, if such a 0-splitting with Gödel number $\leq s$ exists; otherwise, take them to be the immediate extensions of $T_{0,s}(\sigma)$ on $T_{-1,s}$. Note that, as pictured in Figure 1.1, $T_{0,s}$ will consist of two layers, a bottom layer made up of 0-splittings and a top layer which looks like the identity tree. Notice that as s increases the changes

Figure 1.1.

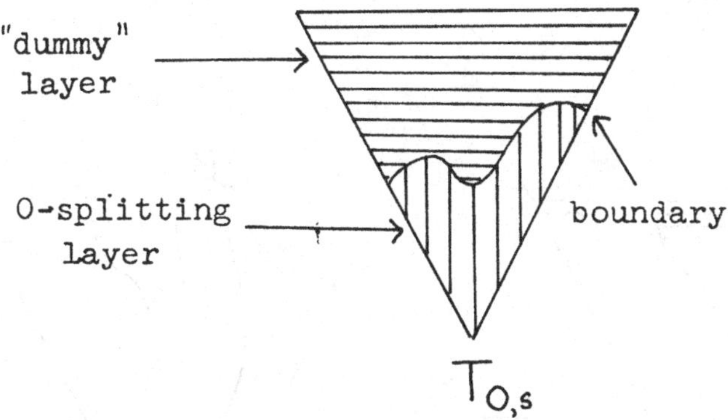

in the tree occur at the boundary between the layers as the bottom layer grows (pushing the boundary up). Once a 0-splitting is put up it stays up. Along some paths the boundary eventually stops and along others it may push "to ∞"

It is thus clear that $\{T_{0,s}\}$ converges (pointwise) to a tree T_0. We can still think of T_0 as consisting of two layers. (See Figure 1.2.) What is important is that the 0-splitting part of T_0 is partial recursive and for any path A on T_0 either every branching along A is a 0-splitting or there is some $\alpha \subsetneq A$ such that α is not 0-splittable on T_0. In particular, this will be true for B and so we can obtain T_0^* satisfying the 0-th minimal degree requirement for B as follows. If every branching along B on T_0 is a 0-splitting then we take T_0^* to be the 0-splitting part of T_0. (This is partial recursive because once a 0-splitting goes up it stays up.) Otherwise, let $\beta \subsetneq B$ be on T_0 and not 0-splittable on T_0 and take T_0^* to be the full subtree of T_0 above β (which is $Fu(Id,\beta)$). See Figure 1.3.

Figure 1.2.

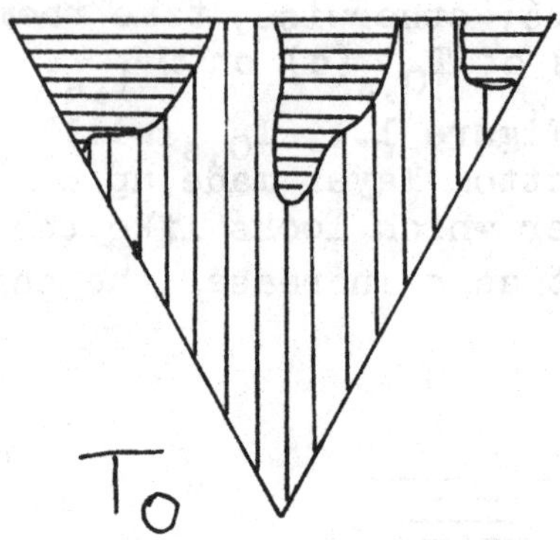

Figure 1.3.

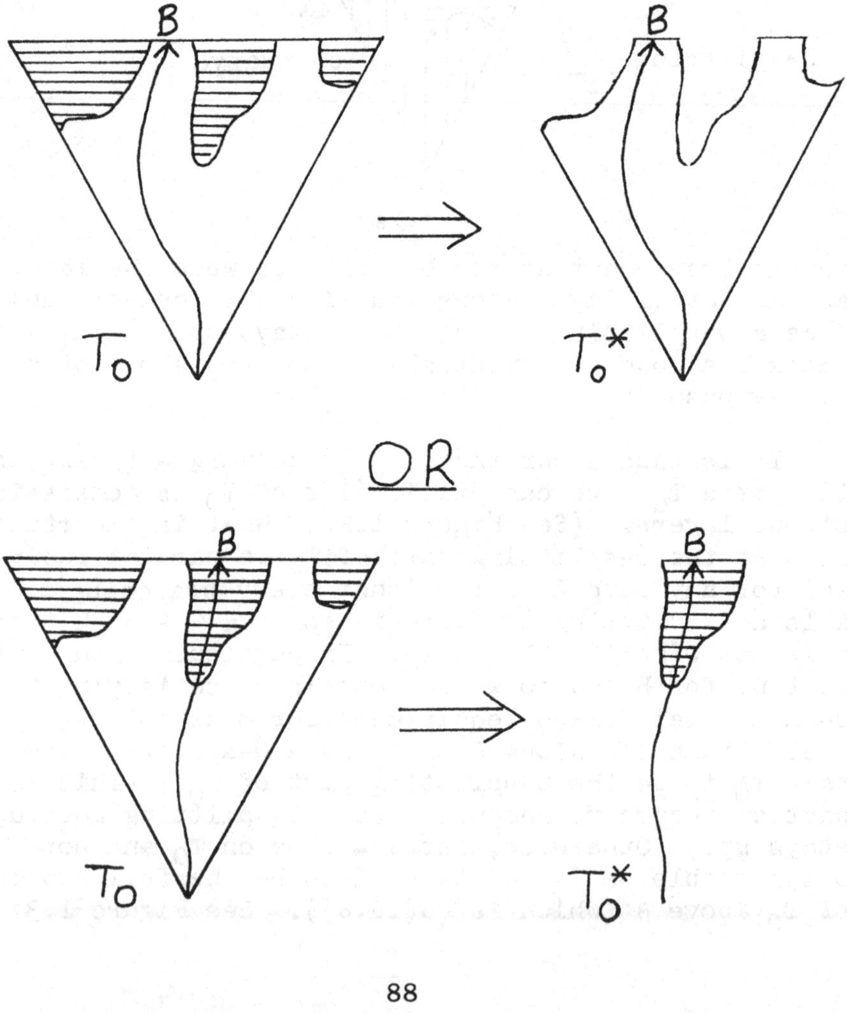

B

T_0

B

T_0^*

OR

B

T_0

B

T_0^*

Now let us consider how to define $T_{1,s}$. The obvious approach is to simply take $T_{1,s}$ to be as much of a 1-splitting subtree of $T_{0,s}$ as we can while limiting search times to s (i.e., just copy the definition of $T_{0,s}$ replacing -1's by 0's and 0's by 1's.) Unfortunately this strategy doesn't work. The problem is that $T_{0,s}$, unlike $T_{-1,s}$, is changing with s. As a result, 1-splittings put up at some stage may have to be taken down again later because one or both components are no longer on T_0. Under the above strategy there would be no guarantee that $\{T_{1,s}\}$ would even converge (let alone converge to a tree ensuring the existence of T_1^*).

The solution to this problem uses a technique which was first used by Friedberg (1958) in order to construct a maximal r.e. set. (The reader familiar with that proof should note the strong similarity between these two constructions. They are in fact almost identical.) As we noted earlier, the changes in $T_{0,s}$ only occur above the 0-splitting layer. Thus, 1-splittings put up below the boundary on $T_{0,s}$ are safe with respect to changes on T_0. Further, if we knew in advance that B was going to lie on the 0-splitting part of T_0 then we could succeed in ensuring the existence of T_1^* having the desired properties by only putting up 1-splittings which lie in the 0-splitting part of T_0. (T_0^* would then be the 0-splitting part of T_0 and T_1^* could be taken to be either a 1-splitting or full subtree of $\underline{T}_0^*$ depending on whether the boundary on T_1 stopped or went to ∞ along B.) Now of course we don't know that B will lie on the 0-splitting part of T_0 so we need an alternate strategy which will succeed in case B lies above the final boundary on T_0. In the alternate strategy we only put up 1-splittings above strings which lie at or above the boundary on $T_{0,s}$. Now some of the 1-splittings put up in applying this strategy may have to be taken down. This will occur if the 1-splitting is put up above some string lying at or above the boundary on $T_{0,s}$ and the boundary later moves beneath this string. However, if the boundary eventually stops along B (if it doesn't then this strategy is irrelevant) then no splitting put up above the point along B where the T_0 boundary stops will ever have to be taken down and so the strategy will succeed.

We thus have two strategies for defining $T_{1,s}$. One works if the boundary on T_0 goes to ∞ along B and the other works it it stops at some point along B. All that remains is to observe that there is no difficulty in applying both strategies simultaneously. We use the "below boundary" strategy to define the part of $T_{1,s}$ lying below the boundary on $T_{0,s}$ and the "above boundary" strategy to define the part of $T_{1,s}$ lying within the upper layer on $T_{0,s}$. This leads to a picture like that in Figure 1.4. In general there will be four layers on $T_{1,s}$: a bottom layer consisting of 1-splittings of 0-splittings; a layer consisting of dummy extensions which are 0-splittings; a layer consisting of 1-splittings of dummy extensions; finally a layer consisting of dummy extensions of dummy extensions.

Figure 1.4.

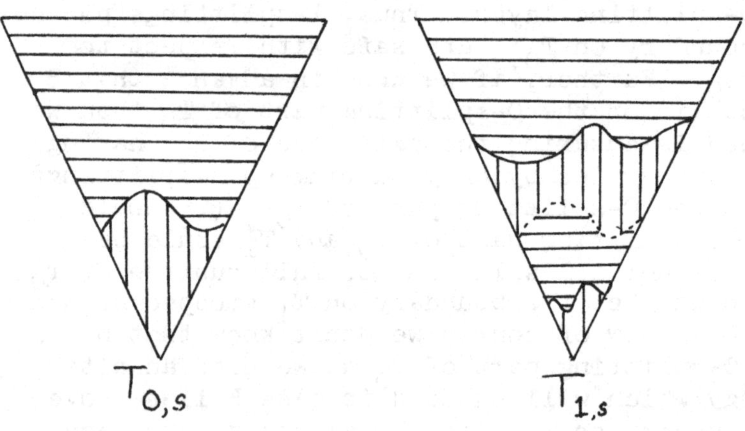

Now just as we have two T_1 strategies corresponding to the two T_0 layers we will have four T_2 strategies corresponding to the four T_1 layers. Each of these strategies will define the part of $T_{2,s}$ lying within the corresponding layer of $T_{1,s}$. The strategy coresponding to the final T_1 layer of B will succeed in ensuring the existence of T_2^*. The four T_2 strategies will give rise to eight T_2 layers leading to eight T_3 strategies giving rise to sixteen layers on T_3 etc.

Though this may appear to be rather complicated, the construction can be formalized very easily using the "e-state" technique of Yates (Yates (1965) and (1970b)). To each string on $T_{e,s}$ we assign an "e-state" which specifies in which of the 2^e possible layers on $T_{e,s}$ the string lies. Formally, an e-state is a binary string of length e+1. The i-th element of the e-state associated with a string δ on $T_{e,s}$ at stage s is 1, if δ is part of an (i-1)-splitting with Godel number $\leq$ s on $T_{i-1,s}$; 0, otherwise. We order e-states lexicographically. (Thus, as we move up through the layers on $T_{e,s}$ the e-states will decrease. See Figure 1.5.)

Figure 1.5.

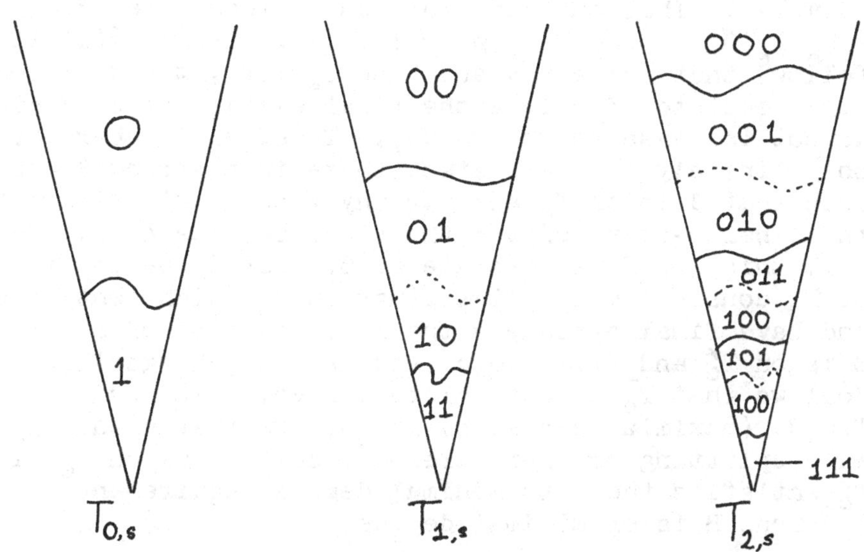

The driving force behind the generation of each $T_{e,s}$ is the desire to maximize the e-states of its branches. This together with a restriction that we not make changes in $T_{e,s}$ which are not of use in attaining this goal is sufficient to ensure that the construction will succeed. More precisely, we impose the following rules on the definition of $T_{e,s}$.

1.6 Rules Governing $T_{e,s}$.

(a) $T_{e,s} \subseteq T_{e-1,s}$.

(b) For all σ, $T_{e,s}(\sigma*0)$ and $T_{e,s}(\sigma*1)$ have the same (e-1)-state at s.

(c) For all $s > 0$ and all σ, if $T_{e,s}(\sigma) = T_{e,s-1}(\sigma)$ and has the same e-state at s as it had at s-1, but either $T_{e,s}(\sigma*0)$ or $T_{e,s}(\sigma*1)$ has changed from what it was at s-1, then $T_{e,s}(\sigma*0)$, $T_{e,s}(\sigma*1)$ have a greater e-state at s as the corresponding strings had at s-1 (i.e., don't make useless changes.)

(d) For all σ, $T_{e,s}(\sigma*0)$ and $T_{e,s}(\sigma*1)$ are defined so as to have maximum possible e-state at s consistent with the other conditions.

(e) There exists a string $\sigma \neq \emptyset$ such that for all sufficiently large s, $T_{e,s}(\emptyset) = T_{e-1,s}(\sigma)$.

Rules 1.6 (c) and (e) and the fact that there are only finitely many e-states for each ensure that for each e and σ, $\{T_{e,s}(\sigma)\}$ and the e-state at s of $T_{e,s}(\sigma)$ converge. Thus the construction defines a set B $= \cup_e T_e(\emptyset) = \lim_s T_{s,s}(\emptyset) = \lim_s \beta_s$. From 1.6(d) it follows that for all γ and δ on T_e, if $\gamma \subseteq \delta$ then the final e-state of γ is $\geq$ the final e-state of δ (e-states do not increase as you go up). Thus, since there are only finitely many e-states, there is a string $\beta \subset B$ such that β is on T_e and for any γ on T_e with $\beta \subseteq \gamma \subset B$, the final e-state of γ = the final e-state of β. We call this the final e-state of B. Let T_e^* be the subtree of T_e consisting of all strings on T_e which contain β and have final e-state = the final e-state of B. Then B is on T_e^* and from 1.6(c) (no useless changes) it follows that T_e^* is partial recursive. Finally, from 1.6(d) (maximize e-states) it follows that either T_e^* is e-splitting or there are no e-splittings on T_e^*, i.e., T_e^* satisfies the e-th minimal degree requirement for B. Thus B is of minimal degree.

2 Minimal Degrees and R.E. Degrees. We next combine the previous construction with an important technique known as r.e. permitting in order to show that every non-recursive r.e. degree bounds a minimal degree. This theorem is due to Yates (1970b). The proof given here is essentially that of Epstein (1975) who used a construction based on this in order to construct minimal complements for r.e. degrees > 0.

2.1 R.E. Permitting. R.E. permitting is a technique which, in its simplest form, is used in order to construct a set B which is recursive in a given non-recursive r.e. set A. The idea is quite simple. Choose some recursive enumeration of A, $\{a_s\}$. This enumeration defines an effective approximation for A, $\{A_s\}$, by taking A_s to be $\{a_i : i \le s\}$. B is constructed as $\lim_s \beta_s$ where $\{\beta_s\}$ is recursive and subject to the following condition

(*) for all s, $lh(\beta_s) = s$ and $\beta_s[a_{s+1}] \subseteq \beta_{s+1}$.

As the reader can easily verify, it follows from (*) that for all x, $B[x] = \beta_{C(x)}[x]$ where C denotes the computation function for $\{A_s\}$, i.e.,

$C(x) = \mu s > x[A_s[x] = A[x]]$.

(Note that this makes use of the fact that once A_s is correct up to some point it stays correct up to that point.) Thus, since C has the same degree as A, it is immediate that $B \le_T A$. (Note that (*) automatically ensures that $\{\beta_s\}$ will in fact converge (pointwise) to a set.)

R.E. permitting affects a construction in the following way. At any stage s and for any x, $\beta_s[x]$ represents our best guess at what we want B[x] to be based on the finite amount of information which can be processed in finitely many stages. At later stages, as we gain more information, we might determine that it would be advantageous to change some part of our approximation to B[x]. In an r.e. permitting construction we will not always be able to make such changes. Specifically, by the stage at which A_s is correct through x, we have become permanently committed to a particular value of B[x]. Thus as $\{A_s\}$ converges to A we become committed to longer and longer initial segments of B. It is the computation function C which determines how much time we have before becoming permanently committed to a given portion of B.

In applying r.e. permitting to our minimal degree construction we encounter the following problem. Suppose that $T_{e,s+1}(\sigma)$ is properly contained in $\beta_s[a_{s+1}]$ and we see a "new" e-splitting of $T_{e,s+1}(\sigma)$ which the rules of our minimal degree construction (1.6) would have us put up on $T_{e,s+1}$. Suppose however that neither branch of this e-splitting is compatible with $\beta_s[a_{s+1}]$ (see Figure 2.2). If we put up this splitting then it will not be possible to define β_{s+1} so as to satisfy (*) <u>and</u> lie on $T_{e,s+1}$.

Figure 2.2.

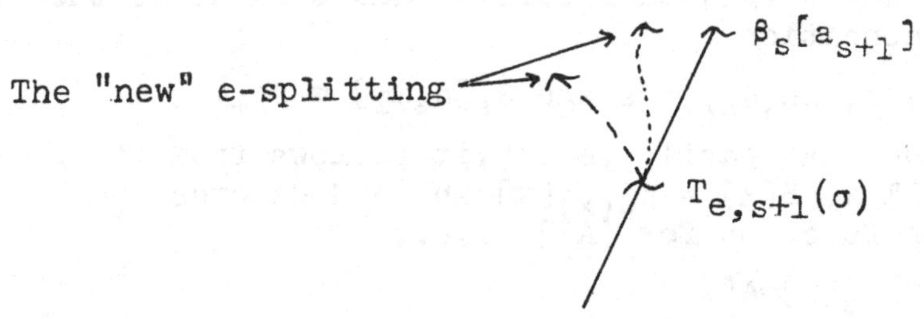

The solution to this problem is simply to not allow such splittings to be put up. More precisely, we add two additional rules to those stated in 1.6.

(i) For all σ, if $T_{e,s+1}(\sigma*0) \neq T_{e,s}(\sigma*0)$ or $T_{e,s+1}(\sigma*1) \neq T_{e,s}(\sigma*1)$ then both $T_{e,s+1}(\sigma*0)$ and $T_{e,s+1}(\sigma*1)$ extend $\beta_s[a_{s+1}]$. (In other words, don't put up a new branching unless <u>both</u> components extend $\beta_s[a_{s+1}]$.)

(ii) $T_{e,s+1}(\emptyset)$ must be compatible with $\beta_s[a_{s+1}]$.

Condition (ii) ensures that (*) will be satisfied and condition (i) ensures that it will be possible to satisfy (ii)

We must show that even with these additional restrictions our construction will still produce a minimal degree. There is clearly no problem in showing that for each e, $\{T_{e,s}\}$ converges pointwise to a total tree T_e. (The additional restrictions only make it harder to change the trees.) Further, the argument that each string on T_e has a final e-state and that the final e-states do not increase along any path remains intact provided that in computing e-states we only count i-splittings which were put up intensionally on $T_{i,s}$. Thus, B will have a final e-state on T_e and so we can define T_e^* just as before. It remains to show that T_e^* satisfies the e-th minimal degree requirement for B.

Suppose for contradiction that every beginning of B is e-splittable on T_e^* but that T_e^* is not an e-splitting tree. Let $\beta \subseteq B$ be on T_e^*. Then, since all strings on on T_e^* have the same final e-state, it was not possible to put up an e-splitting of β on T_e. Why was this not possible? Since β (along with every other string $\subseteq$ B) is e-splittable on T_e^* and any such e-splitting would otherwise be allowed, it must be the case that whenever e-splittings of β having the proper e-1 state (i.e., the final (e-1)-state of B) were found, they were found too late to be of use.

We obtain our contradiction as follows. Let t be such that $\{T_{e,s}\}$ has converged through the level of β and β is in its final e-state by stage t. Let u denote the final e-state of B. Since every beginning of B is e-splittable on T_e^* we can define a recursive function f by

$$f(n) = \mu s > \mathrm{Max}(t,n)\{\ \exists \text{ e-splitting of } \beta, (\sigma,\delta), \text{ such that } \sigma \text{ and } \delta \text{ have Gödel number} < s, \text{ e-state } u \text{ at } s, \text{ and both extend } \beta_s[n+1].$$

We claim that f dominates the computation function for $\{A_s\}$. For suppose that for some n, $A_{f(n)}[n] \neq A[n]$. Consider the first $s > n$ with $a_s \leq n$. Then $\beta_{s-1}[a_s] = \beta_{f(n)}[a_s] \subseteq \beta_{f(n)}[n] \subseteq$ both σ and δ, where σ and δ are as in the definition of f(n). σ and δ have the correct (e-1)-state and are found in time s. Thus there is no reason why σ and δ or some other e-splitting is not put up at stage s. This contaradicts the assumption that no e-splitting of β is put up on T_e. On the other hand, since A is not recursive, no recursive function can dominate the computation function for $\{A_s\}$. We have thus contradicted the assumption that T_e^* does not satisfy the e-th minimal degree requirement for B.

3 <u>Minimal Degrees Below High Degrees</u>. In our final example of a full approximation construction we give another proof that every high degree bounds a minimal degree. The construction used here is based on that used in Cooper's original proof of this theorem (Cooper (1973)). Though this construction is considerably more difficult than the oracle construction of Jockusch (1977) (descrbed in §IIA5), the present construction illustrates an interesting generalization of r.e. permitting to arbitrary Δ_2^0 sets and is likely to be useful in extending other initial segment results

from $\mathcal{D}(\le 0')$ to $\mathcal{D}(\le h)$ for h high.

3.1 <u>Basic Δ_2^0 Permitting</u>. Δ_2 permitting is a generalization of r.e. permitting which can be used in order to construct a set B which is recursive in a given Δ_2 set A. Let A be given and let $\{\alpha_s\}$ be a recursive sequence of strings converging to A. We assume that $lh(\alpha_s) = s$ for all s.

In the simplest form of Δ_2 permitting we construct B recursive in A as follows. Just as in r.e. permitting B is defined as $\lim_s \beta_s$ where $\{\beta_s\}$ is recursive and subject to the following restriction.

(#) For all s and $x \le s$, if $\alpha_s[x] = A[x]$ then $\beta_s[x] = B[x]$.

It is clear that (#) is sufficient to ensure that B is recursive in A. In order to satisfy (#) we must make the following <u>commitment</u> for each s and each $x \le s$:

if $\alpha_s[x] = A[x]$ then for all sufficiently large t, $\beta_t[x] = \beta_s[x]$.

Such a commitment is said to be "binding" if the antecedent is true.

In attempting to honor our commitments we run into a slight problem. Since A will in general not be recursive, there is no effective way of determining at a given stage t, which commitments made prior to t are in fact binding (because we cannot effectively determine whether or not $\alpha_s[x] = A[x]$).

The solution to this problem is quite simple. At stage t we take as binding those commitments made prior to t which <u>appear</u> to be binding on the basis of α_t. Precisely, we require

(##) for all $s < t$ and all x, if $\alpha_s[x] = \alpha_t[x]$ then $\beta_s[x] = \beta_t[x]$.

Since $\{\alpha_s\}$ converges to A it is clear that (##) will guarantee that all truly binding commitments will eventually always be honored and so (#) will be satisfied. Note however that at some stages we will honor commitments which are not in fact binding and fail to honor commitments which are.

In order to keep track of which commitments to honor at each stage it is useful to associate with $\{\alpha_s\}$ a certain partial ordering of stages. For each $s > 0$ let $m(s)$ be the largest $x \geq -1$ such that for some $r < s$, $\alpha_r[x] = \alpha_s[x]$ and let $p(s)$ be the least $r < s$ such that $\alpha_r[m(s)] = \alpha_s[m(s)]$. (In other words, $m(s)$ is the maximum length of agreement between α_s and any α_r, $r < s$, and $p(s)$ is the first stage at which this maximum length of agreement is achieved.) Since $p(s) < s$ for all s, we can define a partial ordering $\prec$ by $t \prec s$ iff for some $n \geq 1$, $t = p^{(n)}(s)$ (where $p^{(1)}(s) = p(s)$ and $p^{(n+1)}(s) = p(p^{(n)}(s))$).

It is useful to picture this partial ordering as a <u>tree</u> of stages with predecessor function p. We will then think of the construction of B as actually being a tree of "potential" constructions corresponding to the different paths on this tree. As we next point out, this tree of stages has only one infinite path and it is the construction corresponding to this one "true" path which actually defines B.

A stage t is said to be <u>true</u> if the length of agreement between α_t and A is > than the length of agreement between α_r and A for all $r < t$. (Equivalently, t is true iff for some $x \geq -1$, $t = \mu s(\alpha_s[x]=A[x])$.) (This terminology was first used by Soare (1976) in connection with r.e. sets.) Certain facts about true stages are easily verified. First, if t is true then the length of agreement between α_t and A is greater than the length of agreement between α_r and α_t for all $r < t$. This implies that if t is true then the commitments which we honor at stage t under (##) are exactly the permanently binding commitments made prior to stage t, i.e., exactly the ones which should be honored. Further, if $t > 0$ and true, then $p(t)$ is the last true stage prior to t and, for all sufficiently large s, $t \prec s$. From these observations it follows that the set of true stages is an infinite path through $\prec$ and is in fact the only infinite path through $\prec$.

Using $\prec$ we can reformulate Δ_2 permitting as follows.

(###) For all s > 0 we have $\beta_{p(s)}[m(s)] \subseteq \beta_s$.

We are thus free to change our approximation to B above m(s) but bound by $\beta_{p(s)}$ below m(s). A simple inductive argument shows that if (###) is satisfied then for all t and s, if t $\prec$ s then $\beta_t[m(t)] \subseteq \beta_s$. Thus, if t is true then $\beta_t[m(t)] \subseteq \beta_s$ for all sufficiently large s and so $B = \bigcup\{\beta_t[m(t)] : t \text{ is true}\}$. Thus, since the set of true stages is clearly recursive in A, it follows immediately from (###) that $\{\beta_s\}$ converges to a set B which is recursive in A (without even appealing to (#)).

3.2 Δ_2 Permitting With Movable Markers. Note that if B is constructed via the form of Δ_2 permitting described in **3.1 then not** only is B recursive in A but B is in fact weak truth-table reducible to A (See Rogers (1967), Exercise 9-45). We achieve the full flexibility of Turing reducibility by the addition of certain "movable markers". We associate with each number x a movable marker M_x which at each stage s of the construction is positioned on some number L(s,x). (The reader who is uncomfortable with the use of mechanical paraphernalia can forget the markers and only consider L.) Condition (#) of 3.1 is then modified to

(†) for all s and x ≤ s, if $\alpha_s[L(s,x)] = A[L(s,x)]$ then $\beta_s[x] = B[x]$.

If, in addition, we require that for each x, {L(s,x)} is bounded (i.e., no marker is pushed unboundedly high during the course of the construction) then it is clear that (†) is still sufficient to ensure that B is recursive in A.

From the point of view of our construction, the purpose of these markers is to give us a certain amount of freedom to delay the stage at which we become finally committed to a particular value of B(x) for a given x. This is accomplished by pushing M_x upward. So long as M_x stays above the length of agreement between α_s and A, we are free to alter the final value of B(x). Eventually, of course, we must allow each M_x to be "captured", i.e., fall below the lenth of agreement between α_s and A, and so make a final commitment with respect to x.

We can formulate this technique in terms of the partial ordering $\prec$ as follow. For each $s > 0$, let $r(s)$ be the largest $x \geq -1$ such that $L(p(s),x) \leq m(s)$ (we let $L(t,-1) = -1$ for all t). We then require

(††) for all $s > 0$ and all $x \leq r(s)$, $\beta_s(x) = \beta_{p(s)}(x)$ and $L(p(s),x) = L(s,x)$.

Thus, for values of $x > r(s)$ we are free to change our approximation to $B(x)$ and reposition M_x. For values of $x \leq r(s)$ we are constrained by the situation at $p(s)$.

It is easily verified by induction that if $t \prec s$ then for all $x \leq r(t)$ we have $\beta_t(x) = \beta_{s}(x)$ and $L(t,x) = L(s,x)$, assuming of course that (††) is satisfied. Applying this in the case that t is true we see that if $\{L(t,x) : t \text{ is true}\}$ is bounded for all x, then $\{\beta_s\}$ will converge to a set B and $B = \cup\{\beta_t[r(t)] : t \text{ true}\}$. Thus, as in 3.1, it will follow that B is recursive in A, simply because the set of true stages is recursive in A.

3.3 <u>The Difference Between R.E. and Δ_2 Permitting</u>. The reader is perhaps wondering why we cannot use Δ_2 permitting to show that <u>every</u> non-recursive degree $\leq 0'$ bounds a minimal degree. (We know this to be false by the initial segment results described in §IA2.) Why can't we use the same technique combined with Δ_2 as we used in combination with r.e. permitting. The problems seem to be the same. We can't put up new e-splittings which are incompatible with $\beta_s[r(s)]$, but can't we argue just as in IIB2 that since the computation function for $\{\alpha_s\}$ is not dominated by any recursive function, we will be able to put up enough e-splittings to ensure that the e-th minimal degree requirement is satisfied?

This argument fails because of an important difference between arbitrary effective approximations and approximations corresponding to enumerations. Notice that if $\{\alpha_s\}$ corresponds to an enumeration of A then for any s, the length of agreement between α_s and A is $\geq$ the length of agreement between α_r and A for any $r < s$. The importance of this is that in r.e. permitting we do not have to worry about the possibility that the approximation to A will lead us to <u>believe</u>

that we are permitted to make a change in the final value of B(x), some x, when in fact we are not. In other words, in r.e. permitting, binding commitments are always honored from the time they are made.

This is not the case with permitting with respect to an arbitrary Δ_2 set. Thus, for example, the length of agreement between α_s and A may drop way down at some stage s leading us to believe that we are permitted to up certain new e-splittings on T_e. At later stages, when the length of agreement comes back up, it would be necessary to take down these branchings in order to satisfy the permitting conditions. One effect of having to take down e-splittings is that even if we could ensure that every branching along B is an e-splitting on T_e there would be no guarantee that the corresponding T_e^* is partial recursive. (How would you know whether a given e-splitting on $T_{e,s}$, even one with the correct final (e-1)-state is permanent or not?) In **fact** the situation is worse. There is no guarantee with an arbitrary Δ_2 set that, even if T_e^* ought to be e-splitting, even a single e-splitting will be put up on T_e. (If it were then it would be possible to show that there are no minimal degrees below 0'!)

3.4 <u>Permitting Relative to a High Degree</u>. The crucial fact about high degrees which allows us to overcome the problems described in 3.3 is the following result of Cooper (1973) (which extends an earlier result of Robinson (1968)).

<u>Lemma 3.4.1</u>. Suppose a is high. Then a contains a set A which has an approximation $\{\alpha_s\}$ whose computation function dominates every recursive function.

Let such an A and $\{\alpha_s\}$ be fixed. We are going to get around the problems described in 3.3 essentially by brute force. Using the fact that the computation function for $\{\alpha_s\}$ dominates every recursive function we are going to arrange things so that **for every** e, if T_e^* ought to be an e-splitting tree then from some point on, we will be permitted to put up <u>every</u> desired e-splitting along B. Though we have the minimal degree construction clearly in mind, we will describe the basic technique of high permitting in a somewhat general setting. We do this partly in the hope that this will be of use in extending other initial segment results to the degrees below an arbitrary high degree.

We suppose that we wish to construct a set B which is recursive in A and satisfies certain requirements $R_0, R_1, \ldots$. We assume that these requirements behave in the following way. During the course of the construction, a given requirement, R_e, may generate certain "requests". Each R_e will generate only finitely many requests per stage. Each such request will be associated with some string. (For example, in the minimal degree requirement, R_e denotes the e-th minimal degree requirement. R_e's requests are requests that certain e-splittings be put up on T_e. The string associated with such a request is the string above which the e-splitting is to be put.) Once made, a request may remain in effect for the remainder of the construction, in which case it is said to be <u>permanently valid</u>, or at some point it may become invalid. At any particular stage certain requests are <u>granted</u> and others are <u>denied</u>. (A particular request may be granted at some stages and denied at others.) We make the following assumptions about the construction and the requirements.

3.4.2 <u>Basic Assumptions About the Requirements</u>.

(a) For any δ,e, and s, at most one R_e request associated with δ is valid at s.

(b) Without violating (††), all valid requests associated with strings $\supseteq \beta_s[r(s)]$ are granted at s.

(c) If a request is valid and granted at stage s then it will be valid and granted at all stages $\succ$ s.

(d) The strings associated with new requests generated at stage s are compatible with $\beta_s[r(s)]$.

(e) For each e the set of permanently valid R_e requests is recursively enumerable.

(f) If $\sigma \subseteq \delta$ and there are permanently valid R_e requests associated with both strings then the R_e request associated with σ was issued at or before the stage at which the permanently valid R_e request associated with δ was issued.

(g) R_e will be satisfied if there is some $\beta \subseteq B$ such that for all δ, $\beta \subseteq \delta \subseteq B$, if there is a (cont'd) permanently valid R_e request associated with δ then this request is eventually always granted.

We can then ensure that $\{\beta_s\}$ will converge to a set B which is recursive in A and satisfies each R_e by using the following rule for positioning the movable markers at stage s.

3.4.3 <u>Rules For Positioning Markers at Stage s</u>. For each $x > r(s)$, position M_x at the least $m \geq x$ such that for all $e \leq x$, if there is no valid R_e request associated with any string $\supseteq \beta_s[x]$ then $m \geq$ total number of valid R_e requests.

The intuition behind 3.4.3 is fairly simple. By 3.4.2 (g), we satisfy a requirement R_e by ensuring that all but finitely many of the permanently valid R_e requests which are associated with strings contained in B are eventually always granted. The difficulty which the Permitting rule (††) creates for us is that by the time a permanently valid R_e request associated with a given string $\delta \subseteq B$ is issued it may be too late to ultimately grant that request (because we have become finally committed to a portion of B beyond δ). One idea which might come to mind is to simply not "allow" $\{\beta_s\}$ to become committed beyond any string β until a valid R_e request associated with some $\delta \supseteq \beta$ is issued. (We can postpone making a final commitment by pushing appropriate markers sufficiently high.) The difficulty with this, of course, is that for some β, there may never be a valid R_e request associated with some $\delta \supseteq \beta$. The solution is to limit the height to which R_e pushes a given marker at a given stage to the total number of valid R_e requests at that stage. The dominance of the computation function for $\{\alpha_s\}$ will ensure that this is enough.

The argument that $\{\beta_s\}$ converges to a set B recursive in A makes no use of the fact that A is high. As noted in 3.2, it suffices to show that for all x, $\{L(t,x) : t \text{ is true}\}$ is bounded. Suppose for contradiction that this is not the case and let y be the least x such that $\{L(t,x) : t \text{ true}\}$ is unbounded. Let $\beta = \bigcup\{\beta_t[r(t)] : t \text{ true}\}$. Then $lh(\beta) = y$ and for all sufficiently large s, $\beta_s[r(s)] \supseteq \beta$. Hence, by conditions 3.4.2 (a) and (d), for any e, all but finitely many R_e requests issued during the construction are associated with strings which are compatible with β. Let m be the total number of R_e requests, $e \leq y$, which are associated with strings which are incompatible with β. Then for any $e \leq y$ and any true stage t, if there are no valid R_e requests associated with strings $\delta \supseteq \beta$ at t, then the total number of valid R_e requests at t is $\leq m + y$. ($lh(\beta) = y$.) Thus under 3.4.3, at no true stage would M_y be pushed higher than $m + y$. This contradicts the assumption that $\{L(t,y) : t \text{ true}\}$ is unbounded. (The intuition behind the argument above is quite simple. We only push M_y when we are permitted to push it. If we were permitted to push it infinitely often then we would have no reason to push it infinitely often.)

The fact that each R_e is satisfied will follow from 3.4.2 (g) and the assumption that the computation function for $\{\alpha_s\}$ dominates every recursive function. By 3.4.2 (g) it suffices to show that there is some $\beta \subseteq B$ such that for all δ, $\beta \subseteq \delta \subseteq B$, if there is a permanently valid R_e request associated with δ then this request is eventually always granted.

We may assume that the set of permanently valid R_e requests is infinite. By 3.4.2(e) the set of such requests is recursively enumerable. Hence we may define a recursive function f by

$f(n) = \mu s$[the number of permanently valid R_e requests issued by stage s and appearing within s steps of some fixed recursive enumeration of the permanently valid requests is $> n$].

(The point is that by stage $f(n)$ at least n many permanently valid requests have been issued by R_e.) Let C be the computation function for $\{\alpha_s\}$ and let m be such that $C(n) > f(n)$ for all $n \geq m$.

We claim that for all $n \geq Max(e,m)$, if $B[n]$ is associated with some permanently valid R_e request then the corresponding request is eventually always granted. Suppose $n \geq Max(e,m)$ and let t be the first true stage with $\alpha_t[L(t,n)] = A[L(t,n)]$. Since $L(t,n) \geq n \geq m$, $C(L(t,n)) > f(L(t,n))$ and so the number of valid R_e

requests at stage t is $> L(t,n)$. Thus, since $r(t) < n$ and $n \geq e$, there musthave been a valid R_e request associated with some string $\delta \supseteq \beta_t[n] = B[n]$. (Otherwise, by 3.4.3 M_n would have been pushed higher than $L(t,n)$.) By 3.4.2(b) all such requests are granted at t and so, by 3.4.2(c) and the fact that t is true, these requests are permanently valid and eventually always granted. By 3.4.2(f) if $B[n]$ is ever associated with a permanent R_e request it must have already been **associated with** such a request by stage t, in which case that request was eventually always granted.

3.5 <u>High Permitting and the Minimal Degree Construction.</u> Applying the method of 3.4 to construct minimal degrees below high degrees using full approximation is fairly straightforward. R_e denotes the requirement that there exist T_e^* satisfying the e-th minimal degree requirement for B. R_e generates requests that certain e-splittings be put up on T_e. The string associated with such a request is the string above which the e-splitting is to be put. There is thus no problem in ensuring that 3.4.2(a),(b),(c),(d), and (f) hold.

One must be slightly careful in defining validity. This should be done in such a way that for each e the permanently valid R_e requests are the R_e requests for e-splittings in (e-1)-states $\leq$ the final (e-1)-state of B. (Formally, one must argue inductively that each R_e is satisfied. Then in making the inductive step from e-1 to e one can assume that B <u>has</u> a final (e-1)-state.) Roughly, the above can be accomplished by ruling invalid at stage s, any requests for e-splittings in (e-1)-states lower than the (e-1)-state of $\beta_s[r(s)]$. Given the above (with the induction argument) it is clear that the set of permanently valid R_e requests is recusively enumerable and so 3.4.2(e) holds.

Finally, by applying 3.4.3 we can ensure that all but finitely many requests for e-splittings along B and in the final (e-1)-state of B are granted. It then follows that B has a final e-state on T_e and so T_e^* can be defined as before and further that T_e^* so defined will satisfy the e-th minimal degree requirement for B.

REFERENCES

Addison, J. (1965). The method of alternating chains. In Theory of Models, North Holland, Amsterdam.

Cooper, S.B. (1971). Doctoral Dissertation, University of Leicester.

Cooper, S.B. (1972a). Degrees of unsolvability complementary between r.e. degrees. Ann. of Math. Logic, 4, no. 1.

Cooper, S.B. (1972b). Jump equivalence of the Δ_2^0 hyperhyperimmune sets. J. Sym. Logic, 37, pp.598-600.

Cooper, S.B. (1973). Minimal degrees and the jump operator. J. Sym. Logic, 38, no. 2.

Epstein, R.L. (1975). Minimal degrees of unsolvability and the full approximation construction. Memoirs of the Am. Math. Soc., no. 162.

Epstein, R.L. (1979). Degrees of unsolvability: structure and theory. Lecture Notes in Math., no. 759, Springer Verlag, Berlin.

Epstein, R.L. (TA). Initial segments of the degrees below 0'. To appear.

Epstein, R.L., Haas, R. & Kramer, R.L. (TA). Hierarchies of sets and degrees below 0'. To appear in Proceedings of the Conference on Mathematical Logic at the University of Connecticut, 1979, Springer Verlag, Berlin.

Ershov, A. (1968). Hierarchies of sets I,II,III. Algebra and Logic, 7, no. 1 and no. 4, 9, no. 1.

Friedberg, R. (1958). Three theorems on recursive enumeration, J. Sym. Logic, 23, pp. 309-316.

Gold (1965). Limiting recursion. J. Sym. Logic, 30, no.1.

Harrington, L. & Shore, R. (1980). Definable degrees. Abstracts of the Am. Math. Soc., 1, no. 1, p. 393.

Jockusch, C. (1969). The degrees of hyperhyperimmune sets. J. Sym. Logic, 34, pp. 489-493.

Jockusch, C. (1977). Simple proofs of some theorems on high degrees. Can. J. Math., XXIX, no. 5.

Jockusch, C. & Posner, D. (1978). Double jumps of minimal degrees. J. Sym. Logic, 43, pp. 715-724.

Jockusch, C. & Posner, D. (TA). Automorphism bases for degrees of unsolvability. To appear.

Jockusch, C. & Soare, R. (1972). Π_1^0 classes and degrees of theories. Trans. Am. Math. Soc.,173, pp. 33-56.

Kleene, S.C. & Post, E. (1954). The upper semilattice of degrees of recursive unsolvability. Ann. of Math., 59, pp. 1108-1109.

Kreisel, G. (1950). Note on arithmetic models for consistent formulae of the predicate calculus. Fund. Math., 37, pp. 265-285.

Lachlan, A. (1966a). The impossibility of finding relative complements for recursively enumerable degrees. J. Sym. Logic, 31, pp. 434-454.

Lachlan, A. (1966b). Lower bounds for pairs of r.e. degrees. Proc. London Math. Soc., 3, pp. 537-569.

Lachlan, A. & Lebeuf, R. (1976). Countable initial segments of the degrees of unsolvability. J. Sym. Logic., 41, pp. 289-300.

Lerman, M. (1977). Automorphism bases for the semilattic of r.e. degrees. Notices of the Am. Math. Soc., 24, A-251.

Lerman, M. (1978). Initial segments of the degrees below 0'. Notices of the Am. Math. Soc.,25, A-506.

Lerman, M. (TA). The degrees of unsolvability. To appear.

Martin, D. (1966). Classes of r.e. sets and degrees of unsolvability. Zeit. f. Math. Logik und Grund. d. Math., 12, pp. 295-310.

Miller, W. & Martin, D. (1968). The degrees of hyperimmune set. Zeit. f. Math. und. Grund. d. Math., 14.

Oxtoby, J. (1971). Measure and Category. Graduate texts in mathematics, Springer Verlag, Berlin.

Posner, D. (1977). High Degrees. Doctoral Dissertation University of California, Berkeley.

Posner, D. (TAa). The upper semilattice of degrees $\leq 0'$ is complemented. To appear in J. Sym. Logic.

Posner, D. (TAb). Minimal degrees and high degrees. To appear.

Posner, D. (TAc). The join theorem for high degrees. To appear.

Posner, D. & Epstein, R. (1978). Diagonalization in degree constructions, J. Sym. Logic, 43, no. 2.

Posner, D. & Robinson, R. (TA). Degrees joining to 0'. To appear in J. Sym. Logic.

Putnam, H. (1965). Trial and error predicates and the solution of a problem of Mostowski. J. Sym. Logic, 30, no. 1.

Robinson, R.W. (1968). A dichotomy of the r.e. sets. Zeit. f. Math. Logik und Grund. d. Math., 14, pp. 339-356.

Robinson, R.W. (1971). Interpolation and embedding in the r.e. degrees. Ann. Math., 93, pp. 285-314.

Rogers, H. (1967). Theory of Recursive Functions and Effective Computability. McGraw-Hill, N.Y.

Sacks, G.E. (1961). A Minimal degree elss than 0'. Bul. Am. Math. Soc., 67.

Sacks, G.E. (1963). Recursive enumerablity and the jump operator. Trans. Am. Math. Soc., 108, pp. 223-239.

Sacks, G.E. (1966). Degrees of Unsolvability, rev. edition. Annals of Math. Studies No. 55., Princeton University press, Princeton, N.J.

Sasso, L. (1970). A cornucopia of minimal degrees. J. Sym. Logic, 35, pp. 383-388.

Sasso, L. (1974). A minimal degree not realizing least possible jump. J. Sym. Logic, 39, pp. 571-573.

Scott, D. & Tennenbaum, S. (1960). On the degrees of complete extensions of arithmetic. Notices of the Am. Math. Soc., 7, pp. 242-243.

Shoenfield, J. (1959). On degrees of unsolvability. Ann. of Math., 69, pp.644-653.

Shoenfield, J. (1966). A theorem on minimal degrees. J. Sym. Logic, 31, no. 4.

Shoenfield, J. (1971). Degrees of Unsolvability. North Holland, Amsterdam.

Shore, R. (1979). The first-order theory of the degrees below 0'. Notices of the Am. Math. Soc., 26, A-617.

Shore, R. (TA). The homogeneity conjecture. To appear in Proc. Nat. Acad. Sci.

Simpson, S. (1977a). Degrees of unsolvability: a survey of results. In Handbook of Math. Logic, North Holland, Amsterdam.

Simpson, S. (1977b). First-order theory of the degrees of recursive unsolvability. Ann. of Math., 105, pp. 121-139.

Soare, R. (1976). The infinite injury priority method. _J. Sym. Logic_, 41, pp. 513-530.

Spector, C. (1956). On degrees of recursive unsolvability. _Ann. of Math._, 69, pp. 644-653.

Thomason, S. (1971). Sublattices of the r.e. degrees. _Zeit. Math. Logik und Grund. d. Math._, 17, pp. 273-280.

Yates, C.E.M. (1965). Three theorems on the degree of recursively enumerable sets. _Duke Math. J._, 37, pp. 461-468.

Yates, C.E.M. (1970a). Initial segments of the degrees of unsolvability, part I: a survey. In _Mathematical Logic and Foundations of Set Theory_. North Holland, Amsterdam.

Yates, C.E.M. (1970b). Initial segments of the degrees of unsolvability, part II: Minimal degrees. _J. Sym. Logic_, 35, no. 2.

Yates, C.E.M. (1976). Banach-Mazur games, comeager sets, and degrees of unsolvability. _Math. Proc. of the Cambridge Phil. Soc._, 79, pp. 195-220.

DEGREES OF GENERIC SETS

Carl G. Jockusch, Jr.
University of Illinois

1. INTRODUCTION

If $\underset{\sim}{a}$ is a degree of unsolvability, let $\mathcal{D}(\leq \underset{\sim}{a})$ denote the set of all degrees $\underset{\sim}{b} \leq \underset{\sim}{a}$. The study of the orderings of such topped initial segments (or principal ideals) $\mathcal{D}(\leq \underset{\sim}{a})$ has been and remains an important part of degree theory. Within this study the investigation of the specific initial segment $\mathcal{D}(\leq \underset{\sim}{0}')$ (see Posner's survey in this volume) and the characterization of the possible order types of arbitrary topped initial segments culminating in Lachlan and Lebeuf (1976) have played a central role. The current paper may be viewed as a study of the ordering of such initial segments $\mathcal{D}(\leq \underset{\sim}{a})$ for almost every degree $\underset{\sim}{a}$, where "almost everywhere" is in the sense of Baire category. Alternatively, it may be viewed as the study of $\mathcal{D}(\leq \underset{\sim}{a})$ when the degree $\underset{\sim}{a}$ is generic, i.e. contains a set of natural numbers which is Cohen-generic for arithmetic. Of course the two points of view are equivalent because the family of generic sets is comeager in $P(\omega)$, and any two generic sets determine the same initial segment of degrees, up to elementary equivalence as partially ordered sets. (Such basic facts about arithmetical forcing and its connection with Baire category are reviewed in §2.)

The results of this paper may also be viewed as illustrative of the power and limitations of the Kleene-Post method of constructing, say, incomparable degrees by making finitely many irrevocable membership decisions at each state. (The connection between the Kleene-Post method and category arguments was pointed out by Myhill (1961).) Strangely enough, though, we show that generic degrees solve Post's problem over lower degrees and use this in connection with a result of Yates about r.e. degrees proved with the priority method to show that $\mathcal{D}(\leq \underset{\sim}{a})$ is not densely ordered (see §5). Finally the results of this paper may be interpreted as dealing with winning strategies in certain Banach-Mazur games (see Oxtoby, 1971 and Yates, 1976).

We assume the reader to be familiar with the basic results of degree theory and of arithmetical forcing. Rogers (1967), Feferman (1965), and Hinman (1969) and (1978) are good references in these areas. Yates (1976) contains a greatly generalized study of applications of the category method and Banach-Mazur games to degrees of unsolvability.

We now summarize the main results of the paper. Assume throughout this paragraph that the degree $\underset{\sim}{a}$ is generic. In §3, we observe that constructions in the literature may be used to show that $\mathcal{D}(\leq \underset{\sim}{a})$ is not a lattice but has many lattices embedded in it. In §4, we paraphrase and give a new proof of a result of Martin (1967) on category to show that every nonzero degree below $\underset{\sim}{a}$ bounds a generic degree. Thus $\mathcal{D}(\leq \underset{\sim}{a})$ has no atoms (minimal degrees), and in fact no nontrivial initial segment of $\mathcal{D}(\leq \underset{\sim}{a})$ is a lattice. Such results make it natural to conjecture that $\mathcal{D}(\leq \underset{\sim}{a})$ consists only of generic degrees (except for $\underset{\sim}{0}$) and is dense. Both of these conjectures are refuted in §5 as a consequence of the result proved there that $\underset{\sim}{a}$ is r.e. in some strictly lower degree. In §6, it is proved that $\mathcal{D}(\leq \underset{\sim}{a})$ has a strong "cupping up" property and that, in $\mathcal{D}(\leq \underset{\sim}{a})$, every nonzero degree bounds a nonzero complemented degree. In §7, a number of open problems are listed.

Almost all of the results of this paper apply not only to generic sets but to the wider class of sets which are generic for n-quantifier arithmetic, for some specified n (usually 1 or 2) depending on the theorem. This added generality is important for obtaining sharp existence theorems because, for each n, there is an n-generic set of degree below $\underset{\sim}{0}^{(n)}$. Moreover, for n = 1, Jockusch & Posner (1978) have shown that every degree $\underset{\sim}{b}$ which satisfies $\underset{\sim}{b}'' > (\underset{\sim}{b} \cup \underset{\sim}{0}')'$ bounds the degree of a 1-generic set. (In constrast, it is shown in §5 that such degrees $\underset{\sim}{b}$ need not bound minimal degrees.) Since the degrees below even a 1-generic degree cannot form a lattice, it follows that if $\mathcal{D}(\leq \underset{\sim}{b})$ is a lattice, then $\underset{\sim}{b}'' = (\underset{\sim}{b} \cup \underset{\sim}{0}')'$. This connection between the ordering of degrees and the jump operation (which does not mention genericity) has been of use in studying global questions concerning automorphisms and definability in the degrees (see Epstein (1979) p. 109; Nerode & Shore (to appear) §5; Shore (to appear).)

Our notation and terminology are quite standard. In particular a <u>string</u> is a mapping from a finite initial segment of $\omega = \{0,1,2,\ldots\}$ into $\{0,1\}$. Letters such as σ, τ, μ are reserved for strings. We say $\tau \supseteq \sigma$ (τ extends σ) if $\tau(n) = \sigma(n)$ whenever $\sigma(n)$ is defined. A subset A of ω <u>extends</u> σ (written $A \supseteq \sigma$) if the characteristic function of A extends σ. Two partial functions are <u>incompatible</u> if there is an argument on which they are defined and unequal. Strings may also be regarded as finite sequences of 0's and 1's. From this point of view, $|\sigma|$ is the length of σ and $\sigma * \tau$ is the string which results from concatenating σ and τ. If S is a set of strings and $A \subseteq \omega$, then A <u>meets</u> S if A extends some string in S. A set S of strings is <u>dense</u> if for every string σ there is a string τ in S which extends σ. We apply notions of recursion theory to strings via Gödel-numbering.

Let $P(\omega)$ be the power set of ω. We give $P(\omega)$ its usual product topology. Thus a set $A \subseteq P(\omega)$ is comeager if there exist dense sets of strings $S_0, S_1, \ldots$ such that every set $A \subseteq \omega$ which meets each S_n is in A.

An <u>operator</u> is a function from a subset of $P(\omega)$ into $P(\omega)$. The <u>partial recursive</u> operators are those which are determined by recursive reduction procedures. The letters Δ, Γ, Φ are used for operators. Associated with each partial recursive operator Δ is a recursive mapping (also denoted Δ) from strings to strings which is consistent in the sense that $\Gamma(\tau) \supseteq \Gamma(\sigma)$ whenever $\tau \supseteq \sigma$. Conversely, each such consistent recursive mapping from strings to strings gives rise to a partial recursive operator. Let $\Phi_0, \Phi_1, \ldots$ be an effective enumeration of all partial recursive operators.

Let $<.,.>$ be a recursive bijection from ω^2 to ω such that $i \leq <i,j>$ for all $i,j \in \omega$. Define $(n)_0$, $(n)_1$ by $n = <(n)_0,(n)_1>$. For $i \in \omega$, $A \subseteq \omega$, let $(A)_i = \{j : <i,j> \in A\}$. For $a,b \in \omega$, let $[a,b) = \{i \in \omega : a \leq i < b\}$. Let W_e be the eth r.e. subset of ω in some standard enumeration of all such sets.

The author wishes to thank D. A. Martin and others for sharing ideas and information with him so generously. This research was supported by the National Science Foundation.

2. PRELIMINARIES ON FORCING, GENERICITY, AND CATEGORY

Our aim is to study the degrees of unsolvability of sets generic for arithmetic in the sense of Feferman (1965). The purpose of this section is to fix notation and review elementary background information about forcing and genericity. We assume the reader already has some acquaintance with these concepts. We also take a look at the many-one degree (m-degree) analogue of our main problem, which is closely related to the inclusion lattice of r.e. sets modulo finite sets.

Let L^* be a language for first order number theory augmented by a unary predicate symbol $\underline{A}$. If φ is a sentence of L^* and $A \subseteq \omega$, then $A \models \varphi$ means that φ is true in the expansion of the standard model of number theory obtained by interpreting $\underline{A}$ as A. The subset of $P(\omega)$ <u>defined</u> by such a sentence φ is $\{A \subseteq \omega : A \models \varphi\}$. A formula φ of L^* having k free numerical variables defines in a similar way a subset of $\omega^k \times P(\omega)$. The relation $\sigma \Vdash \varphi$ (σ forces φ), for strings σ and sentences φ of L^*, is defined as usual by induction on sentences (see Feferman (1965) or Hinman (1969)). Then $A \Vdash \varphi$ (for $A \subseteq \omega$ and φ a sentence of L^*) means that $\sigma \Vdash \varphi$ for some $\sigma \subseteq A$.

Definition 2.1. A set $A \subseteq \omega$ is generic if for every sentence φ of L^*, $A \Vdash \varphi$ or $A \Vdash \neg\varphi$.

The following lemma is standard.

Lemma 2.2.

(i) If A is generic and φ is any sentence of L^*, then $A \Vdash \varphi$ iff $A \models \varphi$.

(ii) For any fixed sentence φ of L^*, $\{\sigma : \sigma \Vdash \varphi\}$ is an arithmetical set of strings.

(iii) The family G of all generic sets is comeager in $P(\omega)$.

Of course (iii) holds because, by the definition of forcing, for any fixed sentence φ the family of strings which force either φ or $\neg\varphi$ is dense.

The next lemma (again standard) gives two characterizations of genericity which do not mention forcing.

Lemma 2.3. The following are equivalent for any set $A \subseteq \omega$:

(a) A is generic

(b) for any arithmetical set S of strings, there is a string $\sigma \subseteq A$ such that either σ is in S or no string $\tau \supseteq \sigma$ is in S

(c) for every comeager arithmetical subset $\mathcal{A}$ of $P(\omega)$, $A \in \mathcal{A}$.

Proof. The result may be easily proved in the order (b) $\Rightarrow$ (a) $\Rightarrow$ (c) $\Rightarrow$ (b). To prove (b) $\Rightarrow$ (a), for a given sentence φ of L^*, let $S = \{\sigma : \sigma \Vdash \varphi\}$. Then S is arithmetical by Lemma 2.2(ii), and if no $\tau \supseteq \sigma$ is in S, then $\sigma \Vdash \neg\varphi$ by definition. To prove (a) $\Rightarrow$ (c), let φ be a formula defining a given comeager $\mathcal{A} \subseteq P(\omega)$. Since the intersection of two comeager subsets of $P(\omega)$ is again comeager, it follows from the Baire category theorem and Lemma 2.2(iii) that any string σ has a generic extension $B \in \mathcal{A}$. Thus by Lemma 2.2(i), no string σ forces $\neg\varphi$. Hence every generic set forces φ and hence belongs to $\mathcal{A}$ by Lemma 2.2(i). To prove (c) $\Rightarrow$ (b), let S be a given arithmetical set of strings. Define $\mathcal{A}$ to be the family of all sets A such that for some $\sigma \subseteq A$ either $\sigma \in S$ or no $\tau \supseteq \sigma$ is in S. Then $\mathcal{A}$ is an arithmetical subset of $P(\omega)$ and $\mathcal{A}$ is comeager. (In fact $\mathcal{A}$ is open and dense.) Applying (c) to S yields (b) immediately.

Of course generic sets are never arithmetical, although there do exist hyperarithmetical generic sets. On the other hand, existence theorems in degree theory almost always yield degrees of arithmetical sets with the property in question. To overcome this discrepancy, we consider as in Hinman (1969) or (1978), a series of weaker notions than genericity. As usual, certain formulas of L^* are inductively classified as Σ^0_n or Π^0_n. In particular, the Π^0_n formulas are just the negations of Σ^0_n formulas, and, for $n \geq 1$, the Σ^0_{n+1} formulas are those obtained from Π^0_n formulas by existential quantification. (We are using the notation Σ^0_n ambiguously to classify formulas of L^* and subsets of $\omega^k \times P(\omega)^\ell$. Of course the two uses are closely related since a subset of $\omega^k \times P(\omega)^\ell$ (with $\ell \leq 1$) is Σ^0_n iff there is a Σ^0_n formula of L^* which defines it.)

<u>Definition 2.4</u>. If $A \subseteq \omega$ and $n \geq 1$, then A is <u>n-generic</u> if for every Σ^0_n sentence φ of L^*, $A \Vdash \varphi$ or $A \Vdash \neg\varphi$.

Analogously to the results mentioned earlier for generic sets, we have the following for n-generic sets. (See Hinman (1969) where Σ^0_n, Π^0_n are denoted $\bigvee_n(\alpha)$, $\bigwedge_n(\alpha)$ respectively.)

<u>Lemma 2.5</u>. Let $n \geq 1$.

(i) If A is n-generic and φ is Σ^0_n or Π^0_n, then

$$A \Vdash \varphi \Leftrightarrow A \models \varphi.$$

(ii) $\{\langle\sigma,\varphi\rangle : \varphi$ is a Σ^0_n sentence and $\sigma \Vdash \varphi\}$ is Σ^0_n and hence recursive in $0^{(n)}$. Similarly, forcing for Π^0_n sentences is Π^0_n.

The following lemma is standard.

<u>Lemma 2.6</u>.

(i) If $n \geq 1$, there is an n-generic set A such that $A \leq_T 0^{(n)}$.

(ii) If A is n-generic, then $A^{(n)} \equiv_T A \oplus 0^{(n)}$.

<u>Proof</u>. Part (i) is immediate since Lemma 2.5(ii) shows that the usual construction of a generic set may be carried out recursively in $0^{(n)}$.

To prove part (ii), let $\varphi(x,\underline{A})$ be a Σ^0_n formula which defines the relation "$x \in A^{(n)}$". Then by Lemma 2.5(i), one has

$$i \in A^{(n)} \Leftrightarrow (\exists\sigma)\ [\sigma \subseteq A \text{ and } \sigma \Vdash \varphi(i,\underline{A})].$$

The bracketed portion is recursive in $A \oplus 0^{(n)}$ by Lemma 2.5(ii) so $A^{(n)}$ is r.e. in $A \oplus 0^{(n)}$. Replacing φ by $\neg\varphi$ above shows that the complement of $A^{(n)}$ is also r.e. in $A \oplus 0^{(n)}$, so

$A^{(n)} \leq_T A \oplus 0^{(n)}$. Since $A \oplus 0^{(n)} \leq_T A^{(n)}$ holds for every A, the proof is complete.

The next result gives a characterization of n-genericity which does not involve forcing. The result for $n = 1$ appeared in D. Posner's dissertation (1977).

Lemma 2.7. Suppose $n \geq 1$ and $A \subseteq \omega$. Then the following two assertions are equivalent.

(a) A is n-generic

(b) for every Σ_n^0 set S of strings, there exists a string $\sigma \subseteq A$ such that either σ is in S or no extension of σ is in S.

Proof. To show (b) ⇒ (a), given φ let $S = \{\sigma : \sigma \Vdash \varphi\}$. To show (a) ⇒ (b), given S let $\mathcal{A} = \{B \subseteq \omega : (\exists\sigma)[\sigma \in S \text{ and } \sigma \subseteq B]\}$. Let φ be a Σ_n^0 sentence of L^* which defines $\mathcal{A}$ and let τ be a string extended by A which forces φ or forces $\neg\varphi$. If $\tau \Vdash \varphi$, then $A \in \mathcal{A}$. If $\tau \Vdash \neg\varphi$, then no extension σ of τ is in S. (Otherwise consider a generic set B which extends σ and get a contradiction since $B \in \mathcal{A}$ and $B \Vdash \neg\varphi$.)

We call a degree $\mathbf{a}$ generic if there is a generic set of degree $\mathbf{a}$. (The n-generic degrees are defined analogously.) If $\mathbf{a}$ is a degree, let $\mathcal{D}(\leq \mathbf{a})$ be the initial segment of degrees $\leq \mathbf{a}$, or more precisely the restriction of the usual partial ordering of degrees to the set of degrees $\mathbf{b} \leq \mathbf{a}$. Our main intent is to study the partial ordering $\mathcal{D}(\leq \mathbf{a})$ when $\mathbf{a}$ is generic. The next proposition shows that the theory of this structure is independent of the choice of the generic degree $\mathbf{a}$.

Proposition 2.8. If $\mathbf{a}$ and $\mathbf{b}$ are both generic degrees, then the structures $\mathcal{D}(\leq \mathbf{a})$, $\mathcal{D}(\leq \mathbf{b})$ are elementarily equivalent.

Proof. Let φ be any sentence of the language of partial orderings. Let $\mathcal{A}$ be the family of all sets $A \subseteq \omega$ whose degree $\mathbf{a}$ is such that $\mathcal{D}(\leq \mathbf{a})$ satisfies φ. A routine argument shows that $\mathcal{A}$ is an arithmetical subset of $P(\omega)$ and hence is definable by some formula ψ of L^*. Choose any string σ such that $\sigma \Vdash \psi$ or $\sigma \Vdash \neg\psi$. Every generic set differs only finitely from some generic set extending σ. Since $\mathcal{A}$ is invariant under Turing equivalence and hence under finite differences, it follows that either every generic set is in $\mathcal{A}$ or no generic set is in $\mathcal{A}$. (Alternatively, this result could be proved by using the 0 - 1 law for category (Oxtoby, (1971)) to show that $\mathcal{A}$ is meager or comeager and applying Lemma 2.3.)

It is not known whether the isomorphism type of $\mathcal{D}(\leq \mathbf{a})$ is independent of $\mathbf{a}$ for generic $\mathbf{a}$. We will point out in §7 that

Theorem 5.1 and a plausible conjecture about initial segments of degrees suggest a negative answer. On the other hand, when Turing degrees are replaced by m-degrees, the corresponding initial segments below generic sets are determined up to isomorphism, as the following proposition shows.

Proposition 2.9. If A is a 1-generic set, the ordering of m-degrees of sets m-reducible to A (other than ϕ and ω) is isomorphic to the inclusion ordering of r.e. sets modulo finite sets.

Proof. Let A be 1-generic. We claim that the required isomorphism may be obtained by mapping, for each recursive function f, the m-degree of $f^{-1}(A)$ to the equivalence class modulo finite sets of $\mathrm{rng}(f)$. (Here $\mathrm{rng}(f)$ denotes the range of f, and we exclude the case where $f^{-1}(A)$ is ϕ or ω.) To show that this mapping gives an isomorphism amounts to showing that for any recursive functions f and g,

$$f^{-1}(A) \leq_m g^{-1}(A) \Leftrightarrow (\mathrm{rng}(f) - \mathrm{rng}(g)) \text{ is finite.}$$

The implication from right to left is trivial and uses only the (tacit) assumption that $g^{-1}(A)$ is neither ϕ nor ω. For the other direction assume that $\mathrm{rng}(f) - \mathrm{rng}(g)$ is infinite. Given a recursive function h it must be shown that $f^{-1}(A) \neq h^{-1}g^{-1}(A)$. Let S be the set of strings σ which make this condition hold, i.e. such that there exists n such that $\sigma(f(n))$ and $\sigma(gh(n))$ are defined and different. Then S is a Σ_1^0 set of strings, and S is dense (i.e. every string has an extension in S) because $\mathrm{rng}(f) - \mathrm{rng}(g)$ is infinite. Since A is 1-generic, there is a string $\sigma \subseteq A$ such that $\sigma \in S$, so $f^{-1}(A) \neq h^{-1}(g^{-1}(A))$. Therefore $f^{-1}(A) \nleq_m g^{-1}(A)$.

If A is n-generic and f is a 1 - 1 recursive function, then $f^{-1}(A)$ is also n-generic, as may be seen quite easily from Lemma 2.7. Since every infinite r.e. set is the range of a 1 - 1 recursive function, the following result now follows from the proof of Proposition 2.9.

Proposition 2.10. If A is n-generic, then every nonrecursive m-degree below that of A has an n-generic representative. The same result holds with "generic" in place of "n-generic."

In §§4 and 5 it will be shown that the analogue of Proposition 2.10 for Turing reducibility is false although a weaker but still useful version of the Turing analogue of Proposition 2.10 is true.

3. CONSEQUENCES OF CLASSICAL CONSTRUCTIONS

It is well known that many classical constructions in recursion theory show that some arithmetical subset of $\mathcal{P}(\omega)$ is comeager and thus give information about generic sets. For instance, the Kleene-Post (1954) construction of incomparable degrees shows that every generic (even 1-generic) set is of the form $A \oplus B$, where A, B are Turing incomparable (Jockusch & Posner (1978), Lemma 2). In the same way, if A is 1-generic, then the degrees of its "components" $(A)_0$, $(A)_1, \ldots,$ are strongly independent in the sense that for no i is $(A)_i$ recursive in $\{<j,k> \in A : j \neq i\}$. It follows as in Sacks (1963) that if $\underset{\sim}{a}$ is 1-generic then every countable partially ordered set can be embedded in $\mathcal{D}(\leq \underset{\sim}{a})$. Similarly it may be shown as in Yates (1976) that if $\mathcal{L}$ is any countable distributive lattice and $\underset{\sim}{a}$ is 1-generic, then there is a lattice embedding from $\mathcal{L}$ into $\mathcal{D}(\leq \underset{\sim}{a})$. Also the argument of Thomason (1970), Theorem 6 shows that if $\underset{\sim}{a}$ is 2-generic, then every finite lattice is lattice embeddable in $\mathcal{D}(\leq \underset{\sim}{a})$. Similarly, it may be shown as in Thomason (1969) that if $\underset{\sim}{a}$ is 1-generic, then the lattice consisting of least and greatest elements and a denumerably infinite set of pairwise incomparable (and hence complementary) elements can be embedded in $\mathcal{D}(\leq \underset{\sim}{a})$. Also it is easy to show, using a slight modification of J. R. Shoenfield's 1959 construction of a non-r.e. degree below $\underset{\sim}{0}'$ that if $\underset{\sim}{a}$ is 1-generic, then there is no nonzero r.e. degree below $\underset{\sim}{a}$.

The Kleene-Post (1954) and Spector (1956) proofs that the degrees do not form a lattice are not category arguments per se since they construct upper bounds to given chains of degrees. Nonetheless, the proof of the following theorem shows that they can be construed as category arguments.

Theorem 3.1. If $\underset{\sim}{a}$ is 1-generic, then $\mathcal{D}(\leq \underset{\sim}{a})$ is not a lattice.

Proof. Let A be a 1-generic set. To prove the theorem we must find sets B, C, each recursive in A such that the degrees of B, C have no greatest lower bound. The set B will be chosen so that $(B)_i = (A)_{3i}$ for all i. The set C will be chosen so that the symmetric difference of $(C)_i$ and $(B)_i$ (denoted $(C)_i \Delta (B)_i$) is finite for each i. (The precise choice of the finite set F_i such that $(C)_i = (B)_i \Delta F_i$ will depend on $(A)_{3i+1}$ and $(A)_{3i+2}$.) It can then be shown (as in Kleene-Post (1954)) that every set recursive in both B and C is recursive in some finite join of $(B)_i$'s, from which it follows easily that the degrees of B and C have no greatest lower bound.

For any set A and any $i < \omega$, let

$$\Delta_i(A) = \{j : \langle 3i+1,j\rangle \in A \ \&\ (\forall k \le j)[\langle 3i+2,k\rangle \in A]\}$$

If A is 1-generic, then A has no infinite r.e. subset, so $\Delta_i(A)$ is finite for each $i < \omega$. Letting $\Delta_i(A)$ play the role of F_i, we now define the functionals Γ, Θ which will map A to B and C, respectively. For any set A, let $\Gamma(A)$, $\Theta(A)$ be the unique sets which satisfy

$$(\Gamma(A))_i = (A)_{3i}$$

$$(\Theta(A))_i = (A)_{3i} \ \Delta\ \Delta_i(A)$$

for all $i < \omega$. Define $\Delta_i(\sigma)$, $\Gamma(\sigma)$, $\Theta(\sigma)$ for strings σ in the obvious way.

Assume again that A is 1-generic, and let $B = \Gamma(A)$, $C = \Theta(A)$. We show that if $\Phi_b(B)$ and $\Phi_c(C)$ are the same total function, then that function is recursive in a finite join of sets of the form $(A)_{3i}$. Since each $(A)_{3i}$ is recursive in $\Gamma(A)$ and in $\Theta(A)$, it will then follow from the strong independence property of the components of A (mentioned earlier in this section) that the degrees of B and C have no greatest lower bound. Let S be the set of strings σ such that $\Phi_b(\Gamma(\sigma))$ and $\Phi_c(\Theta(\sigma))$ are incompatible. S is obviously a recursive set of strings, so there is a string $\nu \subseteq A$ such that $\nu \in S$ or $\tau \notin S$ for all $\tau \supseteq \nu$. If $\nu \in S$, then $\Phi_b(B)$, $\Phi_c(C)$ are incompatible. Assume that $\tau \notin S$ for all $\tau \supseteq \nu$ and that $\Phi_b(B)$, $\Phi_c(C)$ are each total. Call a string ρ <u>admissible</u> if ρ is compatible with the restriction of the characteristic function of B to numbers $\langle i,j\rangle$ with $i \le |\nu|$, $j < \omega$. Clearly the set of admissible strings is recursive in a finite join of $(A)_{3i}$'s. We claim now that $\Phi_b(B;x) = \Phi_b(\rho;x)$ whenever ρ is admissible and both sides are defined. If the claim is false there is a string μ such that $\mu \subseteq C$ and $\Phi_c(\mu)$, $\Phi_b(\rho)$ are incompatible, where ρ is admissible. But given such μ, ρ there is a string σ such $\sigma \supseteq \nu$, $\Gamma(\sigma) \supseteq \rho$, and $\Theta(\sigma) \supseteq \mu$. The construction of such a σ is quite straightforward and we omit the details. Intuitively speaking, σ exists because the requirement $\sigma \supseteq \nu$ leaves $\Gamma(\sigma)$, $\Theta(\sigma)$ independent of each other on arguments $\langle i,j\rangle$ with $i > |\nu|$, and other arguments are taken care of by the assumption that ρ is admissible and $C \supseteq \mu$. (The formal verification uses the facts that $i \le \langle 3i,j\rangle, \langle 3i+1,j\rangle, \langle 3i+2,j\rangle$ for all i and j.) Now any such σ is in S, which contradicts our choice of ν. Thus the claim is proved, and $\Phi_b(B;x)$ can be computed recursively in a finite join of $(A)_{3i}$'s by searching for any admissible ρ with $\Phi_b(\rho;x)$ defined and computing $\Phi_b(\rho;x)$.

4. MARTIN'S CATEGORY THEOREM

D. A. Martin (1967) showed that if $\mathcal{A}$ is a meager set of degrees not containing $\underset{\sim}{0}$, and $\mathcal{A} \cup \{\underset{\sim}{0}\}$ is an initial segment of the degrees, then the upward closure of $\mathcal{A}$ is also meager. Martin never published his proof, but a proof of a generalized version is given by Yates (1976). A typical application of Martin's result is obtained by letting $\mathcal{A}$ be the class of minimal degrees and concluding that the set of degrees which have minimal predecessors is meager. It follows that if $\underset{\sim}{a}$ is generic, then $\mathcal{D}(\leq \underset{\sim}{a})$ has no minimal elements. Another application of Martin's result is obtained by letting $\mathcal{A}$ be the class of non-zero degrees which bound no n-generic degree. By Martin's theorem the upward closure of $\mathcal{A}$ is meager. Since the class of n-generic sets is arithmetical for each n, it follows that for each generic degree a and each n, each nonzero degree $\underset{\sim}{b} \leq \underset{\sim}{a}$ bounds an n-generic degree. Below we prove a sharpening of this result. Our proof is similar in spirit to the proofs in Martin (1967) and Yates (1976), although we believe our proof is somewhat simpler. It is inessential that we have formulated the result in terms of genericity rather than category since Martin's theorem (and indeed Yates' improvement thereof) may be easily deduced from our lemmas, and the formulation here may easily be gotten from the proofs in Martin (1967) and Yates (1976).

<u>Theorem 4.1</u> (Martin). Suppose $n \geq 2$. If $\underset{\sim}{a}$ is an n-generic degree and $\underset{\sim}{0} < \underset{\sim}{b} \leq \underset{\sim}{a}$, then there exists an n-generic degree $\underset{\sim}{c}$ such that $\underset{\sim}{c} \leq \underset{\sim}{b}$. (If $\underset{\sim}{a}$ is generic, $\underset{\sim}{c}$ may also be chosen to be generic.)

<u>Proof</u>. We first narrow down the class of partial recursive operators we need to consider. If Ψ is a partial recursive operator, we say that strings σ_0, σ_1 are <u>Ψ-split</u> if $\Psi(\sigma_0)$ and $\Psi(\sigma_1)$ are incompatible.

<u>Definition 4.2</u>. A partial recursive operator Ψ is called <u>totally splittable</u> if

(i) every string has a Ψ-split pair of extensions, and

(ii) for every string σ and $x \in \omega$ there exists $\tau \supseteq \sigma$ with $\Psi(\tau;x)$ defined.

<u>Lemma 4.3</u>. Suppose that A is 2-generic and $0 <_T B \leq_T A$. Then there is a totally splittable partial recursive operator Ψ such that $\Psi(A) = B$.

<u>Proof</u>. Let Φ be a partial recursive operator such that $\Phi(A) = B$. Let S be the set of strings σ such that either

(i) $(\exists x)(\forall\tau \supseteq \sigma)[\Phi(\tau;x)$ is undefined], or

(ii) σ has no Φ-split pair of extensions.

Then S is a Σ_2^0 set of strings, and A extends no string in S because $\Phi(A)$ is total and nonrecursive. Since A is 2-generic, we may choose by Lemma 2.7 a string $\nu \subseteq A$ such that no $\sigma \supseteq \nu$ is in S. Define Ψ on strings of length at least $|\nu|$ by $\Psi(\sigma * \tau) = \Phi(\nu * \tau)$ whenever $|\sigma| = |\nu|$ and τ is any string. Then $\Psi(A) = B$ because $A \supseteq \nu$, and Ψ is totally splittable because no extension σ of ν is in S.

As in initial segment constructions with admissible triples, we define a <u>tree</u> to be a triple (g,f_0,f_1) of recursive functions such that g is strictly increasing, f_0, f_1 are 0-1 valued and, for each n, the restrictions of the functions f_0, f_1 to the interval $[g(n),g(n+1))$ are incompatible. Fix such a tree $T = (g,f_0,f_1)$. Let $I(n)$ denote the interval $[g(n),g(n+1))$, for $n < \omega$. For $n < \omega$, $i \leq 1$, let $T(n,i)$ be the set of strings of length $g(n+1)$ which are compatible with f_i on $I(n)$. If Ψ is a partial recursive operator, call T <u>strongly Ψ-splitting</u> if for each $n < \omega$ the following two conditions hold:

(a) σ_0 and σ_1 are Ψ-split whenever $\sigma_0 \in T(n,0)$, $\sigma_1 \in T(n,1)$.

(b) $|\Psi(\sigma)| \geq g(n)$ whenever $\sigma \in T(n,0) \cup T(n,1)$.

Note that condition (a) is somewhat stronger than the usual notion of "splitting tree" because $T(n,i)$ may include many strings which are not compatible with either f_0 or f_1 on intervals $I(m)$ for $m < n$. However, the construction of a strongly Ψ-splitting tree in the following lemma is essentially identical to the usual splitting tree construction.

<u>Lemma 4.4</u>. If Ψ is a totally splittable partial recursive operator, then there exists a tree $T = (g,f_0,f_1)$ which is strongly Ψ-splitting.

<u>Proof</u>. The tree T is constructed by an inductive procedure. To start the induction, let τ_0, τ_1 be any Ψ-split pair of strings of the same length ℓ. Let $g(0) = \ell$ and $f_i(x) = \tau_i(x)$ for all $x < \ell$. Assume now that $g(n)$ and the restrictions of f_0, f_1 to the interval $[0,g(n))$ have already been defined We will define two incompatible strings τ_0, τ_1 of the same length ℓ and then let $g(n+1) = g(n) + \ell$, $f_i(g(n)+j) = \tau_i(j)$ for $i \leq 1$, $j < \ell$. Each τ_i is obtained as the final term τ_i^p of an ascending sequence of $2^{g(n)}$ strings, $\tau_i^0 \subseteq \ldots \subseteq \tau_i^p$ where $p = 2^{g(n)} - 1$. Let $\mu_0,\ldots,\mu_p$ be all strings of length $g(n)$.

The strings τ_0^k, τ_1^k are chosen by induction on k so that $\tau_i^k \supseteq \tau_i^{k-1}$, $|\tau_0^k| = |\tau_1^k|$, $|\Psi(\mu_k * \tau_i^k)| \geq g(n)$ for $i \leq 1$, and $\mu_k * \tau_0^k$, $\mu_k * \tau_1^k$ are Ψ-split. Such strings τ_0^k, τ_1^k must exist because for any σ there is no bound on $|\Psi(\tau)|$ for $\tau \supseteq \sigma$, and any pair of strings can be componentwise extended to a Ψ-split pair of strings, as may be easily seen from that fact that Ψ is totally splittable.

If Θ is an operator, a string σ is called <u>Θ-good</u> if Θ maps the set of strings extending σ <u>onto</u> the set of strings extending $\Theta(\sigma)$. The following lemma is the heart of the argument and is essentially the same as Corollary 5.2 of Yates (1976). However, our operator Φ is <u>total</u> (i.e. maps total functions to total functions), whereas the corresponding operators in the proofs of Martin and Yates apparently are not.

<u>Lemma 4.5</u>. If Ψ is a totally splittable partial recursive operator, then there is a total recursive operator Φ such that, if $\Theta = \Phi \circ \Psi$, there is a dense recursive set of Θ-good strings.

<u>Proof</u>. Let $T = (g, f_0, f_1)$ be a strongly Ψ-splitting tree. We will construct Φ so that $\Theta(\sigma;n) = i$ whenever $\sigma \in T(n,i)$, where $\Theta = \Phi \circ \Psi$. The operator Φ is defined by associating with each string μ a string $\Phi(\mu)$. In order to be sure that the domain of $\Phi(\mu)$ is a finite initial segment of ω, we require that $\Phi(\mu;n)$ be defined only when $n < |\mu|$ and $\Phi(\mu;m)$ is defined for each $m < n$. Assuming that these conditions are satisfied, we set $\Phi(\mu;n) = i$ if $\mu \supseteq \Psi(\sigma)$ for any $\sigma \in T(n,i)$. Also we arbitrarily set $\Phi(\mu;n) = 0$ if μ is incompatible with $\Psi(\sigma)$ for each $\sigma \in T(n,0) \cup T(n,1)$. (In all other cases, $\Phi(\mu;n)$ is undefined.) $\Phi(\mu)$ is single-valued because T satisfies condition (a) in the definition of "strong splitting tree." Also Φ is consistent, i.e. $\Phi(\tau) \supseteq \Phi(\sigma)$ whenever $\tau \supseteq \sigma$ as may easily be seen from the consistency of Ψ. To see that Φ induces a total mapping from $P(\omega)$ to $P(\omega)$, observe that $\Phi(\mu;n)$ is defined whenever $|\mu| \geq |\Psi(\sigma)|$ for all σ in $T(m,i)$ for any $m \leq n$, $i \leq 1$. Finally if $\sigma \in T(n,i)$, then $\Phi(\Psi(\sigma);m)$ is defined for $m < n$ since $|\Psi(\sigma)| \geq g(n) \geq |\tau| \geq |\Psi(\tau)|$ whenever $\tau \in T(n,i)$ with $m < n$, $i \leq 1$. From this it follows that $\Phi(\Psi(\sigma);n) = \Theta(\sigma;n) = i$ whenever $\sigma \in T(n,i)$.

Let S be the set of all strings in $T(n,i)$ for any $n < \omega$, $i \leq 1$. S is obviously a dense recursive set of strings, so it suffices to show that every string in S is Θ-perfect. Suppose $\sigma \in S$, say $\sigma \in T(n,i)$. Then $|\Theta(\sigma)| = n + 1$. (To prove this observe that $\Theta(\sigma;n) = i$ since $\sigma \in T(n,i)$ and $\theta(\sigma;n+1)$ is undefined because σ has an extension in $T(n+1,0)$ and an extension in $T(n+1,1)$.) Let τ be any extension of $\Theta(\sigma)$. To show that σ is Θ-perfect we must define $\mu \supseteq \sigma$ with $\Theta(\mu) = \tau$.

Let μ be the unique extension of σ such that $|\mu| = g(|\tau|)$ and $\mu \in T(m,\tau(m))$ whenever $n < m < |\tau|$. It must be shown that $\Theta(\mu;m) = \tau(m)$ for $m < |\tau|$. If $m \leq n$, then $\Theta(\sigma;m)$ is defined, so $\Theta(\mu;m) = \Theta(\sigma;m) = \tau(m)$. If $n < m < |\tau|$, then $\mu \in T(m,\tau(m))$ so $\Theta(\mu;m) = \tau(m)$. Finally $|\Theta(\mu)| = |\tau|$ since $|\mu| = g(|\tau|)$ and $\mu \in T(|\tau| - 1,0) \cup T(|\tau|-1,1)$. This completes the proof of Lemma 4.5.

The next lemma will complete the proof of the Theorem.

<u>Lemma 4.6</u>. If Θ is a partial recursive operator and there is a dense recursive set P of Θ-good strings, then, for any $n \geq 1$, $\Theta(A)$ is total and n-generic whenever A is n-generic.

<u>Proof</u>. To show that $\Theta(A)$ is total, for each n let $S_n = \{\sigma : \Theta(\sigma;n)$ is defined$\}$. Then, for each n, S_n is a dense recursive set of strings. (Given any string ν choose $\sigma \in P$ with $\sigma \supseteq \nu$ and then choose $\tau \supseteq \sigma$ with $|\Theta(\tau)| > n$.) Thus, by Lemma 2.7 and the 1-genericity of A, for each n there is a string $\sigma \in S_n$ with $\sigma \subseteq A$, so $\Theta(A)$ is total. Assume now that A is n-generic in order to show that $\Theta(A)$ is n-generic. Let S be any Σ^0_n set of strings. Let $T = \{\sigma : \sigma \in P$ & $(\exists\tau)[\tau \in S$ & $\Theta(\sigma) \supseteq \tau]\}$. Then T is also Σ^0_n so from Lemma 2.7 and the n-genericity of A we conclude that there is a string ν such that $\nu \subseteq A$ and either $\nu \in T$ or no $\sigma \supseteq \nu$ is in T. If $\nu \in T$, we see from the definition of T that $\Theta(A)$ extends some string $\tau \in S$. Suppose now that no $\sigma \supseteq \nu$ is in T. Since P is dense, we may choose $\rho \supseteq \nu$ with $\rho \in P$. We claim that no string $\tau \supseteq \Theta(\rho)$ is in S. If not then there exists $\tau \supseteq \Theta(\rho)$ with $\tau \in S$ and $\sigma \supseteq \rho$ with $\sigma \in P$ & $\Theta(\sigma) \supseteq \tau$, so $\sigma \in T$, contrary to choice of ν. Thus there exists a string $\lambda \subseteq \Theta(A)$ such that $\lambda \in S$ or no extension of λ is in S. Since S was an arbitrary Σ^0_n set of strings, it follows from Lemma 2.7 it follows that $\Theta(A)$ is n-generic.

The proof of Theorem 4.1 is now complete. Specifically if A is n-generic and $0 <_T B \leq_T A$, we choose by Lemma 4.3 a totally splittable partial recursive operator Ψ such that $\Psi(A) = B$ and then obtain a total recursive operator Φ satisfying the conclusion of Lemma 4.5. Let $C = \Phi(\Psi(A)) = \Phi(B)$. Then $C \leq_T B$ (and in fact $C \leq_{tt} B$ by the totality of Φ and a result of Nerode (Rogers (1967), p. 143)). Finally if $\Theta = \Phi \circ \Psi$ then $C(= \Theta(A))$ is n-generic by Lemma 4.6.

Theorem 4.1 yields an interesting "downward homogeneity" property of $\mathcal{D}(\leq \underset{\sim}{a})$ when $\underset{\sim}{a}$ is 2-generic. Such a property holds in much stronger form in the lattice of r.e. sets (where the sublattice of r.e. subsets of any given infinite r.e. set is isomorphic to the entire lattice) and has a heuristic analogue in the r.e. degrees (where many, but not all, constructions of r.e. degrees with a specified property may be done below any given nonzero r.e. degree.)

The following corollary lists some facts about $\mathcal{D}(\leq \underset{\sim}{a})$ which follow at once from Theorem 4.1 and the results in §3. Further applications of Theorem 4.1 will be made in §§5 and 6.

Corollary 4.7. Suppose the degree $\underset{\sim}{a}$ is 2-generic and $\underset{\sim}{0} < \underset{\sim}{b} \leq \underset{\sim}{a}$. Then $\underset{\sim}{b}$ is not a minimal degree. In fact $\mathcal{D}(\leq \underset{\sim}{b})$ is not a lattice but embeds all finite lattices and countable distributive lattices.

We now comment briefly on analogues of Theorem 4.1 for other reducibilities. Proposition 2.10 and the remarks before it demonstrate that strengthened analogues of Theorem 4.1 which state that every nonzero predecessor of a generic degree is generic hold for one-one and many-one reducibilities. The remarks based on the totalness of Φ at the conclusion of the proof of Theorem 4.1 establish that the exact analogue of Theorem 4.1 for truth-table reducibility holds with the same proof. For the reducibilities "arithmetical in," "hyperarithmetical in," and "constructible in," analogues of Theorem 4.1 for appropriate notions of genericity may be proved using forcing and the method of Theorem 4.1, as remarked in Martin (1967). (Here "constructible in" means constructible by ordinals $< \gamma$, where γ is the ordinal of a countable well-founded model of ZF.) Solovay in fact has shown that every nonzero degree of constructibility below that of a Cohen-generic set contains a Cohen-generic set.

We now consider whether some of the strengthened versions of Theorem 4.1 we have mentioned for other reducibilities also hold for Turing reducibility. Martin has shown [private communication] that the upward closure of a meager set of nonzero degrees need not be meager. His proof in fact shows that every generic degree bounds a nonzero degree which fails to be generic, so that the strengthened analogues of Theorem 4.1 we have mentioned for many-one reducibility and degrees of constructibility fail for Turing degrees. In the following section we use the idea of Martin's construction to demonstrate this failure in concrete form.

5. RELATIVE RECURSIVE ENUMERABILITY OF 1-GENERIC DEGREES

The results of the previous section show that, if $\underset{\sim}{a}$ is 2-generic, then $\mathcal{D}(\leq \underset{\sim}{a})$ has no minimal elements. This suggests that $\mathcal{D}(\leq \underset{\sim}{a})$ might be densely ordered, but the following result (whose proof is based on an idea of D. A. Martin) will imply that $\mathcal{D}(\leq \underset{\sim}{a})$ is not dense for $\underset{\sim}{a}$ 1-generic as well as provide other contrasts with Theorem 4.1.

Theorem 5.1. If $\underset{\sim}{a}$ is 1-generic, there is a degree $\underset{\sim}{b} < \underset{\sim}{a}$ such that $\underset{\sim}{a}$ is r.e. in $\underset{\sim}{b}$.

Proof. Let $p(i,j) = 2^i 3^j$. Let the total recursive operator Φ be defined by:

$$\Phi(A) = \{p(i,j) : i \in A \,\&\, p(i,j) \notin A\}.$$

We claim that if A is 1-generic, then $\Phi(A) <_T A$, and A r.e. in $\Phi(A)$. Certainly $\Phi(A) \leq_T A$. Also if A is immune, then for every i there exists j with $p(i,j) \notin A$. Thus the following holds for all i:

$$i \in A \iff (\exists j)[p(i,j) \in \Phi(A)].$$

Since 1-generic sets are immune, it follows that A is r.e. in $\Phi(A)$.

It remains to show that $A \not\leq_T \Phi(A)$ when A is 1-generic. Roughly this is so because no finite amount of information about $\Phi(A)$ can ever imply that $n \notin A$, if n is not a number in the range of p. The following lemma makes this precise. Given σ, let $\Phi(\sigma)$ be the string which gives the information about $\Phi(A)$ which follows from the information $A \subseteq \sigma$. More precisely, $\Phi(\sigma)$ is the unique string ν of the same length as σ such that $\nu^{-1}(1) = \{p(i,j) : \sigma(i) = 1 \,\&\, \sigma(p(i,j)) = 0\}$.

Lemma 5.2. If ν, σ are strings with $\nu \subseteq \sigma$ and n is any number $\geq |\nu|$ not in the range of p, then there exists a string $\tau \supseteq \nu$ with $\tau(n) = 1$ and $\Phi(\tau) \supseteq \Phi(\sigma)$.

Proof. If $\sigma(n)$ is 1 or undefined, simply let τ be any extension of σ with $\tau(n) = 1$. Now assume that $\sigma(n) = 0$. Let S be the smallest set (with respect to inclusion) such that $n \in S$ and such that $p(i,j) \in S$ whenever $i \in S$, $\sigma(i) = 0$, and $p(i,j) < |\sigma|$. Let τ be the unique string of the same length as σ such that $\tau^{-1}(1) = \sigma^{-1}(1) \cup S$. Then $\tau \supseteq \nu$ since every element of S is $\geq |\nu|$, as may easily be seen by induction on the construction of S, using that $n \geq |\nu|$ at the base step. Clearly $\tau(n) = 1$ since $n \in S$. It remains to show that $\Phi(\tau) \supseteq \Phi(\sigma)$. Let any number $k < |\Phi(\sigma)| = |\sigma|$ be given, in order to show that $\Phi(\sigma;k) = \Phi(\tau;k)$. If k is not in the range of p, then $\Phi(\sigma;k) = \Phi(\tau;k) = 0$. Assume that $k = p(i,j)$.

Case 1. $\Phi(\sigma;p(i,j)) = 0$. Then either $\sigma(i) = 0$ or $\sigma(p(i,j)) = 1$. If $\sigma(p(i,j)) = 1$, then $\tau(p(i,j)) = 1$, so $\Phi(\tau;p(i,j)) = 0$ as required. Assume now that $\sigma(i) = 0$. If $i \notin S$, then $\tau(i) = 0$ so $\Phi(\tau;p(i,j)) = 0$. If $i \in S$, then $p(i,j) \in S$ by the closure properties of S, so $\tau(p(i,j)) = 1$. Again it follows that $\Phi(\tau;p(i,j)) = 0$.

Case 2. $\Phi(\sigma;p(i,j)) = 1$. Then $\sigma(i) = 1$ and $\sigma(p(i,j)) = 0$. Since $\sigma(i) = 1$, clearly also $\tau(i) = 1$. Also $p(i,j) \notin S$ since it is easy to show by induction on the construction of S (using that n is not in the range of p for the base step) that S contains no numbers of the form $p(k,\ell)$ with $\sigma(k) = 1$. Since $\sigma(p(i,j)) = 0$ and $p(i,j) \notin S$, it follows that $\tau(p(i,j)) = 0$. Since $\tau(i) = 1$ and $\tau(p(i,j)) = 0$, we conclude that $\Phi(\tau;p(i,j)) = 1$, completing the proof of the lemma.

To complete the proof of Theorem 5.1, suppose for a contradiction that A is 1-generic and Ψ is a partial recursive operator with $\Psi(\Phi(A)) = A$. Let S be the set of strings σ such that σ is incompatible with $\Psi(\Phi(\sigma))$. Obviously A extends no string in S and S is a recursive set of strings, so by Lemma 2.7 there is a string $\nu \subseteq A$ such that no extension of ν is in S. Choose $n \geq |\nu|$ such that n is not in the range of p and $n \notin A$. Since $\Psi(\Phi(A)) = A$ and $n \notin A$, we may choose $\sigma \supseteq \nu$ with $\Psi(\Phi(\sigma);n) = 0$. By Lemma 5.1 choose $\tau \supseteq \nu$ such that $\tau(n) = 1$ and $\Phi(\tau) \supseteq \Phi(\sigma)$. Then $\Psi(\Phi(\tau);n) = 0$ and $\tau(n) = 1$ so $\tau \in S$. This is a contradiction since $\tau \supseteq \nu$.

Corollary 5.3. If $\underset{\sim}{a}$ is 1-generic, then $\mathcal{D}(\leq \underset{\sim}{a})$ fails to be dense. In fact, if $\underset{\sim}{a}$ is 2-generic, then no nontrivial initial segment of $\mathcal{D}(\leq \underset{\sim}{a})$ is dense.

Proof. Choose $\underset{\sim}{b} < \underset{\sim}{a}$ such that $\underset{\sim}{a}$ is r.e. in $\underset{\sim}{b}$. By relativizing to $\underset{\sim}{b}$ the proof that every nonzero r.e. degree bounds a minimal degree in Yates (1970), obtain a degree $\underset{\sim}{c}$ such that $\underset{\sim}{b} < \underset{\sim}{c} < \underset{\sim}{a}$ and no degree $\underset{\sim}{d}$ satisfies $\underset{\sim}{b} < \underset{\sim}{d} < \underset{\sim}{c}$. This proves the first sentence of the Corollary, and the second follows from this and Theorem 4.1.

It is not known whether $\mathcal{D}(\leq \underset{\sim}{a})$ contains maximal nonunit elements for generic $\underset{\sim}{a}$. However, the following corollary shows that not every degree $\underset{\sim}{b} < \underset{\sim}{a}$ is bounded by such maximal element.

Corollary 5.4. If $\underset{\sim}{a}$ is 1-generic, there is a degree $\underset{\sim}{b} < \underset{\sim}{a}$ such that for every degree $\underset{\sim}{c}$ with $\underset{\sim}{b} \leq \underset{\sim}{c} < \underset{\sim}{a}$ there exists $\underset{\sim}{d}$ with $\underset{\sim}{c} < \underset{\sim}{d} < \underset{\sim}{a}$.

Proof. Let $\underset{\sim}{b}$ be such that $\underset{\sim}{b} < \underset{\sim}{a}$ and $\underset{\sim}{a}$ is r.e. in $\underset{\sim}{b}$. If $\underset{\sim}{b} \leq \underset{\sim}{c}$, then $\underset{\sim}{a}$ is r.e. in $\underset{\sim}{c}$. Relativize to $\underset{\sim}{c}$ the fact that r.e. degrees are never minimal.

Let us call a set A 1-generic over a set C if for every set S of strings which is r.e. in C, there is a string $\sigma \subseteq A$ such that $\sigma \in S$ or no $\tau \supseteq \sigma$ is in S. By relativizing the proof of Theorem 5.1, we see that if A is 1-generic over C, then there exists B such that A is r.e. in B, $B \leq_T A$, and $A \nleq_T B \oplus C$. Also by Post's Hierarchy Theorem, a set is $(n+1)$-generic iff it is 1-generic over $0^{(n)}$.

A degree $\underset{\sim}{b}$ is said to have least possible jump if $\underset{\sim}{b}' = \underset{\sim}{b} \cup \underset{\sim}{0}'$. If $\underset{\sim}{a}$ is 1-generic, then $\underset{\sim}{a}$ has least possible jump by Lemma 2.6. On the other hand we will show in Corollary 5.5 that every 2-generic degree $\underset{\sim}{a}$ bounds a degree $\underset{\sim}{b}$ which does not have least possible jump and hence is not 1-generic. For notational convenience we first extend the notion of having least possible jump as in Jockusch & Posner (1978). For $n \geq 1$, let GL_n be the set of all degrees $\underset{\sim}{a}$ such that $\underset{\sim}{a}^{(n)} = (\underset{\sim}{a} \cup \underset{\sim}{0}')^{(n-1)}$. Then GL_1 consists precisely of the degrees having least possible jump and $GL_n \subseteq GL_{n+1}$ for all n.

Corollary 5.5. Every 2-generic degree a bounds a degree $\underset{\sim}{b} \in GL_2 - GL_1$.

Proof. Let $\underset{\sim}{a}$ be 2-generic. By the discussion after Corollary 5.4 with $C = 0'$, there is a degree $\underset{\sim}{b} < \underset{\sim}{a}$ such that $\underset{\sim}{a}$ is r.e. in $\underset{\sim}{b}$ and $\underset{\sim}{a} \not\leq \underset{\sim}{b} \cup \underset{\sim}{0}'$. Since $\underset{\sim}{a}$ is 2-generic, $\underset{\sim}{a}'' \leq \underset{\sim}{a} \cup \underset{\sim}{0}''$ by Lemma 2.6. Also $\underset{\sim}{a} \leq \underset{\sim}{b}'$ since $\underset{\sim}{a}$ is r.e. in $\underset{\sim}{b}$. Thus $\underset{\sim}{b}'' \leq \underset{\sim}{a}'' \leq \underset{\sim}{a} \cup \underset{\sim}{0}'' \leq \underset{\sim}{b}' \cup \underset{\sim}{0}'' \leq (\underset{\sim}{b} \cup \underset{\sim}{0}')'$. Therefore $\underset{\sim}{b}'' \leq (\underset{\sim}{b} \cup \underset{\sim}{0}')'$, i.e. $\underset{\sim}{b} \in GL_2$. Assume now for a contradiction that $\underset{\sim}{b} \in GL_1$, i.e. $\underset{\sim}{b}' \leq \underset{\sim}{b} \cup \underset{\sim}{0}'$. Then $\underset{\sim}{a} \leq \underset{\sim}{b}' \leq \underset{\sim}{b} \cup \underset{\sim}{0}'$, contrary to the choice of $\underset{\sim}{b}$.

Corollary 5.6. Every 2-generic degree $\underset{\sim}{a}$ bounds a nonzero degree $\underset{\sim}{b}$ which is not 1-generic.

This corollary follows at once from the previous corollary and the fact that 1-generic degrees are in GL_1. It contrasts with Theorem 4.1. The following corollary contrasts with Martin's theorem on category.

Corollary 5.7 (Martin). There is a meager set C of nonzero degrees such that the upward closure of C is non-meager.

Proof. Let C be the class of all degrees $\notin GL_1$. Then C is meager since it is disjoint from the comeager class of 1-generic degrees. The upward closure of C is comeager since, by Corollary 5.3, it contains the comeager class of 2-generic degrees. (Of course, by Corollary 5.6, the class of non-generic degrees (or non 1-generic degrees) also satisfies Corollary 5.7. However, we prefer to use the complement of GL_1 since it is definable in the degrees using the ordering of degrees and the jump operation.)

We remark that, in Theorem 5.1, the set A provides a solution to Post's problem relative to $\Phi(A) = B$. (B' is not recursive in A since $0'$ is not recursive in A whenever A is 1-generic.) This is perhaps of some methodological interest since no use of the priority method is involved in this solution of a relativized Post problem. Another weak relative solution of Post's problem may be obtained without the priority method by showing that if f is any $1-1$ recursive function with nonrecursive range and A is 2-generic, then $A <_T f(A) <_T A'$, so $f(A)$ solves Post's problem over A. It might be objected that these supposedly simple solutions use the sophisticated tool of forcing. However, the use of forcing is inessential and may easily be replaced by standard category arguments or Kleene-Post constructions to get solutions of Post's problem relative to certain sets. Still these simple solutions are of limited interest because, of course, the priority method gives solutions to Post's problem relative to <u>all</u> sets.

As in Jockusch & Posner (1978), let GH_n be the set of degrees $\underset{\sim}{b}$ such that $\underset{\sim}{b}^{(n)} = (\underset{\sim}{b} \cup \underset{\sim}{0}')^{(n)}$. (GL stands for generalized low and GH for generalized high. In particular the degrees in GL_1 (respectively GH_1) below $\underset{\sim}{0}'$ are precisely the degrees usually called low (respectively high).) Clearly $GH_n \subseteq GH_{n+1}$ for all n and $GL_i \cap GH_j = \phi$ for all i and j.

In Jockusch (1977) it was shown that every degree in GH_1 bounds both a 1-generic degree and a minimal degree. In Jockusch & Posner (1978), the first half of this result was improved by showing that every degree not in GL_2 bounds a 1-generic degree, and the question was raised whether the second half could be similarly improved by showing that every degree not in GL_2 also bounds a minimal degree. The following theorem, which raises Theorem 5.1 by one level of the arithmetical hierarchy, will yield a negative answer to this question with the aid of Theorem 4.1. The proof is based on the same idea (due to Martin) as Theorem 5.1 but is somewhat more complicated.

<u>Theorem 5.8</u>. If A is 2-generic, then there is a set $B \leq_T A$ such that A is Σ^0_2 in B but not Π^0_2 in B.

<u>Proof</u>. Let $p(i,j) = 2^i 3^j$ and $q(i,j) = 2^i 5^j$. Given A, let B be the set of all numbers $q(i,j) \in A$ such that $i \notin A$ or $(\forall k < j)[p(i,k) \in A]$.

If A is bi-immune, then A is Σ^0_2 in B since for all i,

$$i \in A \iff (\exists k)(\forall j > k)[q(i,j) \notin B].$$

(If $i \in A$, choose k with $p(i,k) \notin A$. If $i \notin A$, use that $\{j : q(i,j) \in A\}$ is infinite.)

To show that A is not Π^0_2 in B, we assume that A is 2-generic and use a forcing argument.

Let Δ be the partial recursive operator which carries each set A to B as defined above. Given any string σ, let $\Delta^*(\sigma)$ be the partial function which gives the information about B which follows from $A \supseteq \sigma$. More precisely, $\Delta^*(\sigma)$ is the partial function ρ defined as follows: $\rho(q(i,j)) = 1$ iff $\sigma(q(i,j)) = 1$ and $(\sigma(i) = 0$ or $(\forall k < j)[\sigma(p(i,k)) = 1])$. Also $\rho(q(i,j)) = 0$ iff $\sigma(q(i,j)) = 0$ or $(\sigma(i) = 1$ and $(\exists k < j)[\sigma(p(i,k)) = 0])$.

Finally $\rho(x) = 0$ if x is not in the range of q. Thus ρ is a partial recursive function with (infinite) recursive domain. The following lemma is the counterpart of Lemma 5.2.

Lemma 5.9. If σ and ν are strings such that $\sigma \supseteq \nu$ and n is a number such that $n > |\nu|$ and n is not in the range of p or q, there is a string τ such that $\tau \supseteq \nu$, $\tau(n) = 1$, and $\Delta^*(\tau) \supseteq \Delta^*(\sigma)$.

Proof. If $\sigma(n)$ is 1 or undefined, just let τ be any extension of σ with $\tau(n) = 1$. Assume now that $\sigma(n) = 0$. Let S be the smallest set such that $n \in S$ and $p(i,k) \in S$ whenever $i \in S$, $\sigma(i) = 0$, and $p(i,k) < |\sigma|$. It is clear by induction on the construction of S that every element of S is $\geq n$. Let τ be the unique string such that $|\tau| = |\sigma|$ and $\tau^{-1}(1) = \sigma^{-1}(1) \cup S$. Then $\tau \supseteq \nu$ since $\min(S) \geq n > |\nu|$. To show that $\Delta^*(\tau) \supseteq \Delta^*(\sigma)$, write ρ for $\Delta^*(\sigma)$ and consider various cases:

Case 1. $\rho(q(i,j)) = 1$. Then $\sigma(q(i,j)) = 1$, so $\tau(q(i,j)) = 1$.

Subcase 1a. $\sigma(i) = 0$. If $i \notin S$, then $\tau(i) = 0$ so $\Delta^*(\tau,q(i,j)) = 1$ as required. Assume $i \in S$. Then for any $k < j$, $p(i,k) < q(i,j) < |\sigma|$. By construction of S, $p(i,k) \in S$ for all $k < j$. Thus $\tau(p(i,k)) = 1$ for all $k < j$, so again $\Delta^*(\tau;q(i,j)) = 1$ as required.

Subcase 1b. $(\forall k < j)[\sigma(p(i,k)) = 1]$. Then $(\forall k < j)$ $[\tau(p(i,k)) = 1]$, so $\Delta^*(\tau,q(i,j)) = 1$.

Case 2. $\rho(q(i,j)) = 0$.

Subcase 2a. $\sigma(q(i,j)) = 0$. Then $\tau(q(i,j)) = 0$ since S contains no numbers in the range of q. Therefore $\Delta^*(\tau;q(i,j)) = 0$.

Subcase 2b. $\sigma(i) = 1$ and $(\exists k < j)[\sigma(p(i,k)) = 0]$. Then since $\sigma(i) = 1$ and n is not in the range of p, no number in the range of p is in S. Therefore $(\exists k < j)[\tau(p(i,k)) = 0]$, so $\Delta^*(\tau;q(i,j)) = 0$.

Since $\Delta^*(\sigma;x) = \Delta^*(\tau;x) = 0$, when x is not in the range of q, the proof of the lemma is complete.

Observe that $\Delta^*(\sigma;x)$ is defined whenever $x < |\sigma|$. Let $\Delta(\sigma)$ be the string obtained by restricting $\Delta^*(\sigma)$ to $|\sigma|$. The next lemma says roughly that $\Delta^*(\sigma)$ is all the information about $\Delta(A)$ which follows from the information $A \supseteq \sigma$. It has no counterpart in the proof of Theorem 5.1.

Lemma 5.10. If γ and σ are strings such that such γ is compatible with $\Delta^*(\sigma)$, then there exists $\nu \supseteq \sigma$ such that $\Delta(\nu) \supseteq \gamma$.

Proof. Let ν be a string such that $|\nu| \geq |\gamma|$, $\nu \supseteq \sigma$, $\nu(q(i,j)) = \gamma(q(i,j))$ whenever $|\sigma| \leq q(i,j) < |\gamma|$, and $\nu(p(i,k)) = 1$ whenever $|\sigma| \leq p(i,k) < |\gamma|$. Then $|\Delta(\nu)| = |\nu| \geq |\gamma|$, so it suffices to show that $\Delta^*(\gamma)$ and γ are compatible (since $\Delta^*(\nu) \supseteq \Delta(\gamma)$). Let x be any number such that $\Delta^*(\nu)$ and γ are both defined on x. We show that $\Delta^*(\gamma;x) = \gamma(x)$.

Case 1. $\Delta^*(\sigma;x)$ is defined. Then $\gamma(x) = \Delta^*(\sigma,x) = \Delta^*(\nu,x)$ since $\gamma(x)$ and $\Delta^*(\sigma;x)$ are defined, γ is compatible with $\Delta^*(\sigma)$, and $\Delta^*(\nu) \supseteq \Delta^*(\sigma)$ (since $\nu \supseteq \sigma$).

Case 2. $\Delta^*(\sigma,x)$ is not defined. Then $x = q(i,j)$ for some i and j and $|\gamma| > x \geq |\sigma|$. Therefore $\nu(x) = \gamma(x)$ by definition of ν and it remains to show that $\Delta^*(\nu;x) = \nu(x)$, where $x = q(i,j)$.

Subcase 2a. For some $k < j$, $\nu(p(i,k)) = 0$. For this k, by definition of ν, $p(i,k) < |\sigma|$, so $\sigma(p(i,k)) = 0$ since $\nu \supseteq \sigma$. If $\sigma(i) = 1$, then $\Delta^*(\sigma;x) = 0$ by definition of $\Delta^*(\sigma)$, violating the hypothesis of Case 2. Since $i < p(i,k) < |\sigma|$, it follows that $\sigma(i) = 0$, so $\nu(i) = 0$. From $\nu(i) = \nu(p(i,k)) = 0$ it follows from the definition of $\Delta^*(\nu)$ that $\Delta^*(\nu;x) = \nu(x)$.

Subcase 2b. There is no $k < j$ with $\nu(p(i,k)) = 0$. Since $x = q(i,j) > p(i,k)$ for $k < j$, and $x < |\nu|$, it follows that $\nu(p(i,k)) = 1$ for all $k < j$. The definition of $\Delta^*(\nu)$ now yields that $\Delta^*(\nu;x) = \nu(x)$.

It remains to complete the proof of the theorem from the two lemmas. Suppose that A is 2-generic and A is $\Pi_2^{0,B}$. Let f be a recursive function such that, for all i,

$$i \in A \iff W^B_{f(i)} \text{ is infinite .}$$

Let $\varphi(\underline{A})$ be a Π_2^0 formula of number theory which is true of exactly those sets X which satisfy

$$(\forall i)[i \in X \Rightarrow W^{\Delta(X)}_{f(i)} \text{ is infinite}].$$

Since A is 2-generic and φ is true of A, there exists a string ν such that $A \supset \nu$ and $\nu \Vdash \varphi$. Fix such a ν and then choose n so that $n > |\nu|$, $n \notin A$, and n is not in the range of p or q. Next choose s so that s exceeds every element of the finite set $W^{\Delta(A)}_{f(n)}$. Let ψ be a Π_1^0 sentence true of exactly those sets X with s greater than every element of $W^{\Delta(X)}_{f(n)}$. Since ψ is true of A, we may choose $\sigma \supseteq \nu$ such that $\sigma \Vdash \Psi$. Using Lemma 5.9 choose τ so that $\tau \supseteq \nu$, $\tau(n) = 1$, and $\Delta^*(\tau) \supseteq \Delta^*(\sigma)$. Let B be a generic set such that $B \supseteq \tau$. Since $B \supseteq \nu$, φ is true of B and so $W^{\Delta(B)}_{f(n)}$ is infinite. Thus there exist $\mu \supseteq \tau$ and $t \geq s$ so that $\Phi_{f(n)}(\Delta(\mu);t)$ is defined.

(Recall that W_e^X is the domain of $\Phi_e(X,\cdot)$.) Now $\Delta^*(\mu) \supseteq \Delta^*(\tau) \supseteq \Delta^*(\sigma)$ and $\Delta^*(\mu) \supseteq \Delta(\mu)$, so $\Delta(\mu)$ is compatible with $\Delta^*(\sigma)$. Apply Lemma 5.10 with $\gamma = \Delta(\mu)$ to obtain a string λ such that $\lambda \supseteq \sigma$ and $\Delta(\lambda) \supseteq \Delta(\mu)$. Let C be a generic set such that $C \supseteq \lambda$. Then $\Delta(C) \supseteq \Delta(\lambda) \supseteq \Delta(\mu)$, so $\Phi_{f(n)}(\Delta(C);t)$ is defined, i.e. $t \in W_{f(n)}^{\Delta(C)}$. It follows that ψ is false of C. This is a contradiction since $C \supseteq \tau \supseteq \sigma$ and $\sigma \Vdash \psi$.

The following corollary is parallel in statement and proof to Corollary 5.5.

Corollary 5.11. Every 3-generic degree $\underset{\sim}{a}$ bounds a degree $\underset{\sim}{b} \in GL_3 - GL_2$.

At last we obtain the result promised before the statment of Theorem 5.8.

Corollary 5.12. There is a degree $\underset{\sim}{b} \in GL_3 - GL_2$ which bounds no minimal degree.

Proof. Let $\underset{\sim}{a}$ be any 3-generic degree, and let $\underset{\sim}{b}$ be as in Corollary 5.11. If $\underset{\sim}{b}$ bounds a minimal degree, so does $\underset{\sim}{a}$ in contradiction to Corollary 4.7.

The first two conjectures below seem reasonably likely to hold on the basis of the results in this section. The third is a shot in the dark, at least as far as the GH_n cases are concerned.

Conjecture 5.13. If A is n-generic $(n \geq 1)$, then there is a set $B \leq_T A$ such that $A \in \Sigma_n^B - \Pi_n^B$.

Conjecture 5.14. For every $n \geq 1$, every $(n+1)$-generic degree $\underset{\sim}{a}$ bounds a degree $\underset{\sim}{b} \in GL_{n+1} - GL_n$. Also every generic degree bounds a degree $\underset{\sim}{b}$ not in GL_n for any n.

Conjecture 5.15. The class of degrees bounding no minimal degrees contains degrees in each class $GL_{n+1} - GL_n$ and $GH_{n+1} - GH_n$ for $n \geq 1$ as well as degrees not in GH_n or GL_n for any n. (The class certainly contains degrees in GL_1 but does not contain degrees in GH_1 by Jockusch (1977).)

It should be pointed out that Conjectures 5.13 and 5.14 (if true) are quite sharp. The conclusion of 5.13 cannot be true of all $(n-1)$-generic A since by Lemma 2.6 there is an $(n-1)$-generic A which is recursive in $0^{(n-1)}$, and hence Π_n^B for every set B. Similarly the conclusion of 5.14 cannot hold of all n-generic $\underset{\sim}{a}$ since for each n there is an n-generic $\underset{\sim}{a} \leq \underset{\sim}{0}^{(n)}$. Such an $\underset{\sim}{a}$ satisfies $\underset{\sim}{a}^{(n)} = \underset{\sim}{0}^{(n)}$, so every $\underset{\sim}{b} \leq \underset{\sim}{a}$ is in GL_n.

6. CUPPING AND COMPLEMENTATION THEOREMS

A degree $\underset{\sim}{a}$ is said to have the "cupping property" if for every degree $\underset{\sim}{c} \geq \underset{\sim}{a}$ there exists $\underset{\sim}{b} < \underset{\sim}{c}$ with $\underset{\sim}{a} \cup \underset{\sim}{b} = \underset{\sim}{c}$. The Friedberg Completeness Criterion shows that $\underset{\sim}{0}'$ has the cupping property. More generally, by Jockusch & Posner (1978), Theorem 3, every degree $\notin GL_2$ has the cupping property. Also the class of degrees with the cupping property is obviously closed upwards. From the last two remarks and Corollary 5.11 it follows that every 3-generic degree has the cupping property. In the next theorem it is shown directly that every 2-generic degree has the cupping property, and from the proof we deduce that there is a degree $\underset{\sim}{a}$ with $\underset{\sim}{a}' = \underset{\sim}{0}'$ which has the cupping property. We do not know whether every 1-generic degree has the cupping property. However, the existence of a degree without the cupping property follows at once from the fact that the three element chain is isomorphic to an initial segment of the degrees.

Theorem 6.1. Every 2-generic degree has the cupping property.

Proof. For any sets A, C, let

$$\Gamma(A,C) = \{n : 3n \in A \ \&\ (n)_0 \in C\} \cup \{n : 3n \notin A \ \&\ 3n+1 \in A\}.$$

If A is coimmune and $A \leq_T C$, then $C \equiv_T A \oplus \Gamma(A,C)$. The reduction $A \oplus \Gamma(A,C) \leq_T C$ is immediate from the assumption $A \leq_T C$. For the reduction $C \leq_T A \oplus \Gamma(A,C)$ observe that, if $3n \in A$ and $k = (n)_0$, then $k \in C$ iff $n \in \Gamma(A,C)$. Further if A is coimmune then for every k there exists n with $3n \in A$ and $k = (n)_0$, and of course such an n may be A-recursively computed from k. It will be shown that $A \not\leq_T \Gamma(A,C)$ whenever A is 2-generic and C is arbitrary. The proof will then be complete since if $\underset{\sim}{a}$ is 2-generic and $\underset{\sim}{a} \leq \underset{\sim}{c}$, we may choose A to be a 2-generic set of degree $\underset{\sim}{a}$, C to be any set of degree $\underset{\sim}{c}$, and $\underset{\sim}{b}$ to be the degree of $\Gamma(A,C)$. Then $\underset{\sim}{a} \cup \underset{\sim}{b} = \underset{\sim}{c}$ and $\underset{\sim}{b} < \underset{\sim}{c}$ (since $\underset{\sim}{a} \not\leq \underset{\sim}{b}$) as required.

If σ, τ are strings, let $\Gamma(\sigma,\tau)$ be the finite partial function which represents the information about $\Gamma(A,C)$ which follows from $A \supseteq \sigma$, $C \supseteq \tau$. Thus, for instance, $\Gamma(\sigma,\tau;n) = 1$ iff either $\sigma(3n) = 1$ and $\tau((n)_0) = 1$ or $\sigma(3n) = 0$ and $\sigma(3n+1) = 1$. If $\tau((n)_0)$ is defined whenever $\sigma(3n) = 1$, then $\Gamma(\sigma,\tau)$ is a string.

For each e, let S_e be the set of strings σ such that for some $\ell < |\sigma|$ and for every string τ with $|\tau| = \ell$ either

(i) σ is incompatible with $\Phi_e(\Gamma(\sigma,\tau))$
or (ii) $\Phi_e(\nu;3n+2)$ is undefined whenever $\nu \supseteq \Gamma(\sigma,\tau)$ and $3n+2 \geq |\sigma|$.

Clearly S_e is a Π^0_1 set of strings (uniformly in e) and, if A meets S_e, then $A \neq \Phi_e(\Gamma(A,C))$ for every set C. Thus it suffices to show that each S_e is dense. Let a string ρ be given. In order to construct $\sigma' \supseteq \rho$ with $\sigma' \in S_e$, we first associate with ρ a number ℓ which bounds the amount of information about C used to compute $\Gamma(\rho,C)$. Then we construct an increasing chain of extensions of ρ, each associated with a particular τ of length ℓ. The final such extension will be the desired $\sigma' \in S_e$. However, to show that $\sigma' \in S_e$ it is important to know that the same number ℓ also bounds the amount of information about C used to compute $\Gamma(\sigma',C)$. To this end, for any σ we define $Q(\sigma) = \{n : \sigma(3n) = 1\}$ and prove the following lemma.

<u>Lemma 6.2</u>. Suppose ρ, τ are strings such that $\tau((n)_0)$ is defined whenever $\rho(3n) = 1$. Then for any string $\nu \supseteq \Gamma(\rho,\tau)$ there exists $\sigma \supseteq \rho$ such that $Q(\sigma) = Q(\rho)$ and $\Gamma(\sigma,\tau) \supseteq \nu$.

<u>Proof</u>. Let σ be an extension of ρ such that $Q(\sigma) = Q(\rho)$ and such that $\sigma(3n+1) = \nu(n)$ whenever $\nu(n)$ is defined and $\rho(3n+1)$ is undefined. Such a string σ clearly exists. To show that $\Gamma(\sigma,\tau) \supseteq \nu$, let n be given with $\nu(n)$ defined. If $\Gamma(\rho,\tau;n)$ is defined, then $\nu(n) = \Gamma(\rho,\tau;n) = \Gamma(\sigma,\tau;n)$ since $\nu \supseteq \Gamma(\rho,\tau)$ and $\sigma \supseteq \rho$. Therefore $\nu(n) = \Gamma(\sigma,\tau;n)$ as required. Now consider the case where $\Gamma(\rho,\tau;n)$ is undefined. We observe that $\rho(3n+1)$ is undefined. (If $\rho(3n+1)$ is defined, then $\rho(3n)$ is defined. If $\rho(3n) = 0$, then $\Gamma(\rho,\tau,n)$ is defined (as $\rho(3n+1)$). If $\rho(3n) = 1$, then $\tau((n)_0)$ is defined by assumption and again $\Gamma(\rho,\tau,n)$ is defined (as $\tau((n)_0)$).) Since $\nu(n)$ is defined and $\rho(3n+1)$ is undefined, $\sigma(3n+1) = \gamma(n)$. Also $\sigma(3n) = 0$ since otherwise $3n \in Q(\sigma) = Q(\sigma_0)$ and again $\Gamma(\sigma_0,\tau;n)$ is defined (as $\tau((n)_0)$ as above. It follows that $\Gamma(\sigma,\tau;n) = \sigma(3n+1) = \nu(n)$ as required.

We now complete the proof that S_e is dense. Let ρ be a given string. Let $\ell = \max\{1 + (n)_0 : n \in Q(\rho)\}$, where we take $\max \phi = 0$. Then $\ell \leq |\rho_0|$ since $(n)_0 \leq n$ for all n, and ℓ is sufficiently big that the hypothesis of Lemma 6.2 applies to τ whenever $|\tau| = \ell$. Let $\tau_0, \tau_1, \ldots, \tau_q$ $(q = 2^\ell - 1)$ be all strings of length ℓ. Inductively choose strings $\sigma_0 \subseteq \sigma_1 \subseteq \ldots \subseteq \sigma_{q+1}$ with $Q(\sigma_0) = \ldots = Q(\sigma_{q+1})$ as follows. Let σ_0 be ρ as above. Now suppose σ_i has been chosen with $Q(\sigma_i) = Q(\sigma_0)$. If clause (ii) in the definition of S_e holds with $\sigma = \sigma_i$ and $\tau = \tau_i$, let $\sigma_{i+1} = \sigma_i$. Otherwise choose a number $3n + 2 \geq |\sigma_i|$ and a string $\nu \supseteq \Gamma(\sigma_i,\tau_i)$ such that $\Phi_e(\nu;3n + 2)$ is defined. Since $Q(\sigma_i) = Q(\sigma_0)$ and $|\tau_i| = \ell$, Lemma 6.2 applies to σ_i, τ_i. Let $\sigma \supseteq \sigma_i$ be a string such that $\Gamma(\sigma_i,\tau_i) \supseteq \gamma$. Finally let σ_{i+1} be a string which agrees with σ except possibly at $3n+2$ such that σ_{i+1} and $\Phi_e(\Gamma(\sigma,\tau_i))$ are incompatible (at $3n+2$). Then $\sigma_{i+1} \supseteq \sigma_i$ since $3n+2 \geq |\sigma_i|$. Also $\Gamma(\sigma,\tau_i) = \Gamma(\sigma_{i+1},\tau_i)$ since $\Gamma(\sigma,\tau)$ is independent of $\sigma(3n+2)$. Thus σ_{i+1} and $\Phi_e(\Gamma(\sigma_{i+1},\tau_i))$ are incompatible, so (i) in the definition of S_e holds of

σ_{i+1}, τ_i. Once the chain of strings $\sigma_0,\ldots,\sigma_{q+1}$ is constructed, let $\sigma' = \sigma_{q+1}$. Then $\sigma' \supseteq \sigma_0$. Also $\sigma' \in S_e$ since $\sigma' \supseteq \sigma_i$ for $1 \leq i \leq q+1$ and, for given τ, the strings σ satisfying (i) or (ii) are closed under extension.

Corollary 6.3. If $\underset{\sim}{a}$ is a 2-generic degree and $\underset{\sim}{d}$ is any nonzero degree which fails to have the cupping property, then $\underset{\sim}{a}$ and $\underset{\sim}{d}$ form a minimal pair.

Proof. Suppose $\underset{\sim}{a}$ and $\underset{\sim}{d}$ fail to form a minimal pair, and choose a nonzero degree b below both $\underset{\sim}{a}$ and $\underset{\sim}{d}$. By Theorem 4.1 there is a 2-generic degree $\underset{\sim}{c} \leq \underset{\sim}{b}$. Then $\underset{\sim}{c}$ has the cupping property by Theorem 6.1 and so $\underset{\sim}{d}$ must also have the cupping property since $\underset{\sim}{c} \leq \underset{\sim}{d}$.

The next corollary shows that if $\underset{\sim}{a}$ is 2-generic, then $\mathcal{D}(\leq \underset{\sim}{a})$ satisfies the formal sentence which asserts that every nonzero degree has the cupping property.

Corollary 6.4. If $\underset{\sim}{a}$ is 2-generic and $\underset{\sim}{0} < \underset{\sim}{d} < \underset{\sim}{b} \leq \underset{\sim}{a}$, then there exists $\underset{\sim}{c} < \underset{\sim}{b}$ with $\underset{\sim}{d} \cup \underset{\sim}{c} = \underset{\sim}{b}$.

Proof. The degree $\underset{\sim}{d}$ must have the cupping property by Corollary 6.3.

In the proof of Theorem 6.1 we did not use the full force of the assumption that A is 2-generic but rather only that A meets each set in a certain sequence $\{T_n\}$ of dense, uniformly Π^0_1, sets of strings. (Let $T_{2e} = S_e$, $T_{2k+1} = \{\sigma : (\exists n)\ [n < |\sigma|\ \&\ (n)_0 = k\ \&\ \sigma(3n) = 1]\}$.) It thus follows, in the language of Yates (1976), that the class of degrees with the cupping property is $(\mathbb{P},\underset{\sim}{0}')$-comeager and hence $(\mathbb{P},\underset{\sim}{0}')$-abundant. This fact (or straightforward direct constructions) yield many results on the existence of degrees with the cupping property and specified other properties. The following corollary is one such result.

Corollary 6.5. If $\underset{\sim}{d}$ is any degree $\geq \underset{\sim}{0}'$, there is a 1-generic degree $\underset{\sim}{a}$ with the cupping property such that $\underset{\sim}{a}' = \underset{\sim}{d}$.

Also, according to Jockusch & Posner (1978), Lemma 6, for each degree $\underset{\sim}{d} \notin GL_2$ and each sequence of dense sets of strings uniformly recursive in $\underset{\sim}{d} \cup \underset{\sim}{0}'$, there is a degree $\underset{\sim}{a} \leq \underset{\sim}{d}$ which contains a set meeting each set of strings in the sequence. From this one obtains at once the following corollary, which extends Theorem 3 of Jockusch & Posner (1978).

Corollary 6.6. If $\underset{\sim}{d} \notin GL_2$, there is a degree $\underset{\sim}{a} \leq \underset{\sim}{d}$ which is 1-generic and has the cupping property.

If d, c are degrees such that $\underset{\sim}{d} \leq \underset{\sim}{c}$, then $\underset{\sim}{d}$ is said to be complemented in $\mathcal{D}(\leq \underset{\sim}{c})$ if there exists a degree $\underset{\sim}{b}$ such that

$\underset{\sim}{b} \cup \underset{\sim}{d} = \underset{\sim}{c}$ and $\underset{\sim}{b} \cap \underset{\sim}{d} = \underset{\sim}{0}$. We do not know whether every degree $\leq \underset{\sim}{c}$ is complemented in $\mathcal{D}(\leq \underset{\sim}{c})$ when $\underset{\sim}{c}$ is generic. For m-degrees the answer to the corresponding question is "no" by the discussion in §2 and the existence of a nonrecursive r.e. set. However, since every infinite r.e. set has an infinite recursive subset, the discussion in §2 shows that in the m-degrees below that of a generic set, every nonzero elements bounds a nonzero complemented element. Below the corresponding result is established for Turing degrees.

Theorem 6.7. If $\underset{\sim}{c}$ is a 2-generic degree, then for every nonzero degree $\underset{\sim}{a}$ in $\mathcal{D}(\leq \underset{\sim}{c})$ there exists a nonzero degree $\underset{\sim}{b} < \underset{\sim}{a}$ such that b is complemented in $\mathcal{D}(\leq \underset{\sim}{c})$.

Proof. By Theorem 4.1 we observe that it suffices to prove Theorem 6.7 with the additional hypothesis that $\underset{\sim}{a}$ is 2-generic, and thus we use the same binary operator Γ as in the proof of Theorem 6.1. Since $C \equiv_T A \oplus \Gamma(A,C)$ when A is 2-generic and $A \leq_T C$, it would suffice to show that every set recursive in both A and $\Gamma(A,C)$ is recursive under these hypotheses. However, this is not true in general since if $C \equiv_T A$ then $\Gamma(A,C)$ itself is a nonrecursive set recursive in both A and $\Gamma(A,C)$. Instead we prove the weaker result that every set recursive in both $\Delta(A)$ and $\Gamma(A,C)$ is recursive, where $\Delta(A) = \{n : 3n \in A\}$. This will still suffice to prove the theorem, for the argument given that $A \oplus \Gamma(A,C) \equiv_T C$ when $A \leq_T C$ and A is 2-generic actually establishes that $\Delta(A) \oplus \Gamma(A,C) \equiv_T C$ under the same hypotheses. Thus if $\underset{\sim}{a}, \underset{\sim}{c}$ are 2-generic degrees, we may choose 2-generic sets A, C of degree $\underset{\sim}{a}, \underset{\sim}{c}$ respectively, and let $\underset{\sim}{b}, \underset{\sim}{d}$ be the degrees of $\Gamma(A,C)$, $\Delta(A)$ respectively. We have $\underset{\sim}{d} < \underset{\sim}{a}$, $\underset{\sim}{d} \cup \underset{\sim}{b} = \underset{\sim}{c}$, and $\underset{\sim}{d} \cap \underset{\sim}{b} = \underset{\sim}{0}$, so $\underset{\sim}{d}$ is the desired degree below $\underset{\sim}{a}$ which is complemented in $\mathcal{D}(\leq \underset{\sim}{c})$.

The proof that the degrees of $\Delta(A)$ and $\Gamma(A,C)$ form a minimal pair is similar in broad outline to the proof that $A \not\leq_T \Gamma(A,C)$ in Theorem 6.1. However the proof that $A \leq_T \Gamma(A,C)$ was greatly facilitated by the fact that $\Gamma(A,C)$ remains the same if A is changed on numbers of the form $3n + 2$. In fact we could have easily proved in Theorem 6.1 that the degrees of $\Gamma(A,C)$ and $\{n : 3n + 2\} \in A$ form a minimal pair. On the other hand, changes in A which affect $\Delta(A)$ also tend to affect $\Gamma(A,C)$, and this creates an obstacle to the current proof.

We must show that, for any a, b, if $\Phi_a(\Delta(A))$ and $\Phi_b(\Gamma(A,C))$ are the same total function, then that function is recursive. To this end, let $S(a,b)$ be the set of strings σ such that there is a number ℓ such that for every τ of length ℓ at least one of the following clauses holds:

(a) $(\exists x)(\forall \rho \supseteq \sigma)[\Phi_a(\Delta(\rho);x)$ is undefined]

(b) $\Gamma(\sigma,\tau)$ has no Φ_b-split pair of extensions

(c) $\Phi_a(\Delta(\sigma))$ and $\Phi_b(\Gamma(\sigma,\tau))$ are incompatible.

Here $\Delta(\sigma)$ denotes, as usual, the string which describes the information about $\Delta(A)$ which follows from $A \supseteq \sigma$. Clearly if A meets $S(a,b)$ and τ is the string of length ℓ extended by C, then $\Phi_a(\Delta(A))$ is not total, or $\Phi_b(\Gamma(A,C))$ (if total) is recursive, or $\Phi_a(\Gamma(A))$ and $\Phi_b(\Gamma(A,C))$ are incompatible. Also $S(a,b)$ is a Σ^0_2 set of strings. Thus to complete the proof of the theorem it suffices to show that each $S(a,b)$ is dense. To this end let a string η be given in order to construct $\sigma' \supseteq \eta$ with $\sigma' \in S(a,b)$.

Case 1. There is a string $\sigma \supseteq \eta$ and a number x such that $\Phi_a(\Delta(\rho);x)$ is undefined for all $\rho \supseteq \sigma$. Let σ' be any such σ. Clearly $\sigma' \in S(a,b)$ via alternative (a) in the definition of $S(a,b)$.

Case 2. Otherwise. Let ℓ be sufficiently large that $\ell > (n)_0$ whenever $\eta(3n) = 1$. Let $\tau_0,\tau_1,\ldots,\tau_q$ $(q = 2^\ell - 1)$ be all strings of length ℓ. We now construct a sort of b-splitting tree divided into "levels" $L(0), L(1),\ldots,L(q+1)$. Here $L(i)$ is a finite set of strings and is defined by induction on i. Let $L(0) = \{\phi\}$, where ϕ denotes the empty string. Suppose inductively that $L(i)$ has been formed, where $i \leq q$. For each string $\nu \in L(i)$, consider two cases:

Case (i). There exist strings ν_1, ν_2 extending ν such that $\Gamma(\eta,\tau_i) * \nu_1$ and $\Gamma(\eta,\tau_i) * \nu_2$ are Φ_b-split i.e. have incompatible images under Φ_b. Choose two such strings ν_1, ν_2 and say that ν contributes ν_1, ν_2 to $L(i+1)$.

Case (ii). Otherwise. Say that ν contributes ν to $L(i+1)$

Then $L(i+1)$ consists of all strings contributed to it by strings in $L(i)$.

Let $Q(\sigma) = \{n : \sigma(3n) = 1\}$, as in the proof of Theorem 6.1. The proof would be somewhat easier if we could use the failure of Case 1 in constructing σ' to deduce the existence of a string $\sigma \supseteq \eta$ with $Q(\sigma) = Q(\eta)$ such that $\Phi_a(\Delta(\sigma);x)$ is defined for each x which is an argument of any Φ_b-splitting used to construct $L(1),\ldots,L(q+1)$. However, no such σ need exist, although the following lemma gives us a usable substitute. If λ, ρ are strings with $\lambda \supseteq \rho$, λ is called an extension of ρ by 0's if $\lambda = \rho * 0^k$ for some k.

Lemma 6.8. Suppose $\eta, \tau_0, \tau_1, \ldots, \tau_q$ are as above and ν is a given string. Then for every sufficiently long extension λ of $\Delta(\eta)$ by 0's and for every $\mu \supseteq \lambda$ there exists $\sigma \supseteq \eta$ such that $\Delta(\sigma) \supseteq \mu$ and $\Gamma(\sigma, \tau_i) \supseteq \Gamma(\eta, \tau_i) * \gamma$ for $0 \leq i \leq q$.

Proof. We show that the lemma holds for λ whenever λ is an extension of $\Delta(\eta)$ by 0's and $|\lambda| \geq |\Gamma(\eta, \tau_i) * \nu|$ for $0 \leq i \leq q$. Given such a λ and given $\mu \supseteq \lambda$, let σ be an extension of η such that $\sigma(n) = \mu(3n)$ whenever $\mu(3n)$ is defined and $\eta(n)$ is undefined, and such that $\sigma(3n+1) = (\Gamma(\eta, \tau_0) * \nu)(n)$ whenever $\Gamma(\eta, \tau_0) * \nu(n)$ is defined and $\eta(3n+1)$ is undefined. One shows that $\Delta(\sigma) \supseteq \mu$ and $\Gamma(\sigma, \tau_i) \supseteq \Gamma(\eta, \tau_i) * \nu$ for $0 \leq i \leq q$ by considering various cases much as in Lemma 6.2. We omit this routine but tedious verification. One easy but important step in it is the observation that the strings $\Gamma(\eta, \tau_i)$ have the same length for $0 \leq i \leq q$, and so the strings $\Gamma(\eta, \tau_i) * \nu$ have the same length and agree on arguments greater than or equal to the common length of the $\Gamma(\eta, \tau_i)$.

We now return to complete Case 2 of the proof that $S(a,b)$ is dense. Let λ be an extension of $\Delta(\eta)$ by 0's which is so long that Lemma 6.8 applies to λ whenever ν is any string in L_{q+1}. Let μ be an extension of λ such that $\Phi_a(\mu;x)$ is defined for each argument x of the Φ_b-splittings used to construct $L(1), \ldots, L(q+1)$. To obtain μ, first choose $\sigma^* \supseteq \sigma_0$ such that $\Delta(\sigma^*) \supseteq \lambda$. This is possible because $\lambda \supseteq \Delta(\eta)$. Next use the failure of Case 1 to build a chain of extensions of σ^* whose final term ρ is such that $\Phi_a(\Delta(\rho);x)$ is defined for all such x. Then let $\mu = \Delta(\rho)$. Clearly $\mu \supseteq \lambda$ since $\rho \supseteq \sigma^*$ and hence $\mu = \Delta(\rho) \supseteq \Delta(\sigma^*) \supseteq \lambda$. Now define strings δ_i for $0 \leq i \leq q+1$ by induction on i with $\delta_i \in L(i)$ and $\delta_i \subseteq \delta_{i+1}$. Let δ_0 be the empty string. If δ_i has been chosen from L_i and δ_i contributes distinct strings ν_1, ν_2 to L_{i+1}, let δ_{i+1} be ν_j, where ν_j is chosen so that $\Phi_b(\Gamma(\eta, \tau_i)) * \nu_j$ and $\Phi_a(\mu)$ are incompatible. This is possible because, for some x, $\Phi_b(\Gamma(\eta, \tau_i)) * \nu_1$, $\Phi_b(\Gamma(\eta, \tau_i)) * \nu_2$, and $\Phi_a(\mu)$ are all defined at x, and the first two differ. If δ_i contributes only itself to L_{i+1}, let $\delta_{i+1} = \delta_i$. Finally, using Lemma 6.8 with $\nu = \delta_{q+1}$, choose $\sigma' \supseteq \eta$ with $\Delta(\sigma') \supseteq \mu$ and $\Gamma(\sigma', \tau_i) \supseteq \Gamma(\eta, \tau_i) * \delta_{q+1}$ for $0 \leq i \leq q$. We claim that $\sigma' \in S(a,b)$ as required. Consider any string τ_i of length ℓ. If δ_i contributes two strings to $L(i+1)$, then $\Phi_a(\mu)$ and $\Phi_b(\Gamma(\eta, \tau_i) * \gamma_{q+1})$ are incompatible. Since $\Delta(\sigma') \supseteq \mu$ and $\Gamma(\sigma', \tau_i) \supseteq \Gamma(\eta, \tau_i) * \gamma_{q+1}$, it follows that $\Phi_a(\Delta(\sigma'))$ and $\Phi_b(\Gamma(\sigma', \tau_i))$ are incompatible, i.e. (c) in the definition of $S(a,b)$ holds with $\sigma = \sigma'$, $\tau = \tau_i$. If γ_i contributes only itself to L_{i+1}, then $\Gamma(\eta, \tau_i) * \nu_i$ has no Φ_b-split pair of extensions and so, since $\Gamma(\eta', \tau_i) \supseteq \Gamma(\sigma_0, \tau_i) * \nu_{q+1}$, $\Gamma(\sigma', \tau_i)$ has no Φ_b-split pair of extensions. Thus (b) holds with $\sigma = \sigma'$, $\tau = \tau_i$. This concludes the proof of Theorem 6.7.

7. OPEN QUESTIONS

Many questions remain open in the area of this paper, but here we mention a few of those we consider most interesting.

(Q1) If $\mathbf{a}$ is generic, does $\mathcal{D}(\leq \mathbf{a})$ have a maximal nonunit element? (In other words, is every generic degree a minimal cover?)

The answer to the m-degree analogue of (Q1) is affirmative, by the existence of maximal r.e. sets and the discussion in §2. However, the Turing degrees containing such maximal m-degrees are demonstrably not maximal in $\mathcal{D}(\leq \mathbf{a})$.

(Q2) Is there a 1-generic degree which bounds a minimal degree? (Compare Corollary 4.7)

(Q3) If $\mathbf{a}$ is generic, is every element of $D(\leq \mathbf{a})$ complemented in $\mathcal{D}(\leq \mathbf{a})$? (Compare Theorem 6.7.)

(Q4) Are the generic degrees dense?

(Q5). Do there exist generic degrees $\mathbf{a}$, $\mathbf{b}$ with $\mathbf{b} < \mathbf{a}$ and $\mathbf{a}$ r.e. in $\mathbf{b}$? (Compare Theorem 5.1.)

(Q6) If $\mathbf{a}$ is generic, is $\mathcal{D}(\leq \mathbf{a})$ elementarily equivalent to $\mathcal{D}(\leq \mathbf{b})$ for every nonzero degree $\mathbf{b} \leq \mathbf{a}$? (See Proposition 2.8 and Corollary 5.6.)

(Q7) If $\mathbf{a}$ is generic, is the first-order theory of $\mathcal{D}(\leq \mathbf{a})$ decidable? (By Proposition 2.8 this theory does not depend on the choice of the generic $\mathbf{a}$.)

(Q8) Is $\mathcal{D}(\leq \mathbf{a})$ isomorphic to $\mathcal{D}(\leq \mathbf{b})$ whenever $\mathbf{a}$, $\mathbf{b}$ are generic?

Negative answers to (Q7) and (Q8) are implied by certain plausible but presumably extremely difficult conjectures about embedding initial segments below given nonzero r.e. degrees. (Actually the conjectures are needed in relativized form.) For instance, suppose that if $\mathbf{b} < \mathbf{a}$ and $\mathbf{a}$ is r.e. in $\mathbf{b}$, then every finite distributive lattice is isomorphic to an initial segment of the degrees between $\mathbf{b}$ and $\mathbf{a}$. Then Theorem 5.1 and the usual trick for showing the undecidability of the theory of degrees imply that $\mathcal{D}(\leq \mathbf{a})$ is undecidable whenever $\mathbf{a}$ is 1-generic. Similarly, if one may embed every countable distributive lattice with a presentation recursive in $\mathbf{b}$ as an initial segment of the degrees between $\mathbf{a}$ and $\mathbf{b}$ when $\mathbf{b} < \mathbf{a}$ and $\mathbf{a}$ is r.e. in $\mathbf{b}$, then one may answer (Q8) negatively by standard methods. Perhaps such conjectures about initial segments below r.e. degrees could be proved using the methods of Lerman or Epstein to obtain initial segments below $0'$.

REFERENCES

Epstein, R. L. (1979). Degrees of unsolvability, structure and theory. Lecture Notes in Math., vol. 759, Springer Verlag, Berlin, Heidelberg, New York.

Feferman, S. (1965). Some applications of the notions of forcing and generic sets. Fund. Math., 56, pp. 325-345.

Hinman, P. G. (1969). Some applications of forcing to hierarchy problems in arithmetic. Z. Math. Logik Grundlagen Math., 15, pp. 341-352.

Hinman, P. G. (1978). Recursion-Theoretic Hierarchies, Springer Verlag, Berlin, Heidelberg, New York.

Jockusch, C. (1977). Simple proofs of some theorems on high degrees of unsolvability. Can. J. Math. 29, pp. 1072-1080.

Jockusch, C. & Posner, D. (1978). Double jumps of minimal degrees. J. Symbolic Logic, 43, pp. 715-724.

Kleene, S. C. & Post, E. L. (1954). The upper semilattice of degrees of recursive unsolvability. Ann. Math., ser. 2, 59, 379-407.

Lachlan, A. & Lebeuf, R. (1976). Countable initial segments of the degrees of unsolvability. J. Symbolic Logic, 41, pp. 289-300.

Martin, D. A. (1967). Measure, category, and degrees of unsolvability. unpublished.

Myhill, J. (1961). Category methods in recursion theory. Pacific J. Math., 11, pp. 1479-1486.

Nerode, A. & Shore, R. (to appear). Reducibility orderings: theories, definability and automorphisms.

Oxtoby, J. C. (1971). Measure and Category. Springer Verlag, New York, Heidelberg, Berlin.

Posner, D. (1977). High Degrees. Doctoral Dissertation, University of California, Berkeley.

Posner, D. (1980). A survey of non-r.e. degrees $\leq \underset{\sim}{0}'$. this volume.

Rogers, H. (1967). Theory of Recursive Functions and Effective Computability. McGraw-Hill, New York.

Sacks, G. E. (1963). Degrees of Unsolvability. Annals of Mathematics Studies, No. 55, Princeton University Press, Princeton, New Jersey.

Shoenfield, J. R. (1959). On degrees of unsolvability. Ann. Math., 69, pp. 644-653.

Shore, R. (to appear). On homogeneity and definability in the first-order theory of Turing degrees

Spector, C. (1956). On degrees of recursive unsolvability. Ann. Math., 64, pp. 581-592.

Thomason, S. K. (1969). A note on non-distributive sublattices of degrees and hyperdegrees. Canadian J. Math., 22, pp. 569-581.

Yates, C. E. M. (1970). Initial segments of the degrees of unsolvability, Part II. Minimal degrees. J. Symbolic Logic, 35, pp. 243-266.

Yates, C. E. M. (1976). Banach-Mazur games, comeager sets, and degrees of unsolvability. Math. Proc. Camb. Phil. Soc., 79, pp. 195-220.

THE DEGREES OF UNSOLVABILITY: SOME RECENT RESULTS

Manuel Lerman

University of Connecticut

INTRODUCTION

Post initiated the study of the degrees of unsolvability with his 1944 address to the American Mathematical Society and his subsequent paper (Post 1944) based on that address. The paper concentrated on "a very limited portion of a subtheory of the hoped for general theory (of recursive functions)", the recursively enumerable sets and their degrees. The algebraic approach adopted by Post to study this theory has given rise to the field of classical recursion theory.

Post's expressed interest in the theory of recursive functions was to present an intuitive development of the general theory which "stripped of its formalism . . . can be followed, if not indeed pursued, by a mathematician, layman though he be in this formal field". Post connected this theory to unsolvability questions in algebra, an area more familiar to the layman. Connections of recursion theory to algebra have since been actively studied by many researchers. The intuitive approach to recursion theory introduced by Post has been adopted by researchers in the field. The proof techniques which have been developed involve very complex combinatorial arguments which are not readily accessible to most mathematicians. But the results proved by these methods can be used to describe a global theory which is more readily accessible to mathematicians.

Post's paper introduced a problem which became the central problem of recursion theory until its solution by Friedberg (1957b) and Mucnik (1956) in 1956. Post showed that there is a least degree of unsolvability, $\underline{0}$, and a greatest recursively enumerable degree of unsolvability, $\underline{0}'$, and asked whether any other recursively enumerable degrees existed. He defined various classes of recursively enumerable sets, none of whose members had degree $\underline{0}$, and tried to show that there was at least one such class with no members of degree $\underline{0}'$. This approach towards solving Post's problem became known as Post's program.

Results of Post and later researchers showed that all of Post's classes have members of degree $\underline{0}'$. For ten years following the posing of Post's problem, much effort was focused on proving that no intermediate degrees exist, and attempts at constructing intermediate recursively enumerable degrees followed the outline of Post's program.

In 1954, around the time of Post's death, a paper was published by Kleene and Post (1954) collecting results about the degrees of unsolvability. This paper expanded the universe of study from the recursively enumerable degrees of unsolvability to the degrees of unsolvability of arbitrary sets of natural numbers. It was shown that the degrees of unsolvability form an uppersemilattice (a partially ordered set in which every pair of elements has a least upper bound) which is not a lattice. Kleene and Post found many degrees strictly between $\underline{0}$ and $\underline{0}'$, but unfortunately, none of these degrees was recursively enumerable, and so did not serve to solve Post's problem. The technique introduced by Kleene and Post to construct degrees intermediate between $\underline{0}$ and $\underline{0}'$ was effectivized in Friedberg 1957a and Mucnik 1956 two years later and combined with a new technique, the priority argument, to construct intermediate recursively enumerable degrees.

The Kleene & Post paper began the study of degrees of unsolvability of arbitrary sets of integers, an algebraic structure which is the focal point of this paper. Much is known about the global theory of the degrees of unsolvability, but little is known about the global theory of the recursively enumerable degrees, Post's original domain of study. Local structure theorems will be listed without proof in this paper, and used to sketch proofs of theorems about global properties of the degrees of unsolvability. The global properties which will be discussed are decidability, homogeneity, automorphisms, and definability, a combination of logical and algebraic properties.

The degrees of unsolvability are obtained as follows. Consider two subsets A and B of N, the set of natural numbers. (An alternate presentation considers functions from N^k into N for $k \in N$ instead of sets, but no new degrees are obtained by doing so). An <u>algorithm with B oracle</u> is a program for a digital computer which allows, in addition to the standard programming instructions, an instruction of the form "if $x \in B$ do the set of instructions in location n; otherwise perform the set of instructions in location m". During the course of computation, when such an instruction is encountered, the B oracle is asked " is $x \in B$?" and yields a correct answer. This answer determines how the computation is to proceed. Define $A \leq_T B$ if there is an algorithm with B oracle such that given a computer programmed with this algorithm and given input x to the computer, the computer outputs 1 if $x \in A$ and 0 if $x \notin A$. $\leq_T$ is an equivalence relation on the set of all subsets of N, whose equivalence classes are the degrees of unsolvability, $\underline{D}$. $\leq_T$ induces a partial ordering, $\leq$, on $\underline{D}$, and given $\underline{a},\underline{b} \in \underline{D}$, $\underline{a} \cup \underline{b}$, the least upper bound of $\underline{a}$ and $\underline{b}$, is the degree of the set $A \oplus B = \{2n:n \in A\} \cup \{2n+1:n \in B\}$ for any $A \in \underline{a}$ and $B \in \underline{b}$. Thus $\mathcal{D} = \langle \underline{D}, \leq \rangle$ is an uppersemilattice.

The jump operator, ', is a natural function from $\underline{D}$ to $\underline{D}$ taking the degree $\underline{d}$ to $\underline{d}'$, the greatest degree which is recursively enumerable in $\underline{d}$, i.e., $\underline{d}'$ is the greatest degree of a set which is the range of a function recursive in a set of degree $\underline{d}$. The structure $\mathcal{D}' = \langle \underline{D}, \leq, ' \rangle$ is also studied in this paper.

It will be convenient to use interval notation for segments of $\underline{D}$ and $\underline{D}'$. Thus for each $\underline{a},\underline{b} \in \underline{D}$, $\underline{D}[\underline{a},\underline{b}] = \{\underline{d} \in \underline{D} : \underline{a} \leq \underline{d} \leq \underline{b}\}$. Open and half-open interval notation is similarly adopted. Also, for each $\underline{a} \in \underline{D}$, $\underline{D}[\underline{a}, \infty) = \{\underline{d} \in \underline{D} : \underline{a} \leq \underline{d}\}$. $\underline{D}[\underline{a}, \infty)$ will sometimes be referred to as the <u>cone</u> of degrees with <u>vertex</u> $\underline{a}$. The passage from $\underline{D}[\underline{a},\underline{b}]$ to $\mathcal{D}[\underline{a},\underline{b}]$ or $\mathcal{D}'[\underline{a},\underline{b}]$, etc., is defined in the usual way. Thus a topology is induced on $\underline{D}$ by taking the open intervals as a base for the topology, giving rise to the concepts of local theorems and global theorems. A <u>local theorem</u> is one whose content can naturally be restricted to a bounded interval of $\underline{D}$, while a <u>global theorem</u> is one whose content fully encompasses $\underline{D}$.

Another important tool for studying $\mathcal{D}$ is the generalized high/low hierarchy. This hierarchy classifies a degree according to how close its jump comes to attaining the highest or lowest possible value. Thus for $n \geq 0$, the class of generalized high_n sets, GH_n, is defined by

$$GH_n = \{\underline{d} \in \underline{D} : \underline{d}^{(n)} = (\underline{d} \cup \underline{0}')^{(n)}\}$$

and for $n \geq 1$, the class of generalized low_n sets, GL_n, is defined by

$$GL_n = \{\underline{d} \in \underline{D} : \underline{d}^{(n)} = (\underline{d} \cup \underline{0}')^{(n-1)}\}$$

where the superscript (n) denotes the nth iterate of the jump operator. The class of generalized intermediate degrees, GI, consists of those degrees not in GH_n or GL_n for any n.

Information about some local properties of $\mathcal{D}$ can be found in Jockusch 1980 and in Posner 1980. More exhaustive studies of $\mathcal{D}$ can be found in Lerman 1982, Epstein 1979, and Simpson 1978. The reader is referred to Soare 1978 for a survey of the recursively enumerable degrees. The first section of this paper lists some local results which are later used to obtain global results.

The next four sections are devoted to decidability, homogeneity, automorphisms, and definability, respectively.

1. LOCAL STRUCTURE THEOREMS

For a fixed language, a rough analysis of the complexity of a theorem is given by its logical complexity in terms of the number of alternations of quantifiers. To talk about $\mathcal{D}$ the language used is the predicate calculus with a binary relation symbol which is interpreted as $\leq$ in $\mathcal{D}$. Note that $\cup$ is definable from $\leq$. If the statement of the theorem also involves the jump operator, a unary function symbol is added to the language and interpreted as the jump operator in $\mathcal{D}'$.

The easiest place to start is with the lowest complexity of sentences in the language for $\mathcal{D}$. Algebraically, the questions asked at this level are those of $\aleph_0$ universality for partially ordered sets, i.e.,"which partially ordered sets can be embedded in $\mathcal{D}$?" Kleene and Post (1954) began with this question. In fact, they investigated the $\aleph_1$-universality of $\mathcal{D}$ by studying $\mathcal{D}$ in terms of which countable partially ordered sets can be embedded into it.

1.1 Theorem (Kleene & Post 1954): Every countable partially ordered set can be embedded into $\mathcal{D}$.

Theorem 1.1, as stated, has the form of a global result. However, as with all theorems stated in this section, an upper bound can easily be obtained for all degrees mentioned in the theorem, so each of these theorems is really a local theorem in disguise.

$\aleph_2$-universality is easily ruled out for $\mathcal{D}$. Since each $d \in \underline{D}$ can have only countably many predecessors, the order type of ω_1+1, the second uncountable ordinal, is a partially ordered set of order-type $\aleph_1$ which cannot be embedded in $\mathcal{D}$. Nevertheless, Sacks 1966 gives a condition for partially ordered sets of cardinality $\aleph_1$ which guarantees embeddability in $\mathcal{D}$.

Following $\aleph_1$-universality, the theorems of the next complexity class are essentially extension theorems, i.e., statements that theorems of the following form are true (or false): "Let $P \subseteq P^*$ be finite partially ordered sets and let $f: P \to \underline{S} \subset \underline{D}$ be an embedding of P onto a subset of $\underline{D}$. Then there is a finite set $\underline{S}^* \subset \underline{D}$ such that $P^* \simeq \underline{S}^*$ via f^*, $\underline{S} \subseteq \underline{S}^*$ and for all $x \in P$, $f(x) = f^*(x)$." It was such a theorem proved by Kleene and Post 1954 which was used to show that the degrees do not form a lattice whose order relation is induced by $\leq$. This fact is also a corollary of a later theorem of

Spector (1956) which classifies the countable ideals of $\mathcal{D}$.

1.2 Theorem (Spector's exact pair theorem): Let $\underline{C}$ be a countable set of degrees and let $\underline{I}(\underline{C})$ be the ideal generated by $\underline{C}$, i.e., $\underline{I}(\underline{C}) = \{\underline{d} \in \underline{D} : \text{for some finite } \underline{F} \subseteq \underline{C}, \underline{d} \leq \cup \underline{F}\}$. Then there are $\underline{a},\underline{b} \in \underline{D}$ such that $\underline{I}(\underline{C}) = \{\underline{d} \in \underline{D} : \underline{d} < \underline{a} \;\&\; \underline{d} < \underline{b}\}$.

Another extension theorem of Kleene and Post (1954) has the following useful corollary.

1.3 Theorem: Let $P \subseteq P^*$ be finite lattices such that no element of $P^* - P$ lies below any element of P. Let $f{:}P \to \underline{S} \subset \underline{D}$ be an isomorphism. Then there is a set $\underline{S}^* \subseteq \underline{D}$ and an isomorphism $f^*{:}P^* \to S^*$ such that for all $x \in P$, $f^*(x) = f(x)$.

The restriction placed in the hypothesis of Theorem 3.3 on the elements of $P^* - P$ is necessary, a fact which follows from initial segments results which we state later.

The next theorem, due to Friedberg (1956a), characterizes the range of the jump operator.

1.4 Friedberg's jump theorem: Let $\underline{d} \in \underline{D}$ be given such that $\underline{d} \geq \underline{0}'$. Then there is an $\underline{a} \in \underline{D}$ such that $\underline{a}' = \underline{a} \cup \underline{0}' = \underline{d}$.

Friedberg's jump theorem was generalized by Selman (1972). Easier proofs can be found in Jockusch 1974 and Jockusch & Solovay 1977.

1.5 Corollary: Let $n \geq 1$ and $\underline{d} \in \underline{D}$ be given such that $\underline{d} \geq \underline{0}^{(n)}$. Then there is an $\underline{a} \in \underline{D}$ such that $\underline{a}^{(n)} = \underline{a} \cup \underline{0}^{(n)} = \underline{d}$.

An initial segment result is one of the following form: "Let U be an uppersemilattice with least and greatest elements. Then there is a $\underline{d} \in \underline{D}$ such that $U \simeq \mathcal{D}[\underline{0},\underline{d}]$. (The condition that U have a greatest element is unnecessary, and is only introduced for the sake of simplicity.) It is clear that if the $\underline{S}$ in Theorem 1.3 is an initial segment of $\mathcal{D}$, then the theorem is false if the restriction placed on elements of $P^* - P$ is relaxed. It will be a consequence of Theorem 1.7 below that $\underline{S}$ can indeed be chosen to be an initial segment of $\underline{D}$.

Initial segments were first investigated by Spector who proved the following theorem (Spector 1956).

1.6 Theorem: There is a $\underline{d} \in \underline{D}$ such that $\underline{d} \neq \underline{0}$ and for all $\underline{c} \in \underline{D}$, if $\underline{c} < \underline{d}$ then $\underline{c} = \underline{0}$.

Restated, Theorem 1.6 asserts the existence of a minimal degree, or that the two element linearly ordered set is isomorphic to an initial segment of $\mathcal{D}$. Spector's theorem stimulated a great deal of work dealing with initial segments, culminating with the following theorems.

1.7 Theorem (Lachlan 1968): Every countable distributive lattice with least (and greatest) element is isomorphic to an initial segment of $\mathcal{D}$.

1.8 Theorem (Lerman 1971): Every finite lattice is isomorphic to an initial segment of $\mathcal{D}$.

1.9 Theorem (Lachlan & Lebeuf 1976): Every countable upper-semilattice with least (and greatest) element is isomorphic to an initial segment of $\mathcal{D}$.

Recent results announced by Rubin (1979a, 1979b) embed certain uncountable lattices as initial segments of $\mathcal{D}$.

Although Theorem 1.9 implies both Theorem 1.7 and Theorem 1.8, its full power is rarely needed to obtain global properties of $\mathcal{D}$. These theorems have been mentioned since almost all known global properties of $\mathcal{D}$ can be obtained from them.

The last set of structure theorems deals with minimal covers. For $\underline{a},\underline{b} \in \underline{D}$, we say that $\underline{a}$ is a minimal cover for $\underline{b}$ if $\underline{a}$ is a minimal degree in $\mathcal{D}[\underline{b},\infty)$. $\underline{a} \in \underline{D}$ is a minimal cover if $\underline{a}$ is a minimal cover for some $\underline{b} \in \underline{D}$.

1.10 Theorem (Jockusch & Soare 1970): For all $n \in N$, $\underline{0}^{(n)}$ is not a minimal cover.

Jockusch and Soare used Theorem 1.10 together with a weaker version of Theorem 1.11 to compare the arithmetical degrees with all the degrees. The version of Theorem 1.11 which they used depended on the set-theoretical axiom of Borel determinateness. This assumption was later eliminated by Jockusch to obtain the following theorem.

1.11 Theorem (Jockusch 1973): There is a cone of minimal covers.

A sharper version of Theorem 1.11 bounds the vertex of the cone referred to in that theorem. Let $\mathcal{O}$ be the degree of a complete Π^1_1 set.

1.12 Theorem (Harrington & Kechris 1975): $\mathcal{O}$ is the vertex of a cone of minimal covers.

There are many local theorems which have not been mentioned. Those which have been mentioned were chosen either for historical reasons or because they are referred to later in the paper to prove global results.

2. DECIDABILITY

Given a first order algebraic theory, questions arise dealing with the complexity of that theory. We would like to know whether or not the theory is decidable, and if not, we would like to determine the degree of that theory, and the largest natural fragment of the theory which is decidable.

For the theory of $\underline{\mathcal{D}}$ (Th($\underline{\mathcal{D}}$)), the language $\mathcal{L}$ used is the pure predicate calculus with one binary relation symbol, $\leq$, to be interpreted as the order relation on $\underline{D}$. For the theory of $\underline{\mathcal{D}}'$ (Th($\underline{\mathcal{D}}'$)), the language $\mathcal{L}'$ used is $\mathcal{L}$ together with a unary function symbol to be interpreted as the jump operator on $\underline{D}$.

The undecidability of Th($\underline{\mathcal{D}}$) (hence also of Th($\underline{\mathcal{D}}'$)) was first proved by Lachlan.

2.1 <u>Theorem</u> (Lachlan 1968): Th($\underline{\mathcal{D}}$) is undecidable.

<u>Proof</u>: By Ershov & Taitslin 1963, the theory of distributive lattices (hence of countable distributive lattices) is undecidable. Given a sentence σ in the language $\mathcal{L}$ (a language which is suitable for the theory of distributive lattices), form the sentence σ^*:

$$\forall a\ (\{b : b \leq a\} \text{ is a lattice } \rightarrow \sigma^a)$$

where σ^a is σ with quantifiers restricted to elements $\leq a$. Note that σ^* is expressible in $\mathcal{L}$. Furthermore, by Theorem 1.7, σ is true of all distributive lattices if, and only if σ^* is true in $\underline{\mathcal{D}}$. Hence Th($\underline{\mathcal{D}}$) is undecidable.

Given the undecidability of Th($\underline{\mathcal{D}}$), we would like to find the degree of Th($\underline{\mathcal{D}}$) in order to decide how complex that theory is. Early questions asked were: "Are Th($\underline{\mathcal{D}}'$) and Th($\underline{\mathcal{A}}'$) the same?", "Are Th($\underline{\mathcal{D}}$) and Th($\underline{\mathcal{A}}$) the same?", where $\underline{\mathcal{A}}$ is the uppersemilattice of arithmetical degrees and $\underline{\mathcal{A}}'$ is obtained from $\underline{\mathcal{A}}$ by adding the jump operator. (A degree $\underline{d}$ is arithmetical if $\underline{d} \leq \underline{0}^{(n)}$ for some $n \in N$.) A negative answer to the first question follows almost immediately from Theorem 1.2, as was noted by Jockusch and Soare (1970). For the sentence which asserts that there is an ideal which is closed under the jump operator is true of $\underline{\mathcal{D}}$ but not of $\underline{\mathcal{A}}$. Jockusch and Soare (1970) also showed that the second question has a negative solution if one assumes Borel determinateness as a hypothesis. This last assumption was then eliminated by Jockusch.

2.2 Theorem (Jockusch 1973): $Th(\mathcal{D})$ and $Th(\underline{\mathcal{A}})$ are not elementarily equivalent.

Proof: By Theorem 1.10, no arithmetical degree is the vertex of a cone of minimal covers. By Theorem 1.11, there is a degree which is the vertex of a cone of minimal covers.

The degree of $Th(\mathcal{D})$ was finally characterized by Simpson.

2.3 Theorem (Simpson 1977): $Th(\mathcal{D})$ is recursively isomorphic to second order arithmetic.

By second order arithmetic, we mean the theory of N under addition and multiplication, where quantifiers are allowed to range over subsets of N as well as elements of N. Simpson proved Theorem 2.3 by directly coding a model of arithmetic into $\mathcal{D}$ and using Theorem 1.2 to quantify over subsets of N. (Exact pairs a,b of Theorem 1.2 can be used to pick out any subset of N in the model of arithmetic coded into $\mathcal{D}$.) Simpson also proved some new initial segments theorems which enabled him to code models of arithmetic into $\mathcal{D}$. A subsequent proof by Nerode and Shore (1979, 1980) exhibits such a coding which is obtained in a simpler but more indirect manner, but uses only Theorem 1.7.

Sentences of $\mathcal{L}$ may be placed in prenex normal form and then classified according to the number of alternations of quantifiers. Thus an $\exists_n$ ($\forall_n$) formula is one which begins with an existential (universal) formula and has n-1 alternations of quantifiers. Even when a theory is undecidable, questions of interest in that theory may be of a low enough level of complexity to be decidable. For example, it follows from Theorem 1.1 that the $\exists_1$ theory of $\mathcal{D}$ is decidable. This result can be extended to the $\forall_2$ (and hence $\exists_2$) level.

2.4 Theorem (Lerman 1982, Shore 1978): The $\forall_2$ theory of $\mathcal{D}$ is decidable.

Sketch of proof: A straightforward analysis of a typical $\forall_2$ sentence, reduces its truth to the following extension question: Given a finite lattice P, finite lattices $\{P_i : i \leq m\}$ with embeddings $f_i : P \to P_i$ for $i \leq m$, and an isomorphism g of P onto $\underline{S} \subset \underline{D}$, when is it possible that there is a lattice $\underline{S}_i \subset \underline{D}$ for some $i \leq m$ such that $\underline{S}_i$ extendes S, $\underline{S}_i$ is isomorphic to P_i via g*, and for all $x \in P$, $g(x) = g^*(f_i(x))$? By Theorem 1.8, it is necessary that there be an $i \leq m$ such that no element of $P_i - f_i(P)$ lies below any element of $f_i(P)$. And if such an i exists, then the existence of S_i follows from Theorem 1.3. This gives a recursive way to determine the truth of the original sentence when interpreted in $\mathcal{D}$.

Schmerl (see Lerman 1982 for a proof) has shown that Theorem 2.4 is best possible. He exhibits a recursive class of $\forall_3$ sentences of $\mathcal{L}$ which is undecidable.

2.5 Theorem (Schmerl): The $\forall_3$ theory of $\mathcal{D}$ is undecidable.

Questions similar to those asked in this section can be asked of other structures of degrees, such as $\mathcal{D}[\underline{0},\underline{0}']$. Some results along these lines are mentioned in Posner 1980. A question raised in Jockusch & Soare 1970 deals with the proof of Theorem 2.2 and may have a bearing on definability results within the degrees.

Question: Is $\underline{0}^{(\omega)}$ the vertex of a cone of minimal covers? Is there a cone of minimal covers whose vertex is not above $\underline{0}'$?

3. HOMOGENEITY

Relativization is a pervasive phenomenon of degree theory. All sentences of $\mathcal{L}$ which were true of $\mathcal{D}$ seemed also to be true of $\mathcal{D}[\underline{a},\infty)$ for all $\underline{a} \in \underline{D}$. This motivated Rogers (1967) to formulate the homogeneity problems: Is $\mathcal{D} \cong \mathcal{D}[\underline{a},\infty)$ for all $\underline{a} \in \underline{D}$? Is $\mathcal{D}' \cong \mathcal{D}'[\underline{a},\infty)$ for all $\underline{a} \in \underline{D}$? Variations of these problems in which $\cong$ is replaced by $\equiv$ (elementary equivalence) were also examined.

Progress towards solving the homogeneity problems was first made by Feiner.

3.1 Theorem (Feiner 1970): $\mathcal{D}' \not\cong \mathcal{D}[\underline{0}^{(6)},\infty)$.

Proof: Since the language $\mathcal{L}'$ is being used, it suffices to show that $\mathcal{D}[\underline{0},\underline{0}^{(2)}] \not\cong \mathcal{D}[\underline{0}^{(6)},\underline{0}^{(8)}]$. Feiner constructed a linearly ordered set P which had a presentation of degree $\underline{0}^{(6)}$ but no presentation of degree $\le \underline{0}^{(5)}$. By Theorem 1.7 (an earlier result of Hugill 1969 was, in fact, used by Feiner) P is isomorphic to an initial segment $\mathcal{D}[\underline{0}^{(6)},\underline{d}]$ of $\mathcal{D}[\underline{0}^{(6)},\underline{0}^{(8)}]$. (The lower bound of $\underline{0}^{(6)}$ is obtained by relativizing Theorem 1.7. The upper bound of $\underline{0}^{(8)}$ easily follows from the proof of Theorem 1.7.) As noted by Yates (1970b), Turing reducibility is a relation which is recursive in $\underline{0}^{(3)}$. It follows that for all $\underline{a},\underline{b} \in \underline{D}$, $\mathcal{D}[\underline{a},\underline{b}]$ is $\underline{b}^{(3)}$ presentable. Hence there is no $\underline{c} \le \underline{0}^{(2)}$ such that $\mathcal{D}[\underline{0},\underline{c}] \cong \mathcal{D}[\underline{0}^{(6)},\underline{d}]$.

Improvements to Theorem 3.1 by getting degrees $\underline{c} < \underline{0}^{(6)}$ such that $\mathcal{D}' \not\cong \mathcal{D}'[\underline{c},\infty)$ were obtained by Yates (1970a), Nerode & Shore (1980), Epstein (1979), and Shore (1980a) where the best such result appears.

The methods introduced by Simpson (1977) to classify the degree of $\mathrm{Th}(\mathcal{D})$ were powerful enough to yield a negative solution to another homogeneity problem. Simpson's methods yielded some results about definability in $\mathrm{Th}(\mathcal{D}')$ which showed that there were many degrees definable in $\mathrm{Th}(\mathcal{D}')$. Furthermore, for any sufficiently large such degree $\underline{d}$, Simpson produced a sentence of $\mathrm{Th}(\mathcal{D}')$ which was not in $\mathrm{Th}(\mathcal{D}'[\underline{d},\infty))$. The next theorem then followed.

3.2 Theorem (Simpson 1977): There is a $\underline{d} \in \underline{D}$ such that $\mathcal{D}' \not\equiv \mathcal{D}'[\underline{d},\infty)$.

The remaining homogeneity problems were settled by Shore. Shore began with the following theorem on fixed points of isomorphisms.

3.3 Theorem (Shore 1980b): Let $\phi : \mathcal{D} \stackrel{\simeq}{\to} \mathcal{D}[\underline{b},\infty)$, and let $\underline{c} = \phi^{-1}(\underline{b}^{(2)}) \cup \underline{0}^{(2)}$ and $\underline{a} \geq \underline{c}^{(5)} \cup (\phi(\underline{c}))^{(5)}$. Then $\phi(\underline{a}) = \underline{a}$.

Theorem 3.3 was then used to show that for certain $\underline{b} \in \underline{D}$, $\mathcal{D} \not\equiv \mathcal{D}[\underline{b},\infty)$.

3.4 Theorem (Shore 1980b): Let $\underline{b}$ be the vertex of a cone of minimal covers in $\mathcal{D}[\underline{0}^{(2)},\infty)$. Then $\mathcal{D} \not\equiv \mathcal{D}[\underline{b},\infty)$.

Proof: Suppose that ϕ is an isomorphism taking $\mathcal{D}$ to $\mathcal{D}[\underline{b},\infty)$ for $\underline{b}$ as in the hypothesis of the theorem. A contradiction is derived. Let $\underline{c} = \phi^{-1}(\underline{b}^{(2)}) \cup \underline{0}^{(2)}$. Let L be a distributive lattice which is $\underline{c}^{(2)}$ presentable such that for any degree $\underline{d}$ of a presentation of L as a partially ordered set, $\underline{d}^{(2)} \geq \underline{c}$. By Lachlan's proof of Theorem 1.7, L is isomorphic to an initial segment of $\mathcal{D}[\underline{0},\underline{c}]$. Hence L must also be isomorphic to an initial segment of $\mathcal{D}[\underline{0},\phi(\underline{c})]$. Since $\mathcal{D}[\underline{0},\phi(\underline{c})]$ is $(\phi(\underline{c}))^{(3)}$ presentable, it follows that $\underline{c} \leq (\phi(\underline{c}))^{(5)}$. Since ϕ is an isomorphism into $\mathcal{D}[\underline{b},\infty)$, $\phi(\underline{c}) = \underline{b}^{(2)} \cup \phi(\underline{0}^{(2)}) \leq (\underline{0}^{(2)})^{(2)}$, So $\underline{c} \leq \phi(\underline{0}^{(2)})^{(7)}$, Let $\underline{a} = \underline{c}^{(5)} \cup (\phi(\underline{c}))^{(5)}$. Since $\underline{a} \geq \underline{b}$, $\underline{a}$ is the vertex of a cone of minimal covers in $\mathcal{D}[\underline{0}^{(2)},\infty)$. Also, $\underline{a} \leq \phi(\underline{0}^{(2)})^{(12)}$ so $\underline{a}$ is arithmetical over $\phi(\underline{0}^{(2)})$. Thus by Theorem 3.3, $\phi(\underline{a}) = \underline{a}$ is the vertex of a cone of minimal covers in $\mathcal{D}[\phi(\underline{0}^{(2)}),\infty)$, an impossibility by the relativized version of Theorem 1.10 which implies that no degree arithmetical over $\phi(\underline{0}^{(2)})$ can be the vertex of a cone of minimal covers in $\mathcal{D}[\phi(\underline{0}^{(2)}),\infty)$.

The Nerode & Shore (1980) proof of Theorem 2.3 also led to definability results. Shore (1980a) noted that there is a formula ψ of $\mathrm{Th}(\mathcal{D})$ with one free variable such that for any $\underline{b},\underline{c} \in \underline{D}$, if $\mathcal{D}[\underline{c},\infty) \models \psi(\underline{b})$ then $\mathcal{D}[\underline{b},\infty) \simeq \mathcal{D}[\underline{0}^{(7)},\infty)$. This formula tries to say that $\mathcal{D}[\underline{b},\infty) \simeq \mathcal{D}[\underline{0}^{(7)},\infty)$, a statement which can be made in second order arithmetic. The translation of formulas of second order arithmetic into $\mathrm{Th}(\mathcal{D})$ introduces new parameters, so is only a faithful translation if the isomorphism type which it tries to characterize is a cone whose vertex is a sufficiently large degree, $\underline{0}^{(7)}$ being sufficiently large.) Hence if $\underline{b}$ is chosen as in Theorem 3.4, the non-isomorphism proof of Theorem 3.4 is immediately converted into a non-elementary equivalence proof. This idea is used by Shore to resolve the last homogeneity problem.

3.5 Theorem (Shore 1980a): There is a $\underline{b} \in \underline{D}$ such that $\mathcal{D} \not\equiv \mathcal{D}[\underline{b},\infty)$.

The homogeneity problems can be modified to ask for conditions on degrees $\underline{a},\underline{b}$ which guarantee that $\mathcal{D}[a,\infty) \simeq \mathcal{D}[\underline{b},\infty)$. Similar conditions are sought for other homogeneity problems. The first such results were obtained by Yates.

3.6 Theorem (Yates 1970a): If $\mathcal{D}' \simeq \mathcal{D}'[\underline{a},\infty)$ then $\underline{a}' \leq \underline{0}^{(6)}$.

Martin (a proof can be found in Yates 1970b) showed that under the assumption of projective determinaeness (PD), the following theorem holds.

3.7 Theorem (PD): There is an $\underline{a} \in \underline{D}$ such that for all $\underline{b} \geq \underline{a}$, $\mathcal{D}[a,\infty) \equiv \mathcal{D}[\underline{b},\infty)$.

The proof of Theorem 3.7 remains valid for $\mathcal{D}'$ replacing $\mathcal{D}$. Improvements to Theorem 3.6 were obtained by Jockusch & Solovay (1977), Richter (1977), and finally by Shore (1980a) who proved the following result.

3.8 Theorem (Shore 1980a): If $\mathcal{D}' \simeq \mathcal{D}'[\underline{b},\infty)$ then $\underline{0}^{(3)} = \underline{b}^{(3)}$.

An exact classification of those pairs of degrees $<\underline{a},\underline{b}>$ such that $\mathcal{D}'[\underline{a},\infty) \simeq \mathcal{D}'[\underline{b},\infty)$ (or the counterpart pairs for the other homogeneity problems) remains to be found. In fact, it is unknown whether there is such a pair $<\underline{a},\underline{b}>$ with $\underline{a} \nleftrightarrow \underline{b}$ if isomorphism is sought, and proofs of the existence of such pairs when elementary equivalence is sought depend on set-theoretical assumptions.

4. AUTOMORPHISMS

At the same time that Rogers formulated the homogeneity problems, he asked whether there are any automorphisms of the degrees other than the identity. This question has not yet been answered. Research aimed at answering this question has taken two directions; fixed points and automorphism bases.

The fixed point approach aims to show that any automorphism

of $\mathcal{D}$ or $\mathcal{D}'$ has many fixed points. The first such result for $\mathcal{D}'$ was obtained by Jockusch & Solovay (1977) and improved by Richter.

4.1 Theorem (Richter 1977): Let ϕ be an automorphism of $\mathcal{D}'$ and let $\underline{a} \geq \underline{0}^{(3)}$ be given. Then $\phi(\underline{a}) = \underline{a}$.

Proof: It is first shown that if $\mathcal{D}'[\underline{a},\infty) \simeq \mathcal{D}'[\underline{b},\infty)$ then $\underline{a}^{(2)} \leq \underline{b}^{(3)}$. Fix a lattice L which has a presentation of degree $\underline{a}^{(2)}$ as a partially ordered set, such that if $\underline{d}$ is the degree of any presentation of L, then $\underline{d} \geq \underline{a}^{(2)}$. Such lattices are constructed in Richter 1977, and have been alluded to earlier in this paper. By the proof of Theorem 1.9, there is a degree $\underline{c} \geq \underline{a}$ such that $\underline{c}^{(2)} = \underline{a}^{(2)}$ and $L \simeq \mathcal{D}[\underline{a},\underline{c}]$. (Epstein 1979 enables us to choose L to be distributive, and so use Theorem 1.7 instead of Theorem 1.9). Let $\underline{d} = \phi(\underline{c})$. Then $\mathcal{D}[b,d] \simeq L$ and $\underline{d}^{(2)} = \underline{b}^{(2)}$. Since $\mathcal{D}[\underline{b},\underline{d}]$ is $\underline{d}^{(3)}$ presentable, $\underline{a}^{(2)} \leq \underline{d}^{(3)} = \underline{b}^{(3)}$.

Fix $\underline{a} \geq \underline{0}^{(3)}$ and let $\underline{b} = \phi(\underline{a})$. By Corollary 1.5, there is a degree $\underline{c}$ such that $\underline{c}^{(4)} = \underline{c} \cup \underline{0}^{(4)} = \underline{a}$. Let $\underline{d} = \phi(\underline{c})$. Since ϕ is an automorphism of $\mathcal{D}'$, $\underline{d}^{(3)} = \underline{d} \cup \underline{0}^{(3)} = \underline{b}$. By the preceding paragraph, $\underline{d} \leq \underline{c}^{(3)}$. Hence $\underline{d} \cup \underline{0}^{(3)} \leq \underline{c}^{(3)}$, i.e., $\underline{a} \leq \underline{b}$. Using ϕ^{-1} in place of ϕ, a similar proof yields $\underline{b} \leq \underline{a}$. Hence $\underline{a} = \underline{b} = \phi(\underline{a})$.

An improvement of Theorem 4.1 was obtained by Nerode and Shore.

4.2 Theorem (Nerode and Shore 1980). Let ϕ be an automorphism of $\mathcal{D}$. Then there is an $\underline{a} \in \underline{D}$ such that $\phi(\underline{b}) = \underline{b}$ for all $b \geq a$. Furthermore, if $\phi(\underline{0}') = \underline{0}'$ then $\phi(\underline{b}) = \underline{b}$ for all $\underline{b} \geq \underline{0}^{(3)}$.

The other approach to studying automorphism of $\mathcal{D}$ was initiated by Lerman (1977), following lines suggested by Nerode to study automorphism of the lattice of recursively enumerable sets, and pursued by Shore (1977). This is the approach of automorphism bases.

4.3 Definition: A subset $\underline{B} \subseteq \underline{D}$ is an automorphism base for $\mathcal{D}$ ($\mathcal{D}'$) if the identity automorphism is the only automorphism ϕ of $\mathcal{D}$ ($\mathcal{D}'$) such that $\phi(\underline{b}) = \underline{b}$ for all $\underline{b} \in \underline{B}$.

Thus the behavior of an automorphism on an automorphism base completely determines its behavior on $\underline{D}$. It is hoped that small automorphism bases will be found, adding to our understanding of automorphisms. Jockusch and Posner have found a number of automorphism bases for $\mathcal{D}$, among which are the following.

4.4 Theorem (Jockusch & Posner 1980): The following subsets of $\underline{D}$ are automorphism bases for $\mathcal{D}$:

(i) $GH_{n+1} - GH_n$ for all $n \geq 0$.

(ii) $GL_{n+1} - GL_n$ for all $n \geq 1$.

(iii) GL_1

(iv) $\underline{M} = \{d : d$ is a minimal degree$\}$.

Furthermore, "almost every" cone of degrees is an automorphism base.

Frequently, the proof that a set $\underline{B}$ is an automorphism base for $\mathcal{D}$ proves a stronger fact, namely, that $\underline{B}$ generates $\underline{D}$ under $\cup$ and $\cap$ (greatest lower bound whenever it exists).

Rogers' original question about the existence of non-trivial automorphisms of $\mathcal{D}$ can be refined:

4.5 Questions: How many automorphisms does $\mathcal{D}(\mathcal{D}')$ have? Does $\mathcal{D}(\mathcal{D}')$ have a countable automorphism base?

A countable automorphism base for $\mathcal{D}(\mathcal{D}')$ would produce a non-trivial upper bound on the number of automorphisms of $\mathcal{D}(\mathcal{D}')$.

5. DEFINABILITY

Until this point, we have looked at algebraic properties of $\mathcal{D}$ and $\mathcal{D}'$ and decidability questions about their theoreis. We now investigate which degrees or relations on degrees are definable over $\mathcal{D}$ and/or $\mathcal{D}'$.

The first definability were obtained by Jockusch and Simpson and deal with $\mathcal{D}'$. The definitions given were natural definitions.

5.1 Theorem (Jockusch & Simpson 1975): $\mathcal{O}$ is definable over $\mathcal{D}'$. Also, the following relations are definable over $\mathcal{D}'$:

(i) $\underline{a}$ is arithmetical in $\underline{b}$.

(ii) $\underline{a}$ is hyperarithmetical in $\underline{b}$.

(iii) $\underline{b}$ is the hyperjump of $\underline{a}$.

(iv) $\underline{a}$ is ramified analytical in $\underline{b}$.

Simpson's classification of Th($\mathcal{D}'$) and Th($\mathcal{D}$) gave an effective translation of sentences of second order arithmetic into $\mathcal{L}'$ and $\mathcal{L}$ for those theories. For formulas, however, this translation introduces parameters into the translations in $\mathcal{L}'$ and $\mathcal{L}$. Simpson was still able to get the following additional definability results by this method.

5.2 Theorem (Simpson 1977): If $0^{\#}$ exists, then $0^{\#}$ is definable over $\mathcal{D}'$. The following relations are definable over $\mathcal{D}'$:

(i) $\underline{a}$ is constructible from $\underline{b}$.

(ii) $\underline{a}$ is Δ^1_n in $\underline{b}$, $n \geq 1$.

Simpson noted that by Theorem 5.2 (i), there is a sentence of $\mathrm{Th}(\mathcal{D}')$ which is equivalent to the set-theoretical axiom $V = L$. Hence $\mathrm{Th}(\mathcal{D}')$ is not absolute.

The proof of Simpson's theorem by Nerode and Shore (1979,1980) also produced definability results by translating second order arithmetic into $\mathcal{L}$.

5.3 Theorem (Nerode & Shore 1980): The following relations are definable over $\mathcal{D}'$.

(i) $\underline{a} = \underline{b}^{(\omega)}$.

(ii) $\underline{a} = \underline{b}^{\#}$.

The best results on definability follow from a theorem of Shore 1980a. Instead of $\mathcal{D}'$, Shore uses the structure $\mathcal{D}_p = \langle \underline{D}, \leq, \underline{0}' \rangle$ and proves the following:

5.4 Theorem (Shore 1980a): Any relation $R(\underline{x}_1,\ldots,\underline{x}_n)$ on degrees $\geq \underline{0}^{(3)}$ is definable over $\mathcal{D}_p$ if and only if it is definable in second order arithmetic.

Theorem 5.4 is proved by following a translation of second order arithmetic into $\mathcal{L}$, and observing that for degrees $\geq \underline{0}^{(3)}$, only the parameter $\underline{0}'$ remains. In particular, Shore obtains the following corollary.

5.5 Corollary: $\underline{0}^{(3)}$ is definable over $\mathcal{D}_p$.

Two important questions in definability theory are still unresolved.

5.6 Question: Is $\underline{0}'$ definable over $\mathcal{D}$?

5.7 Question: Is the jump operator definable over $\mathcal{D}$?

A positive answer to Question 5.6 would allow the replacement of $\mathcal{L}_p$ with $\mathcal{D}$ in Theorem 5.4. A positive answer to Question 5.7 would imply that $\mathcal{D}$ and $\underline{\mathcal{D}}'$ are interchangeable structures for the study of most global properties of degrees.

The preparation and presentation of this paper were partially supported by the National Science Foundation under grant number MCS 78-01849, and the British Logic Colloquium.

REFERENCES

Epstein, R. L. (1979). Degrees of Unsolvability: Structure and Theory. Springer-Verlag, Berlin, Heidelberg, New York.

Ershov, Y. Taitslin, M.A. (1963). The undecidability of certain theories. Algebra i Logik, 2, no. 5,pp. 37-41.

Feiner, L. (1970). The strong homogeneity conjecture. J. Symbolic Logic, 35, pp. 375-7.

Friedberg, R. M. (1957a). A criterion for completeness of degrees of unsolvability. J. Symbolic Logic, 22, pp. 159-60.

Friedberg, R. M. (1957b). Two recursively enumerable sets of incomparable degrees of unsolvability. Proc. Nat. Acad. Sci. U.S.A., 43, pp. 236-8.

Harrington, L. & Kechris, A. (1975). A basis result for Σ^0_3 sets of reals with an application to minimal covers. Proc. Amer. Math. Soc., 53, pp. 445-8.

Hugill, D. F. (1969). Initial segments of Turing degrees. Proc. London Math. Soc., 19, pp. 1-16.

Jockusch, C. G., Jr. (1973). An application of Σ^0_4 determinateness to the degrees of unsolvability. J. Symbolic Logic, 38, pp. 293-4.

Jockusch, C. G., Jr. (1974). Review of Selman (1972). Math. Reviews, 47, no. 3155, p. 549.

Jockusch, C. G., Jr. (1980). Degrees of generic sets. This volume.

Jockusch, C. G., Jr. & Posner, D. (1980). Automorphism bases for degrees of unsolvability. To appear.

Jockusch, C. G., Jr. & Simpson, S. G. (1975). A degree-theoretic definition of the ramified analytical hierarchy. Ann. Math. Logic, 10, pp. 1-32.

Jockusch, C. G., Jr. & Soare, R. I. (1970). Minimal covers and arithmetical sets. Proc. Amer. Math. Soc., 25, pp. 856-9.

Jockusch, C. G., Jr. & Solovay, R. (1977). Fixed points of jump preserving automorphisms of degrees. Israel J. Math., 26, pp. 91-4.

Kleene, S. C. & Post, E. L. (1954). The upper semi-lattice of degrees of recursive unsolvability, Ann. Math., 59, pp. 379-407.

Lachlan, A. H. (1968). Distributive initial segments of the degrees of unsolvability. Zeit. f. Math. Logik und Grund. der Math., 14, pp. 457-72.

Lachlan, A. H. & Lebeuf, R. (1976). Countable initial segments of the degrees of unsolvability, J. Symbolic Logic, 41, pp. 289-300.

Lerman, M. (1971). Initial segments of the degrees of unsolvability. Ann. Math., 93, pp. 365-89.

Lerman, M. (1977). Automorphism bases for the semilattice of recursively enumerable degrees, Notices Amer. Math. Soc., 24, no. 77T-E10, p. A-251.

Lerman, M. (1982) Structure Theory for the Degrees of Unsolvability. Ω series, Springer-Verlag, Berlin, Heidelberg, New York. To appear.

Mucnik, A. A. (1956). Negative answer to the problem of reducibility of the theory of alogorithms. Dokl. Acad. Nank SSSR, 108, 194-7.

Nerode, A. & Shore, R. A. (1979). Second order logic and first order theories of reducibility orderings. In The Kleene Symposium, ed. J. Barwise, H. J. Keisler & K. Kunen. North-Holland Publishing Co., Amsterdam, New York. To appear.

Nerode, A. & Shore, R.A. (1980). Reducibility orderings: Theories, Definability and Automorphisms. Ann. Math. Logic, To appear.

Posner, D. (1980) A survey of non-r.e. degrees $\leq \underline{0}'$. This volume.

Post, E. L. (1944). Recursively enumerable sets of positive integers and their decision problems. Bull. Amer. Math. Soc., 50, pp. 284-316.

Richter, L. J. (1970). Degrees of unsolvability of models. Doctoral dissertation, University of Illinois at Urbana-Champaign.

Rogers, H., Jr. (1967). Theory of Recursive Functions and Effective Computability. McGraw-Hill, New York, N.Y.

Rubin, J. M. (1979a). The existence of an ω_1 initial segment of Turing degrees. Notices Amer. Math. Soc., 26, no. 79T-A168, p. A-425.

Rubin, J. M. (1979b). Distributive uncountable initial segments of the degrees of unsolvability. Notices Amer. Math. Soc., 26, no. 79T-E74, p. A-619.

Sacks, G. E. (1966). Degrees of Unsolvability. Ann. Math. Studies, no. 55, Princeton University Press, Princeton, N.J.

Selman, A. L. (1972). Applications of forcing to the degree theory of the arithmetic hierarchy. Proc. London Math. Soc., 25, pp. 586-602.

Shore, R. A. (1977). Determining automorphisms of the recursively enumerable sets. Proc. Amer. Math. Soc., 65, pp. 318-25.

Shore, R. A. (1978). On the $\forall\exists$ sentences of α-recursion theory. In Generalized Recursion Theory II, ed. J. Fenstad et. al., pp. 331-53, North-Holland Publishing Co., Amsterdam, New York.

Shore, R. A. (1980a). On homogeneity and definability in the first order theory of Turing degrees. J. Symbolic Logic, to appear.

Shore, R. A. (1980b). The homogeneity conjecture. Proc. Nat. Acad. Sci. U.S.A., to appear.

Simpson, S. G. (1977). First order theory of the degrees of recursive unsolvability. Ann. Math., 105, pp. 121-39.

Simpson, S. G. (1978). Degrees of unsolvability: A survey of results. In Handbook of Mathematical Logic, ed. J. Barwise, pp. 631-52. North-Holland Publishing Co., Amsterdam, New York.

Soare, R. I. (1978). Recursively enumerable sets and degrees. Bull. Amer. Math. Soc., 84, pp. 1149-81.

Soare, R. I. (1983) Recursively Enumerable Sets, Ω series, Springer-Verlag, Berlin, Heidelberg, New York. To appear.

Spector, C. (1956). On degrees of recursive unsolvability. Ann. Math., 64, pp. 581-92.

Yates, C. E. M. (1970). Initial segments of the degrees of unsolvability Part I: A survey. In Mathematical Logic and Foundations of Set Theory, ed. Y. Bar-Hillel, pp. 63-83. North-Holland Publishing Co., Amsterdam, New York.

Yates, C. E. M. (1972). Initial segments and implications for the structure of degrees. In Conference in Mathematical Logic-London 1970, ed. W. Hodges, pp. 305-35. Lecture Notes in Mathematics, no. 255, Springer-Verlag, Berlin, Heidelberg, New York.

SOME CONSTRUCTIONS IN α-RECURSION THEORY

Richard A. Shore
M.I.T. and Cornell University

1 PRELIMINARIES

In these lectures we will try to give some idea of the new problems that confront a recursion theorist trying to ply his trade in the realm of admissible ordinals. Of course we will also try to illustrate some of the methods that have been devised to overcome these problems. As you have already learned from the lectures on classical recursion theory, the recursion theorist's stock in trade consists to a great extent of priority arguments. We will therefore first consider one of these constructions for α-r.e. degrees.

To set the stage we will sketch the basic facts and definitions relevant to the α-r.e. degrees. An ordinal α is admissible iff L_α satisfies Σ_1-replacement. In this case it is fairly straightforward to set up a recursion theory with α or L_α as the domain of discourse, $\Sigma_1(L_\alpha)$ as α-r.e., $\Delta_1(L_\alpha)$ as α-recursive and members of L_α as the α-finite sets (equivalently α-recursive and bounded). This recursion theory proves all the same basic results as does CRT with similar proofs (e.g. enumeration, s-m-n and recursion theorems). [A key point is that L_α has a Σ_1 well ordering]. We thus have α-recursive pairing functions and listings $\{D_u\}_{u<\alpha}$ of canonical indices for α-finite sets and $\{W_e\}_{e<\alpha}$ of the α-r.e. sets together with a simultaneous α-recursive enumeration $W_{e,s}$. (Our conventions are that $D_u \subseteq u$ and $W_{e,s} \subseteq s$.)

Our basic object of study here is relative computability and the ordering of α-degrees. We start with a definition of a multi-valued function.

1.1 Definition

$\{e\}^A(x) = y \iff (\exists u,v)[D_u \subseteq A \;\&\; D_v \subseteq \bar{A} \;\&\; \langle u,v,x,y\rangle \in W_e]$.
We call $\langle u,v,x,y\rangle$ the computation of $\{e\}^A(x) = y$ and y is its output.

Now for the reducibilities.

1.2 Definition

$B \leq_{w\alpha} A \iff \{e\}^A = c_B$ for some e. (c_B is the characteristic function of B and we will abuse notation by writing B for its own characteristic function.)

1.3 Definition

$B \leq_\alpha A \iff \{e\}^A = c_B^*$ for some e where

$$c^*_B(u) = \begin{cases} 0 & \text{if } D_u \subseteq A \\ 1 & \text{if } D_u \subseteq \overline{A} \\ 2 & \text{otherwise} \end{cases}$$

We approximate $\{e\}^A(x)$ as follows:

1.4 Definition

$\{e\}^A_s(x) = y \iff (\exists u,v)[D_u \subseteq A \;\&\; D_v \subseteq \overline{A} \;\&\; \langle u,v,x,y\rangle \in W_{e,s}]$. When we are constructing a sequence A_s of α-recursive sets (to get an α-r.e. $A = \bigcup A_s$) we will modify this definition to make it single valued by always taking the least convergent computation:

$$\{e\}^{A_s}_s(x) = y \iff (\exists u,v)[D_u \subseteq A \;\&\; D_v \subseteq \overline{A} \;\&\; \langle u,v,x,y\rangle \in W_{e,s} \;\&\; \neg\exists \langle u',v',y'\rangle < \langle u,v,y\rangle [D_{u'} \subseteq A \;\&\; D_{v'} \subseteq \overline{A} \;\&\; \langle u',v',x,y'\rangle \in W_{e,s}]].$$

Note that if $\{e\}^{A_s}_s(x) = y$ then only information below s about A was used in the computation.

The α-degrees are of course the equivalence classes of subsets of α under $\leq_\alpha$. One very useful fact about the α-r.e. degrees is that they contain representatives that behave well with respect to these approximations.

1.5 Fact (Sacks)

Every α-r.e. degree contains a regular α-r.e. set A, i.e. one such that $A \cap \beta$ is α-finite for every β.

Introductions to and surveys of aspects of α-recursion theory may be found in Lerman (1978), Shore (1977) and (1978a) and Simpson (1974).

2 α-FINITE INJURY PRIORITY ARGUMENTS AND THE SACKS SPLITTING THEOREM

Let us begin with the first priority argument from Soare's lectures on CRT: For each non-recursive r.e. C there is a simple r.e. A with $C \not\leq A$. Our first problem would be how to translate this into α-recursion theory. Even given the reducibility notions there are several possible versions of simplicity. This is actually a serious problem for the study of the α-r.e. sets as a lattice and we suggest Lerman (1978) for a discussion of these questions. For degree theoretic purposes we only need to worry about the unbounded sets so we take as our positive requirements the

$$P_e : W_e \text{ unbounded in } \alpha \Rightarrow W_e \cap A \neq \emptyset.$$

The negative requirements are as before

$$N_e : C \neq \{e\}^A.$$

(Note that this actually guarantees that $C \not\leq_{w\alpha} A$).

Suppose we now mimic the construction of A in CRT and try to verify that it succeeds. Unfortunately what goes wrong is, as is so frequently the case, so self-evident in CRT that it is not even dignified with a number. Although it is still true that each P_e acts at most once, we cannot conclude that each N_e is injured only boundedly often (and so that I_e is α-finite). The point is that although, for any $e \geq \omega$, the set $\{e' < e \mid P_{e'} \text{ injures } N_e\}$ is bounded (by e) it is only α-r.e. and not in general α-recursive. Thus the stages at which N_e is injured could be unbounded in α. (For example $\alpha = \omega_1^{ck}$, and have $P_{e'}, e' < \omega$, act at stage $|e'|$ for $e' \in \mathcal{O}$).

The solution to this problem of course is to never let it arise. We want every α-r.e. subset of e to be α-finite, but can we do this and still list all the α-r.e. sets? Well there is clearly some bound on the ordinals available to us. Let $\alpha^* = \mu\beta$ (there is an α-r.e. non-α-finite subset of β). If $\alpha^* = \alpha$ we have no worries so suppose $\alpha^* < \alpha$. As frequently happens in recursion theory the cause of our dismay is also the source of our salvation: Let $B \subseteq \alpha^*$ be α-r.e. but not α-finite. We now choose an α-recursive f which is a 1-1 enumeration of B with domain some $\beta \leq \alpha$. If $\beta < \alpha$ the admissibility of α would imply that B is α-finite. Thus $\beta = \alpha$ and f gives us our desired α-recursive listing of α-r.e. sets by ordinals $e < \alpha^*$:

$$W^*_{e,s} = \begin{cases} W_{x,s} & \text{if } (\exists x < s)[f(x) = e] \\ \emptyset & \text{otherwise} \end{cases}$$

$$W^*_e = \bigcup_{s<\alpha} W^*_{e,s}.$$

For the rest of this section we will write W_e and $W_{e,s}$ for the starred versions. We also adjust the definition of $\{e\}^A$, $\{e\}^A_s$ and $\{e\}^{A_s}_s$ to use this new indexing $e < \alpha^*$ as well. It is now easy to verify that I_e is α-finite for every $e < \alpha^*$. (We postpone the arguments for a bit, however.)

It would now be possible to verify Lemmas 2.3 and 2.2 of Soare's notes to show that the negative requirements succeed. The next problem would come in checking that the positive requirements also succeed. Consider P_e. We know that for each $e' < e$ the restraint function reaches a limit $r(e') < \alpha$ but this does not tell us that $R(e) = \bigcup_{e' \leq e} r(e')$ is less than α. The point here is that the map $r(e')$ is only Π_1 and so Σ_1-admissibility does not suffice to bound it.

Again we would like to shorten our listings so that this problem cannot arise. To this end we let $\gamma = \mu\lambda$ [there is an unbounded Π_1, or equivalently Σ_2, map $\lambda \to \alpha$]. Unfortunately simple cardinality considerations show that there is no hope of indexing the α-r.e. sets by ordinals below γ. (For example, there are $\alpha > \aleph_1$ with $\gamma = \omega$). The solution then must be somehow more subtle. We really only want our requirements to be indexed by γ and so we must make one requirement do the work of many. Consider therefore any block B of reduction procedures. We wish to guarantee by a single requirement that $\{e\}^A \neq C$ for any $e \in B$. We will therefore try to preserve new computations giving more information about C from any $e \in B$. As before, however, it is crucial to also preserve disagreements when we can. Of course, if we have a "guaranteed" disagreement between $\{e\}^A$ and C we do not accept further computations from e. As we shall see, increasing the scope of the negative requirements in this way will not interfere too much with the verification that they all succeed.

To implement this blocking of reduction procedures we must split α^* up into γ many pieces. If g is any unbounded Σ_2 map from γ into α it is easy to see that $h = f\circ g$ is an unbounded Σ_2 map from γ into α^*. [If $\mathrm{rg}\, f\circ g \subseteq \beta < \alpha$ then $f^{-1}\restriction\beta$ is an unbounded α-recursive function as $\mathrm{dom}\, f^{-1} \cap \beta$ is an α-r.e. and so α-finite subset of β]. The usual approximation to a Σ_2 function gives us an α-recursive $h(e,s)$ with $\lim_{s\to\alpha} h(e,s) = h(e)$:

$$h(e,s) = y \iff \exists u\, [(\forall v<s)\phi(u,v,e,y)\ \&\ (\forall \langle y',u'\rangle < \langle y,u\rangle)\ (\exists v<s)\neg\phi(u',v,e,y')]$$

where $h(e) = y \iff \exists u \forall v \phi(u,v,x,y)$. Note that for each $e < \alpha^*$ there is in fact a stage s_e such that $\forall e' \leq e\ \forall s > s_e[h(e',s) = h(e')]$. At the expense of minor adjustments $(h'(e) = \bigcup_{e'<e} h(e'),\ h'(e,s) = \bigcup_{e'<e,s'<s} (e',s'))$ we can assume that $h(e)$ and $h(e,s)$ are monotonic and continuous as well. (This is for convenience only).

We can now describe our auxiliary functions and construction. For the same effort we go directly to the Sacks splitting theorem for α-r.e. sets.

2.1 Theorem (Shore (1975))

If B and C are regular α-r.e. sets with C non-α-recursive then we can find regular α-r.e. A_0 and A_1 such that $A_0 \cup A_1 = B$, $A_0 \cap A_1 = \emptyset$ and $C \not\leq_\alpha A_i$ for $i = 0,1$.

2.1.1 Proof. We let B_s and C_s be given by one-one α-recursive enumerations of B and C. We will give an enumeration A_s of A satisfying the one positive requirement

$$P: x \in B_{s+1}-B_s \Rightarrow x \in A_{i,s+1}-A_{i,s} \text{ for exactly one } i,$$

and for each $i < 2, \delta < \gamma$ the negative requirement

$$N_{i,\delta}: \ C \neq \{e\}^{A_i} \text{ for any } e < h(\delta).$$

We define our auxiliary functions as follows:

$$u_i(e,x,s) = \begin{cases} \mu z < s(\{e\}_z^{A_i,s}(x) \text{ is defined)} \\ \text{undefined if no such } z \text{ exists.} \end{cases}$$

$$D_i(\delta,s) = \{e<h(\delta,s) \mid (\exists x<s)[\{e\}_s^{A_i,s}(x) = 0 \ \& \ C_s(x) = 1]\}.$$

(These are the dormant e's in the $h(\delta)^{th}$ block at stage s.)

$$\ell_i(\delta,s) = \mu x(\neg(\exists e<h(\delta,s))[e \notin D_i(\delta,s) \ \& \ \{e\}_s^{A_i,s}(x)=C_s(x)).$$

We now seek to preserve all the relevant computations

$$r_i(\delta,s) = \bigcup\{u_i(e,x,s) \mid e \in D_i(\delta,s) \text{ and } x \text{ is the witness for this fact minimizing } u_i(e,x,s)\}$$

$$\bigcup\{u_i(e,x,s) \mid x<\ell_i(\delta,s) \text{ and } e \text{ is the witness for this fact minimizing } u_i(e,x,s)\}$$

$$R_i(\delta,s) = \bigcup_{\delta'\leq\delta} r_i(\delta',s).$$

2.1.2 Construction: Say $x \in B_{s+1}-B_s$. We put x into A_{1-i} where $\langle\delta,i\rangle$ is the least $\langle\delta',i'\rangle$ such that $x \leq R^{i'}(\delta',s)$ if one exists and $i = 1$ otherwise.

Note that all claims except $\{e\}^{A_i} \neq C$ are immediate.

We nust now verify by simultaneous induction on $\langle\delta,i\rangle$ for $\delta<\gamma, i = 0,1$ that

(1) $I_{i,\delta} = \{x \mid (\exists s)[x \in A_{i,s+1}-A_{i,s} \ \& \ x \leq R^i(\delta,s)]\}$ is α-finite.

(2) $\lim R_i(\delta,s) = R_i(\delta) < \alpha$.

and

(3) $\{e\}^{A_i} \neq C_i$ for $e < h(\delta)$.

Suppose that (1) - (3) hold for $\langle\delta',i'\rangle < \langle\delta,i\rangle$.

1) As $\delta < \gamma$ and the maps from $\delta' < \delta$ to the stage by which $R_{1-i}(\delta',s)$ is constant is Σ_2 its range on δ is bounded say by $s_0 < \alpha$. (It is total by inductive assumption (2)).

As $R_{1-i}(\delta') \leq s_0$ for every $\delta' < \delta$ it is clear that $I_{i,\delta} \subseteq s_0$. (For all $x > s_0$, $<\delta,i>$ would be the least requirement that could be injured by x and so $R_i(\delta,s)$ would keep it out of A_i and so out of $I_{i,\delta}$). Thus $I_{i,\delta}$ is bounded. The only way an element $x \leq s_0$ can enter $I_{i,\delta}$ is for x to be enumerated in $B_{s+1}-B_s$. As B is regular $B \cap s_0$ is α-finite and so entirely enumerated by some stage $s_1 \geq s$. After stage s_1, no x can enter $I_{i,\delta}$. Thus $I_{i,\delta}$ is α-recursive as well as bounded and so α-finite.

2) By the above argument no computation protected by $r_i(\delta,s)$ is ever ruined for any $s \geq s_1$. We may as well assume that $h(\delta,s) = h(\delta)$ for $s \geq s_1$. We now see that $D_i(\delta,s)$ is a non-decreasing subset of $h(\delta)$. Let

$$D_i(\delta) = \{e < h(\delta) \mid (\exists s \geq s_1)\, [e \in D_i(\delta,s)]\}.$$

It is an α-r.e. subset of $h(\delta) < \alpha^*$ and so α-finite. Thus there is an $s_2 \geq s_1$ such that $D_i(\delta,s) = D_i(\delta)$ for $s \geq s_2$. Of course the contributions to $r_i(\delta,s)$ for $s \geq s_2$ from the e's in $D_i(\delta)$ are then bounded by s_2, non-increasing and so eventually constant. As we already have $R_i(\delta',s)$ constant for $\delta' < \delta$, increases in $R_i(\delta,s)$ can only come from increases in $\ell_i(\delta,s)$ and $R_i(\delta,s)$ can be unbounded only if $\lim \ell_i(\delta,s) = \alpha$. If this were true however we could compute C α-recursively for a contradiction: To see if $x \in C$ wait until one gets an $s \geq s_2$ with $\ell_i(\delta,s) > x$. Then $x \in C$ iff $x \in C_s$ for if x entered C after stage x the witness e that shows that $x < \ell_i(\delta,s)$ would enter $D_i(\delta,s)$ for a contradiction. Thus $\ell_i(\delta,s)$ is eventually constant and so therefore is its contribution to $r_i(\delta,s)$.

3) By the argument above $\ell_i(\delta,s)$ is eventually constant. Say $\ell_i(\delta,s) = \ell_i(\delta)$ for each $s \geq s_3 \geq s_2$. If $\{e\}^{A_i} = C$ then clearly $e \notin D_i(\delta)$. Moreover there is a stage $s \geq s_3$ by which $\{e\}_s^{A_{i,s}}(\ell_i(\delta)) = \{e\}^{A_i}(\ell_i(\delta)) = C(\ell_i(\delta)) = C_s(\ell_i(\delta))$. At such a stage s we would have $\ell_i(\delta,s) > \ell_i(\delta)$ for a contradiction.

An interesting companion result to this splitting theorem has recently appeared. Yang (1979) shows that there are α-r.e. sets which cannot be split into two others of the same α-degree.

3. A CONE OF WELL ORDERED α-DEGREES

In this section we will discuss a problem for which the recursion theorist should be a set theorist. We will prove a theorem of S. Friedman (1981) that shows (among other things) that in contrast to CRT one cannot in general relativize results such as those of section 2. It also gives a striking global difference between the ordering of α-degrees for some α and those for ω (at least with some set theoretic assumptions).

3.1 Theorem

(V = L) If α is a singular cardinal of cofinality $\kappa > \omega$ then the non-hyperregular α-degrees are well ordered. (A subset $A \subseteq \alpha$ (or its α-degree) is said to be non-hyperregular if, for some e, $\{e\}^A$ is a function mapping some $\gamma < \alpha$ unboundedly into α).

3.2 Remark

If A is non-hyperregular it is routine to get a function $t_A: \kappa \to \alpha$ which is monotonic, continuous, cofinal in α and α-recursive in A. (Start with any unbounded function f α-recursive in A and let $t_A(\delta) = \bigcup_{\delta' < \delta} f(\delta')$. As α is a cardinal and $\delta < cf(\alpha)$ there is no problem in computing all of these values for f).

Given two non-hyperregular sets A and B we want to compare the rates of growth of the complexity of their initial segments to order them:

3.3 Definition

$g_A(\delta) = \mu\beta(A \cap t_A(\delta) \in L_\beta)$.

3.4 Claim

If $B \not\leq_\alpha A$ then there is a club, i.e. a closed unbounded set, $C \subseteq \kappa$ contained in $\{\delta | g_A(\delta) < g_B(\delta)\}$.

[Note that this claim suffices to prove the theorem: first if $B \not\leq_\alpha A$ and $A \not\leq_\alpha B$ then there would be clubs C_1 and C_2 on which $g_A(\delta) < g_B(\delta)$ and $g_B(\delta) < g_A(\delta)$ respectively. As $C_1 \cap C_2 \neq \emptyset$ we would have a contradiction. If there were $A_0 > A_1 > A_2 \ldots$ then we would have clubs C_i with $g_{A_{i+1}}(\delta) < g_{A_i}(\delta)$. This would give an infinite descending chain of ordinals $g_{A_i}(\delta)$ for $\delta \in \bigcap C_i$ (which is non-empty) for a contradiction].

3.4.1 Proof of claim: If not $S = \{\delta | \lim(\delta) \ \& \ g_B(\delta) \leq g_A(\delta)\}$ is stationary in κ by definition. When $g_B(\delta) \leq g_A(\delta)$ though we can label $B \cap t_B(\delta)$ by an ordinal below $t_A(\delta)$. We do this by projecting all of $L_{g_A}(\delta)$ onto $t_A(\delta)$: Let P_δ: $L_{g_A}(\delta) \to t_A(\delta)$ be one-one and onto. ($g_A(\delta) < t_A(\delta)^+$ since each subset of $t_A(\delta)$ is constructed before the next cardinal as $V = L$). We can clearly choose P_δ such that it and its inverse P_δ^{-1} are uniformly α-recursive in A as both g_A and t_A are α-recursive in A. Thus if $\delta \in S$, P_δ is defined on $B \cap t_B(\delta)$ i.e. $P_\delta(B \cap t_B(\delta)) < t_A(\delta)$. As δ is a limit ordinal and t_A is monotonic and continuous we in fact see that for every $\delta \in S$, $P_\delta(B \cap t_B(\delta)) < t_A(\gamma)$ for some $\gamma < \delta$. As S was assumed stationary, Fodor's theorem tells us that there is a single δ_0 such that $\{\delta | P_\delta(B \ \ t_B(\delta)) < t_A(\delta_0)\}$ is stationary (and so unbounded) in κ.

The point of all of this was to get unboundedly many initial segments of B labelled below some single $t_A(\delta_0)$ via the functions P_δ. Once we have this we can get the set of ordinals doing this coding as an α-finite set. Using this set together with the P_δ we can then recover all initial segments of B to get $B \leq_\alpha A$. Formally we let

$$T = \{\langle\delta,\eta\rangle | \delta \in S \;\&\; \eta < t_A(\delta_0) \;\&\; P_\delta(B \cap t_B(\delta)) = \eta\}.$$

$T \subseteq n \times t_A(\delta_0)$ and so is α-finite. [V = L implies that every bounded subset of a cardinal α is α-finite.] We can now compute $B \cap \beta$ by finding a $\langle\delta,\eta\rangle \in T$ with $P_\delta^{-1}(\eta)$ a set with supremum above β as by the definition of $T, P_\delta^{-1}(\eta) = B \cap t_B(\delta)$.

4 MINIMAL PAIRS OF α-R.E. DEGREES

We will now consider the problem of building a minimal pair $\underset{\sim}{a}_0, \underset{\sim}{a}_1$ of α-r.e. degrees. Again we try to follow the path outlined in Soare's paper on CRT: We want one side or the other to always hold the correct computation. Our first problem is that even if each time we destroy a computation we also preserve its mate, we cannot get our desired conclusion. The point is that we can alternate between the two, destroying first one side and then the other. If we do this cofinally in some $\lambda < \alpha$ then at stage λ we will have no correct computations left. We can prevent such an unbounded alternation by guaranteeing that the needed computations of $\{e\}_s^A(x)$ are destroyed only finitely often. To accomplish this we will impose new restraints. They will however be α-finite and so not overly inimical to the success of our positive requirements.

The second problem is that we cannot employ the guessing strategy to make the negative restraints drop back simultaneously. There will in general be too many (i.e., not α-finite) possible guesses. The solution here is to note that there is no real problem (or rather it has already been solved). As the computation of $\{e\}^{A_i}(x)$ can be destroyed only finitely often, it is fairly easy to see that if the lengths of agreements for e reach new maximums at stages cofinal in some $\lambda < \alpha$ then they reach a new maximum at stage λ. Thus the stages at which the restraint imposed by N_e is 0 form a closed set. If this happens only boundedly often then the corresponding restraint is bounded. For each of the others however we get a club on which its restraint is 0. Intersecting them all gives us a club on which they are all zero simultaneously. So if we can uniformly bound the bounded ones the positive requirements will have no trouble succeeding.

Thus we see that we again have cofinality problems (for the bounding) and projectum problems (for picking out the e's with bounded restraint). They are however more serious than in the case of the finite injury arguments. The set of e's whose restraint function is bounded is only Σ_2 . Thus we should index our r.e. sets by a listing of length ρ such that any Σ_2 subset of $\delta < \rho$ is α-finite. We can do this at the expense of substituting a Σ_2 projection function and an α-recursive approximation to it for the α-recursive one of §2. The bounding problem is again one of a Σ_2 cofinality function (from e to the bound on its restraint). Unfortunately it is not always possible to interweave these constraints. Unlike α^*, ρ need not have the same Σ_2-cofinality as α. [We can have $\rho = \aleph_1$, but $cf(\alpha) = \sigma 2cf(\alpha) = \omega$.] Ultimately this is the source of a problem we do not yet know how to solve. For now, though, let us set up the auxiliary functions for the construction.

To begin we let ρ be the Σ_2 projectum of α i.e., $\rho = \mu\delta$ (there is a Σ_2 subset of δ which is not in L_α). As usual by Jensen (1972) there is a partial Σ_2 function f mapping a subset of ρ one-one onto α. We let $f(e,s)$ be an α-recursive approximation to f as in §2. (Thus if $e \in \text{dom } f$ then $\lim f(e,s) = f(e)$. If $e \notin \text{dom } f$, $f(e,s)$ need not converge). We could now reindex the reduction procedures and their approximations but prefer to write $\{f(e,s)\}$ instead of $\{e\}$ when required. We can now define for $e < \rho$ the length of agreement functions for the sets A_i that we will construct as in §4 of Soare's lectures.

$$\ell(e,s) = \bigcup\{x \mid \forall y < x[\{f(e,s)\}_s^{A_0,s}(y) = \{f(e,s)\}_s^{A_1,s}(y)\}$$

$$m(e,s) = \bigcup\{\ell(e,t) \mid t \leq s\} .$$

Our restraint function is like that for the highest priority requirements in CRT as we need no guessing here. An additional clause is needed however to reflect the fact that we are only approximating the Σ_2 indexing of procedures.

$$r(e,s) = \begin{cases} 0 \text{ if } \ell(e,s) = m(e,s) \text{ or if} \\ \qquad \lim_{t \to s} f(e,t) \neq f(e,s) \\ \bigcup\{t < s \mid \ell(e,t) = m(e,t)\} \text{ otherwise} \end{cases}$$

$$R(e,s) = \bigcup\{r(e',s) \mid e' \leq e\}.$$

The positive requirements will also be indexed below ρ so that $P_{i,e}$, for example, tries to guarantee that if $f(e)$ is defined and $W_{f(e)}$ is unbounded then $W_{f(e)} \cap A_i \neq \emptyset$. One new constraint is needed because of this approximation procedure. When $f(e,s)$ changes we do not allow $P_{i,e}$ to later injure any computation $\{c\}^{A_i}(y)$ for $c,y < s$. (Note that this replaces line 11 p. 414 of Shore (1978) which as phrased does not quite accomplish what was intended.) For convenience we will now assume that $\alpha^* = \alpha$ and show how to eliminate the assumption later.

4.1 Construction

Stage s. For $i = 0,1$ and each $e < s$ for which $P_{i,e}$ is not satisfied (i.e., $W_{f(e,s)} \cap A_{i,s} = \emptyset$) see if there is an $x \in W_{f(e,s),s}$ such that $x > R(e,s)$. We also require that putting x into A_i does not injure any current computation of $\{c\}^{A_i,s}_s(y)$ for $c,y < \bigcup(\{t<s|$ some element was put into A_i at stage t to satisfy some

$P_{e',i}$ with $f(e',t) \leq f(e,s)\} \cup \{t < s|\ f(e,t) \neq \lim_{t'\to t} f(e,t')\})$.

If there is such an x choose one with $\langle i,f(e,s)\rangle$ minimal and put it into A_i.

4.2 Clubs

It is clear from the definition of the restraint functions that $r(e,s) < r(e,s')$ implies that $r(e,t) = 0$ for some t between s and s'. Thus either lim inf $r(e,s) = 0$ or lim sup $r(e,s) < \alpha$. The key point is that $S_e = \{s|\ r(e,s) = 0\}$ is a club: Suppose S_e is unbounded in λ. If $\lim_{t\to\lambda} f(e,t) \neq f(e,\lambda)$ then $r(e,\lambda) = 0$ as required. Otherwise $f(e,t)$ is constant on a final segment of λ. Our new preservation requirements guarantee that, for any c and y, if a computation of $\{c\}^{A_i,s}_s(y)$ is injured to satisfy some $P_{i,e'}$ at a stage t' once reestablished it can be later reinjured at $t'' > t'$ only for some $P_{i,e''}$ with $f(e'',t'') < f(e',t')$. Thus no such computation is ever injured more than finitely often. So if $\{e\}^{A_i,s}_s(y)$ is defined cofinally in λ it is constant on a final segment. Thus

$m(e,\lambda) = \bigcup\{s < \lambda|m(e,s)\} = \{s < \lambda|\ell(e,s)\} = \ell(e,\lambda)$ and $r(e,\lambda) = 0$ as required.

4.3 The positive requirements

Suppose $f(e) = \lim f(e,s)$ exists and $W_{f(e)}$ is unbounded. Let s_o be such that $f(e,s) = f(e,s_o) = f(p)$ for every $s \geq s_o$. As $\alpha^* = \alpha$ the set of e's such that we ever act to satisfy a $P_{i,e'}$ at some stage s_d with $f(e',s_d) = d < f(e)$ is α-finite. Thus there is a bound $s_1 \geq s_o$ on all such stages s_d. Our extra requirement on finding an acceptable $x \in W$ can therefore only protect computations $\{c\}^{A_i,s}_s(y)$ $f(e)$ with $c,y < s$. By the argument in 4.2 any such computation can be injured only finitely often. Thus the set of all such computations corresponds to a Σ_1 subset of $s_1 \times s_1 \times w$ (The final entry is the number of times the computation has been previously injured.) Again as $\alpha^* = \alpha$ this set is α-finite and so therefore is the restraint imposed by our extra requirement. Let $\delta_o \geq s_1$ be a bound on this restraint.

For the basic requirement that $x > R(e,s)$ we consider the set $U = \{ e' \leq e|\lim\inf r(e',s) = 0\}$.

It is a Π_2 subset of $e < \rho$ and so α-finite as is $B = \{e' \leq e | \liminf r(e',s) \neq 0\}$. As we have seen, if $e' \in B$, $\limsup r(e',s) < \alpha$. Moreover, the map from e' to $r(e') = \limsup r(e',s)$ is Σ_2. Thus if $\rho \leq \sigma 2cf(\alpha)$, $\bigcup\{r(e',s) \mid e' \in B\}$ is bounded by some $\delta_1 < \alpha$. On the other hand $S_{e'}$ is an α-recursive club for each $e' \in U$ and so $S = \bigcap \{S_{e'} | e' \in U\}$ is a club. Thus if $x \in W_{f(e),s}$ is greater than δ_o and δ_1 and $s \in S$ we put x into A_o if $A_{o,s} \cap W_{f(e)} \neq \emptyset$. At the next such stage s' we put an $x \in W_{f(e)}$ into A_1 if $A_{1,s} \cap W_{f(e)} \neq \emptyset$.

4.4 The negative requirements

Suppose that $f(e) = \lim f(e,s)$ exists and that $\{e\}^{A_o} = \{e\}^{A_1} = g$ is a total function. We must show that g is α-recursive. Consider first the set $D = \{e' < e \mid \lim f(e',s)$ exists$\}$. It is Σ_2 and so α-finite. If $\rho \leq \sigma 2cf(\alpha)$ then we can find a stage s_o such that $f(e',s) = f(e',s_o)$ for all $e' \in D$. We can then find an $s_1 \geq s_o$ by which $W_{f(e'),s_1} \cap A_{i,s_1} \neq \emptyset$ for all $e' \in D$ such that $W_{f(e)} \cap A_i \neq \emptyset$. To compute $g(x)$ we claim it suffices to find a stage $s > s_1$ at which $\ell(e,s) = m(e,s) > x$ and $f(e',s) \neq \lim_{t \to s} f(e',t)$ for $e' < e, e' \notin D$. Indeed we claim that $g(x) = \{e\}_s^{A_i,s}(x)$ for such an s. Our extra requirement guarantees that no $P_{i,e'}$ of higher priority with $e' \notin D$ can ever injure any future computation of $\{e\}^{A_i}(x)$. Our choice of s_1 guarantees that no $P_{i,e'}$ of higher priority with $e' \in D$ can ever act again. Thus only lower priority requirements can injure computations $\{e\}^{A_i}(x)$ at any stage $t \geq s$. The basic restraints now guarantee that at most one of these computations is ever injured at any such later stage. As our new requirements also make sure that such injuries can occur only finitely often we see that for every $t \geq s$,

$$\{e\}^{A_i,t}(x) = \{e\}_s^{A_i,s}(x) \text{ for at least one of } i = 0,1.$$

Thus $\{e\}_s^{A_i,s}(x) = g(x)$ as required.

All that remains is to argue that such a stage s actually exists. Note that $\{s | f(e',s) \neq \lim_{t \to s} (e',t)\}$ is clearly a club. As we have seen in 4.2 that $\{s | \ell(e,s) = m(e,s)\}$ is closed it suffices to show that $\ell(e,s)$ is unbounded. If not then the argument of 4.2 shows that there is a least β such that $\exists t_o \forall t > t_o (\ell(e,t) < \beta)$. Let t_o be the required witness for β. If $\beta = \delta + 1$ we can find a stage $t_1 > s_1, t_o$ such that $\{e\}^{A_i,t}(\delta) = \{e\}^{A_i}(\delta)$ for every $t > t_1$. By the miminality of β there is a $t > t_1$ at which $\ell(e,t) = \delta$. But the definition of ℓ would then show that $\ell(e,t) \geq \delta + 1$ for our contradiction. Finally if β is a limit ordinal the sets $\{s | \ell(e,s) \geq \delta\}$ for $\delta < \beta$ are α-recursive and unbounded. Again by the argument of 4.2 they are closed and so have a point $t > t_0$ in common. Thus $\ell(e,t) \geq \beta$ for our contradiction.

We have thus proved the following:

4.5 Theorem (Shore (1978))

If $\sigma 2p(\alpha) \leq \sigma 2cf(\alpha)$ and $\alpha^* = \alpha$ then there is a minimal pair of α-r.e. degrees.

By modifying this construction we can cover the cases done in Lerman & Sacks (1972) which include $\alpha^* < \alpha$. We will only briefly sketch the changes needed and refer to the paper of Lerman & Sacks for unexplained terminology. We use δ = the tame Σ_2 projectum of α for indexing our requirements with the usual α-recursive approximation $f(e,s)$ convergent on intial segments. The major change in the construction is in our extra requirement on elements going into the A_i: Suppose that $P_{i,e}$ put an element into A_i at stage s. For each $e' < \delta$ let $s' \geq s$ be the first stage (if one exists) such that $\ell(e',s') = m(e',s')$ & $f(e',s) = f(e',t)$ for every t between s and s'. We never let any $P_{j,e''}$ with $e'' > e$ put an element less than s' into A_j ($j = 0,1$) at any stage $s'' \geq s'$. It is also convenient to add on the usual requirements to make the A_i hyperregular as in Sacks & Simpson (1972) or Shore (1975).

One can now argue that the new requirements impose only α-finitely much restraint on $P_{i,e}$ even though α^* may be less than α. They do however suffice to prevent infinite alternations once things have settled down and so to guarantee that a final segment of S_e is closed for each e. The major change in the verification that the $P_{i,e}$ succeed is to consider a stage s well above the new restraint by which f has settled down through e, in which $W_{f(e)}$ is unbounded and which is of α-cofinality $> e$. Given such a stage the cofinality assumption gives a uniform bound on the basic restraint imposed by all the $e' < e$ which have $r(e',t)$ bounded in s. The others again have $r(e',s) = 0$ on a club in s and so $P_{i,e}$ succeeds before stage s. The key point then is the existance of such a stage s. The assumption needed here is that there is an α-cardinal greater than e.

To see that the negative requirements succeed note that we can now wait until all $P_{i,e'}$ of higher priority have stopped acting completely. Thus the usual argument that one side or the other holds the computation suffices as before.

We can in this way prove the following:

4.6 Theorem (Lerman & Sacks (1972).

If for every $e < \delta = t\sigma 2p(\alpha)$ there is an α-cardinal greater than e then there is a minimal pair of α-r.e. degrees.

The case that remains open is when $\sigma 2cf(\alpha) < \sigma 2p(\alpha) =$

greatest cardinal of $\alpha < \alpha^* = \alpha$. (Even here these techniques suffice when there is an unbounded set of sufficiently nice ordinals below α. See Maass(1980)).

To close we would like to thank the Natural Science Foundation for its support under grant MCS 77-04013 and the British Logic Colloquim for bringing us to the Leeds meeting.

REFERENCES

Friedman, S.D. (1981). Negative Solutions to Post's Problem II, to appear.

Jensen, R.B. (1972). The fine structure of the constructible hierarchy, Ann. Math. Logic, 4, pp. 229-308.

Lerman, M. & Sacks, G.E. (1972). Some minimal pairs of α-recursively enumerable degrees, Ann. Math. Logic, 4 pp. 415-442.

Maass, W.(1980) On minimal pairs and minimal degrees in higher recursion theory, to appear.

Sacks, G.E. & Simpson. S.G. (1972). The α-finite injury method, Ann. Math. Logic, 4, pp. 343-367.

Shore, R.A. (1975). Splitting an α-recursively enumerable set, Trans. Am. Math. Soc., 204, pp. 65-78.

Shore, R.A. (1977). α-Recursion Theory, in Handbook of Mathematical Logic, ed. J. Barwise, pp. 653-680, North-Holland, Amsterdam.

Shore. R. A. (1978). Some more minimal pairs of α-recursively enumerable degrees, Zeitschr. f. Math. Logik and Grundlagen de Math., 24 pp. 409-418.

Shore, R.A. (1978a). On the $\forall\exists$ sentences of α-recursion theory, in Generalized Recursion theory II, eds. Fenstad et al., pp. 331-353, North Holland, Amsterdam.

Simpson, S.G. (1974). Degree theory on admersible ordinals, in Generalized Recursion theory, eds. J. Fenstad and P. Hinman, pp. 165-194, North-Holland, Amsterdam.

THE RECURSION THEORY OF THE CONTINUOUS FUNCTIONALS

Dag Normann, Leeds and Oslo

INTRODUCTION

Classical or ordinary recursion theory has been subject to many generalizations and extensions. When we generalize a mathematical theory we want to give the old concepts a new meaning similar to the original one but within a new context. In generalized recursion theory this mainly means to generalize the concepts of finite, computation and reduction-procedure. There have been several successful generalizations of recursion theory, e.g. admissible recursion theory and recursion in normal higher type objects. There are also axiomatic approaches to a notion of 'general recursion theory'. Fenstad [5] is a good introduction to this area.

When we extend a mathematical theory we want to see if the old concepts are meaningful in a wider context. The main concept of recursion theory is that of an algorithmic procedure. Elsewhere in this volume (Tucker [20]) there is a survey of finite algorithmic procedures over general algebraic structures, a typical example of extended recursion theory.

In [11] Kleene extended the notion of an algorithm to arbitrary objects of finite type. The pure finite types are defined as follows:

$Tp(0) = \omega$ = the natural numbers

$Tp(n+1) = {}^{Tp(n)}\omega$ = the set of total maps from $Tp(n)$ to ω.

He described indices e denoting algorithms for functions e operating on finite lists of functionals of finite type. The algorithms are described using nine schemata S1 - S9.

It can be and is discussed whether S1 - S9 gives a true extension of classical recursion theory, since computations no longer are finite. Moreover there are alternative ways to extend classical recursion theory to a hierarchy of functionals (Platek [19], Kleene [13]). But it is generally agreed that Kleene's S1 - S9 gives an important analysis of the concept of an algorithm.

In the late fifties Kleene [12] and Kreisel [14] discovered an alternative to the functionals of higher type, the countable or continuous functionals. Kleene regarded them as elements of his hierarchy of total functionals and he showed that they are closed under his notion of computability, S1 - S9, Kreisel defined them as equivalence-classes of certain equivalence-relations. Common for both definitions is that the functionals are globally described by a countably amount of information coded in the associates, and they are locally described by a finite amount of information.

In this paper, we will define the continuous functionals and survey some of the results concerning their recursion theory. We will only give a few elementary proofs. For detailed proofs we refer to the original papers, or to Normann [17] .

BASIC DEFINITIONS AND RESULTS

First we will give a simultaneous definition of the continuous functionals, their associates and a topology for them.

<u>Definition</u>

$Ct(0) = \omega$

$Ct(1) = {}^{\omega}\omega$ with its usual topology

$Ct(2) = \{F:Ct(1) \to \omega;\ \ F \text{ is continuous}\}$

Let $\alpha:\omega \to \omega$. α <u>is an associate</u> for $F:Ct(2) \to \omega$ if

<u>i</u> $\forall \beta \in {}^{\omega}\omega \exists n\ \alpha(\overline{\beta}(n)) > 0$

<u>ii</u> $\forall \beta \in {}^{\omega}\omega \forall n[\alpha(\overline{\beta}(n)) > 0 \Rightarrow \alpha(\overline{\beta}(n)) = F(\beta)+1]$

<u>iii</u> $\forall \beta \in {}^{\omega}\omega\ \{n;\alpha(\overline{\beta}(n)) = 0\}$ is an initial segment of ω.

It is easy to show that $F:Ct(1) \to \omega$ is continuous if and only if F has an associate.

Let As(2) be the set of associates for type-2 objects and let $\rho_2:As(2) \to Ct(2)$ be the map sending an associate to the corresponding functional. Let T_2 be the topology on Ct(2) defined as the finest topology making ρ_2 continuous, i.e.

$O \in T_2 \Leftrightarrow \rho_2^{-1}[O]$ is open in As(2)

By induction on $k > 2$ we define $Ct(k)$, $As(k)$, ρ_k and T_k as follows

$Ct(k) = \{\psi : Ct(k-1) \to \omega; \psi \text{ is continuous}\}$

Let $\alpha : \omega \to \omega$. α is an associate for ψ: $Ct(k-1) \to \omega$ if

<u>i</u> $\forall \beta \in As(k-1) \exists n\, \alpha(\overline{\beta}(n)) > 0$

<u>ii</u> $\forall \beta \in As(k-1) \forall n[\alpha(\overline{\beta}(n)) > 0 \Rightarrow \alpha(\overline{\beta}(n)) = \psi(\zeta_{k-1}(\beta)+1]$

<u>iii</u> $\forall \beta \in {}^{\omega}\omega\{n; \alpha(\overline{\beta}(n)) = 0\}$ is an initial segment of ω

It is easy to show that $\psi : Ct(k-1) \to \omega$ is continuous if and only if ψ has an associate.

The rest of the definition is completely analogous to the case $k = 2$.

In this survey we will not need the precise definition of S1 - S9-computations. One of the main schemes is S8 for functional application: given an algorithm for computing $\Phi(\phi)$ for all $\phi \in Ct(k)(Tp(k))$ we uniformly get an algorithm for computing $\Psi(\Phi)$ from Ψ where $\Psi \in Ct(k+2)(Tp(k+2))$. S8 is actually a relativized version of this. Another essential scheme is S9 which automatically gives the enumeration property.

Kleene [12] showed that if $\psi : Ct(k) \to \omega$ is S1 - S9-computable relative to a finite list $\underset{\sim}{\psi}$ of continuous functionals then $\psi \in Ct(k+1)$. Moreover he showed that all computable functionals have recursive associates. The idea behind the proof is to construct a "universal associate" for all computations and the proof shows that the value of a computation is decided by finite bits of any set of associates for the arguments involved.

Kleene's algorithms work directly on the functionals, but there is another interesting notion of an algorith working uniformly on the associates.

<u>Definition</u>

$\psi \in Ct(k)$ is <u>recursive</u> if ψ has a recursive associate.

$\psi \in Ct(k)$ is <u>recursive in</u> $\phi \in Ct(t)$ if there is a recursive function mapping any associate for ϕ onto an associate for ψ.

There is a notion of partial recursiveness too which we will not consider here, see e.g. Feferman [4] or Hyland [10].

The results of Kleene mentioned above show that the relation "S1 - S9-computable in" is a subrelation of "recursive in". We will now mention some results in chronological order and the first shows that "S1 - S9-computable in" is a proper subrelation of "recursive-in". We will later discuss a couple of general methods.

If $f \in {}^{\omega}\omega$ we let $C_f = \{g:\omega\to\omega:\forall n\, g(n)\le f(n)\}$. C_f will be compact so any continuous F will be uniformly continuous on C_f.

If σ is a sequence of length n of natural numbers let

$$B_\sigma = \{g:\bar{g}(n) = \sigma\}$$

Let

$$\Phi(F,f) = \mu n \text{ (if } \sigma \text{ has length n then F is constant on } C_f \cap B_\sigma)$$

Φ is called the <u>fan functional</u> . We may code Φ as a functional of type 3 and we then have

THEOREM 1 (Tait unpublished, see Gandy, Hyland [6] or Normann [17])

Φ has a recursive associate but Φ is not Kleene-computable in any element of Ct(2).

It was immediately realised that a functional F in Ct(2) is uniformly computable in any of its associates and if F is computable in some $\alpha \in {}^{\omega}\omega$ then F has an associate recursive in α. The same considerations are valid for "recursive in". If we in a uniform way could compute an associate for F from F it is easy to see that all type 3 functionals will be computable in any of their associates, so by Tait's result there is no such uniform way. Indeed there is no way at all because Hinman [8] showed the following

THEOREM 2

There is a continuous functional of type 2 which is not recursively equivalent to any $\alpha:\omega \to \omega$.

It is then of course not S1 - S9-equivalent to any $\alpha:\omega \to \omega$.

These two results have later been improved by the following theorems:

THEOREM 3 (Gandy-Hyland [6])

There is a recursive functional $\Gamma \in$ Ct(3) such that Γ is not computable in Φ and any $F \in$ Ct(2) (where Φ is the fan-functional).

THEOREM 4 (Dvornickov [3])

For each $k > 2$ there is a $\Phi \in$ Ct(k) such that Φ is not recursively equivalent to any $\phi \in$ Ct(k-1).

THEOREM 5 (Normann [16])

For each $k > 3$ there is a recursive functional Δ in Ct(k) such that Δ is not S1 - S9-computable in any $\phi \in Ct(k-1)$.

For the rest of this paper we will mainly be interested in S1 - S9 computability. Among other things we will be looking at the following two sets:

<u>Definition</u>

Let $\psi \in Ct(k)$ (or Tp(k))

a 1-section (ψ) = 1-sc(ψ) = $\{\alpha:\omega\to\omega;\ \alpha$ is computable in $\psi\}$

b 2-envelope (ψ) = 2-en(ψ) = $\{A\subseteq {}^{\omega}\omega\ ;A$ is semicomputable in $\psi\}$

where a set is semicomputable if it is the domain of a partial computable functional.

THE MODULUS OF A SEQUENCE

Many results concerning the continuous functionals have been obtained by investigating approximations to phenomena. This was used by one of the pioneers, namely T. Grilliot, see [7] , in investigating the computational power of these functionals. Among other things he proved

THEOREM 6

Let $F:{}^{\omega}\omega\to\omega$. Then F is continuous if and only if $\forall\alpha\in{}^{\omega}\omega$ (^2E is not computable in F,α) where

$$^2E(\beta) = \begin{cases} 0 \text{ if } \forall n\beta(n) = 0 \\ 1 \text{ if } \exists n\beta(n) > 0 \end{cases}$$

This theorem has later been generalized by J. Bergstra in [2]:

THEOREM 7

Let $\psi:Ct(k)\to\omega$. Then $\psi\in Ct(k+1)$ if and only if $\forall\phi\in Ct(k)$ (^2E is not computable in ψ,ϕ).

We will give a quick sketch of Grilliot's proof of Theorem 6:

Assume that F is discontinuous. Let $\{\alpha_i\}_{i\in\omega}\to\alpha$ be such that $F(\alpha)$ is not the limit of $\{F(\alpha_i)\}_{i\in\omega}$ W.l.o.g. we may assume

<u>i</u> $\forall i\ F(\alpha_i)\neq F(\alpha)$

<u>ii</u> $\forall i\ \overline{\alpha}_i(i) = \overline{\alpha}(i)$

by picking up a subsequence if necessary.

Let β be given. Let

$$\nu_\beta(n) = \begin{cases} \alpha_i(n) \text{ if } i \le n \text{ is minimal such that } \beta(i) > 0 \\ \alpha_{n+1}(n) \text{ otherwise} \end{cases}$$

It is easy to see that

$$\exists i\beta(i) > 0 \Rightarrow \nu_\beta = \alpha_i \text{ for the least such } i \Rightarrow F(\nu_\beta) \neq F(\alpha)$$

$$\forall i\beta(i) = 0 \Rightarrow \nu_\beta = \alpha \Rightarrow F(\nu_\beta) = F(\alpha)$$

and we may compute

$$^2E(\beta) = \begin{cases} 0 \text{ if } F(\nu_\beta) \neq F(\alpha) \\ 1 \text{ if } F(\nu_\beta) = F(\alpha) \end{cases}$$

The trick in this proof has developed into a method based on the following notion

Definition

Let $\{\psi_n\}_{n\in\omega}$ be a sequence from Ct(k). We call Φ a modulus-function for $\{\psi_n\}_{n\in\omega}$ if

$$\forall\phi \in Ct(k-1)\ \forall m,n > \Phi(\phi)\ (\psi_m(\phi) = \psi_n(\phi))$$

Remarks

a In the proof above $\gamma(n) = n + 1$ is a modulus for the sequence $\{\alpha_i\}_{i\in\omega}$

b $\{\psi_n\}_{n\in\omega}$ will have a modulus-function if and only if it converges pointwise. It can be shown that $\{\psi_n\}_{n\in\omega}$ is convergent in the sense of T_k if and only if it has a modulus in Ct(k).

Grilliot called a functional $F: {}^\omega\omega \to \omega$ effectively discontinuous if F is discontinuous at some α computable in F. In [7] he also showed

F is effectively discontinuous if and only if ^2E is computable in F.

One of his main tricks was to compute a modulus for $\{F(\alpha_n)\}_{n\in\omega}$ from $F, \{\alpha_n\}_{n\in\omega}$ and a modulus for $\{\alpha_n\}_{n\in\omega}$ provided that $F(\lim\alpha_n) = \lim F(\alpha_n)$.

Wainer [21] iterated this trick. Given an F that is not effectively discontinuous he gave uniformly a primitive recursive approximation $f(n,e,\vec{x})$ to a computation $\{e\}(F,\vec{x})$ and computed moduli for $\{f(n,e,\vec{x})\}_{n\in\omega}$ whenever $\{e\}(F,\vec{x})$ has a value. A consequence was the following

COROLLARY (Normann-Wainer [18])

If F is not effectively discontinuous then there is a $G \in Ct(2)$ such that

$$1\text{-}SC(F) = 1\text{-}SC(G)$$

There are other consequences that will be mentioned later.

In Normann-Wainer [18] Wainer's proof was extended to arbitrary continuous functionals:

THEOREM 8

There is a primitive recursive f and a partial recursive M such that whenever $\{e\}(\vec{\psi}) \simeq k$ then

<u>i</u> $\lim f(n,e,\vec{\psi}) = k$

<u>ii</u> $M(e,\vec{\psi})$ takes a value and $\forall n \geq M(e,\vec{\psi})\ f(n,e,\vec{\psi}) = k$

For each continuous ψ let $f_\psi(n,e,x) = f(n,e,x\psi)$ whenever this makes sense.

COROLLARY

$1\text{-}SC(\psi)$ is recursively generated by its r.e.f_ψ-degrees.

<u>Proof</u>

Let α be computable in ψ, $\alpha(x) = \{e\}(x,\psi)$. Let $\beta(x) = M(e,x,\psi)$ Then β is computable in ψ. Let

$$D = \{(n,x) ; \exists m > n f_\psi(m,e,x) \neq f_\psi(n,e,x)\}$$

D is r.e.(f_ψ). D is recursive in β so D is computable in ψ. Moreover α is recursive in D,f_ψ by $\alpha(x) = f_\psi(n,x)$ for the least n such that $\neg D(n,x)$.

Modulus functions have also been used in constructing interesting continuous functionals of type 2. In [1] Bergstra constructed functionals F_x^y to any pair of r.e. sets W_x,W_y such that from a modulus α for the canonical approximation of W_x we may use F_x^y to compute W_y, while F_x^y is computable at those α which are not moduli for W_x. By coding such functionals F_x^y together in an effective way one may produce many interesting examples of continuous type-2 functionals among which we mention

Bergstra [1]: There is an F such that $\mu\text{-}1\text{-}SC(F) \neq 1\text{-}SC(F)$

Normann [15]: There is an F such that $1\text{-}SC(F) \in \Pi^1_1 \setminus \Sigma^1_1$

Normann-Wainer [18]: If F is not effectively discontinuous then there is a continuous G such that $1\text{-}SC(F) = 1\text{-}SC(G)$

THE PROJECTIVE HIERARCHY

It is easy to show that the set As(k) will be Π^1_{k-1} and a closer analysis of computations due to J. Bergstra [1] shows:

If $\psi \in Ct(k+2)$ $(k \geq 1)$ and α is an associate for ψ then

i $\quad 2\text{-en}(\psi) \subset \Pi^1_k(\alpha)$

ii $\quad 1\text{-}SC(\psi) \in \Pi^1_k(\alpha)$

which gives a bound on the complexity of sections and envelopes. In order to show that this bound is the best possible we need a closer tie-up with the projective hierarchy. The method of proof goes back to Kreisel [14].

LEMMA

Uniformly in k there is a computable sequence $\{\phi^k_i\}_{i\in\omega}$ from Ct(k) such that $\{\phi^k_i; i\in\omega\}$ is topologically dense in Ct(k).

COROLLARY

If $R \subset Ct(k_1) \times Ct(k_2)$ is computable in continuous elements and if

$$\forall\psi \in Ct(k_1)\, \exists\phi \in (Ct(k_2)R(\psi,\phi)$$

then there is a continuous map $\Phi: Ct(k_1) \rightarrow Ct(k_2)$ such that

$$\forall\psi \in Ct(k_1)R(\psi,\Phi(\phi))$$

<u>Proof</u>

Let $\Phi'(\psi) = \mu n R(\psi,\phi^{k_2}_n)$

and let

$$\Phi(\psi) = \phi^{k_2}_{\Phi'(\psi)}$$

Kreisel called this principle the Quantifier Free Axiom of Choice (Q.F.A.C.) and by the aid of this principle he used the continuous functionals to give a constructive interpretation of formulas of analysis. We will use Q.F.A.C. to show a connection between the continuous functionals and the projective hierarchy.

Definition

For each $\psi \in Ct(k)$ let $h_\psi(i) = \psi(\phi_i^{k-1})$ $(k\geq 1)$. Let $H_k = \{h_\psi ; \psi \in Ct(k)\}$.

It is easy to show that H_k is Π^1_{k-1} for $k > 1$.

THEOREM 9

a $k > 1$: Let $A \subset {}^\omega\omega$ be Σ^1_{k-1}. Then there is a recursive relation S such that

i $\alpha \in A \Rightarrow \forall h \in H_k \exists n S(\alpha(n), \overline{h}(n))$

ii $\alpha \notin A \Rightarrow \exists \psi \in Ct(k)$ (ψ is computable in α and $\forall n \neg S(\overline{\alpha}(n), \overline{h}(n))$)

b $k \geq 1$: Let $B \subset {}^\omega\omega$ be Π^1_k. Then there is a recursion relation R such that

$$\alpha \in B \Leftrightarrow \forall h \in H_k \exists n R(\overline{\alpha}(n), \overline{h}(n))$$

Proof

The proof is by induction on $k \geq 1$. b,k = 1 is well known and for each k > 1 a => b is just a matter of coding an arbitrary α as an element in H_k. (Note that $\psi \rightsquigarrow h_\psi$ commutes with the standard pairing

$$\langle\psi_1,\psi_2\rangle(\phi) = \langle\psi_1(\phi), \psi_2(\phi)\rangle).$$

So we must show b,k => a,k+1.

Let A be Σ^1_k. By b,k there is a recursive relation R such that

$$\alpha \notin A \Leftrightarrow \forall\psi \in (Ct(k) \exists n R(\alpha,\psi,n)$$

Let $\Phi_\alpha(\psi) = \mu n R(\alpha,\psi,n)$. Then

$$\alpha \notin A \Rightarrow \forall \psi \in Ct(k) R(\alpha,\psi,\Phi_\alpha(\psi))$$

so

$$\begin{aligned}\alpha \notin A &\Leftrightarrow \exists \Phi \in Ct(k+1)\ \forall \psi \in Ct(k) R(\alpha,\psi,\Phi(\psi)) \\ &\Leftrightarrow \exists \Phi \in Ct(k+1)\ \forall n R(\alpha,\phi_n^k,\Phi(\phi_n^k))\end{aligned}$$

so

$$a \in A \iff \forall \Phi \in Ct(k+1)\ \exists n \neg R(\alpha, \phi^k_n, \Phi(\phi^k_n))$$

It is easy to re-write this in the form of a, k + 1 using Φ_α in proving ii.

COROLLARIES

Let $k > 1$

a As(k) is complete Π^1_{k-1}

b H_k is not $\underset{\sim}{\Sigma}^1_{k-1}$

c If $A \subset {}^\omega\omega$ is Π^1_k then A is semicomputable in ${}^{k+2}O$ (the constant zero functional of type k+2).

Proofs

We leave a and b for the reader, but let us prove c. Let $A \subset {}^\omega\omega$ be Π^1_k. Then there is a recursive relation T such that

$$\alpha \in A \iff \forall \psi \in Ct(k)\ \exists n R(\alpha, \psi, n).$$

Let Φ_α be as above. Then Φ_α is total if and only if $\psi \in A$ so

$$\alpha \in A \iff {}^{k+2}O(\Phi_\alpha) \text{ takes a value}$$

which shows that A is semicomputable.

Together with the results of Bergstra mentioned earlier, we see that this gives a characterization of the 2-envelopes of continuous functionals.

Another corollary is that all projective ordinals (lengths of projective well-founded relations) are computable in some continuous functionals. The proof will appear in a forthcoming paper of H. Vogel.

This connection between the projective hierarchy and the continuous functionals was one of the fundaments of the proof of Theorem 5 and it has also been used to give an alternative to Theorem 4.

From Theorem 8, its corollary and a result of J. Bergstra mentioned above we know

THEOREM 10

Let $K \geq 3, \psi \in Ct(k)$. Then

i 1-SC(ψ) is $\Pi^1_{k-2}(f_\psi)$

ii 1-SC(ψ) is recursively generated by its r.e. f_ψ-degrees modulo f_ψ.

As an application of Theorem 9 there is a converse to this result

THEOREM 11 (Normann-Wainer [18])

Let $k \geq 3$. Let $A \subset {}^\omega\omega$ be a set, $f \in A$ a function such that

i A is closed under recursion in finite lists from A

ii A is $\Pi^1_{k-2}(f)$

iii modf:A is recursively generated by the r.e. degrees in A

Then there is a $\Phi \in Ct(k)$ computable in the jump of f such that

$A = 1\text{-}Sc(\Phi)$.

The proof is too long to be given here. It makes essential use of part a of Theorem 9.

FINAL REMARKS

It must be admitted that the degree-structure of the continuous functionals is not yet well understood. If we use "recursive in" as our basic notion we know from Dvornickov's result (Theorem 4) that there are new degrees at each type. These degrees are in between the old ones since all functionals are recursive in some $\alpha: \omega \to \omega$.

Problems

1. Is the degree-structure of Ct(1) an elementary substructure of of the degree-structure of $\langle Ct(k)\rangle_{k\in\omega}$?

2. Are they elementary equivalent?

3. Is there a canonical or natural extension of the jump-operator to $\langle Ct(k)\rangle_{k\in\omega}$?

We can ask the same questions for S1 - S9 reductions. We may then also look at the degree-structure of say the recursive functionals.

We think that problems of this kind deserve attension and clearly the solution to any of these problems will increase our understanding of algorithmic procedures in general.

We do not claim in this survey to have done justice to everyone who works, or who has worked, within this area of recursion theory. We have for instance completely omitted important structural investigations due to Hyland [9] and others (see [9] for further references). We also avoided possible applications of the structure and its theory, since we wanted to survey results concerning recursion-theory itself. We think that the recursion theory of the continuous functionals is rich and deserves investigation, and that this investigation should use methods different from those normally used in generalized recursion theory. A consequence will be renewed insight into various aspects of computations and algorithms.

REFERENCES

1. J. Bergstra, Computability and continuity in finite types, Utrecht, 1976.

2. J. Bergstra, The continuous functionals and ${}^2E$, in J.E. Fenstad, R.O. Gandy & G.E. Sacks (eds) Generalized Recursion Theory II, North-Holland, 1978, pp 39-54.

3. S.G. Dvornickov, On c-degrees of everywhere defined functionals (in Russian) Logica i Algebra, 18, 1979. pp 32-46

4. S. Feferman, Inductive schemata and recursively continuous functionals, in R.O. Gandy & J.M.E. Hyland (eds) Logic Colloquium '76, North-Holland, 1977.

5. J.E. Fenstad, General Recursion Theory, Springer Verlag, 1980.

6. R.O. Gandy & J.M.E. Hyland, Computable and recursively countabl functionals of higher type, in R.O. Gandy & J.M.E. Hyland (eds) Logic Colloquium '76, North Holland, 1977, pp 907-938.

7. T. Grilliott, On effectively discontinuous type-2 objects, J.S.L. 36, 1977, pp 245-248.

8. P.G. Hinman, Degrees of continuous functionals, J.S.L. 38, 1973, pp 393-395.

9. J.M.E. Hyland, Filter-spaces and continuous functionals, Ann. Math. Log. 16, 1979, pp 101-143.

10. J.M.E. Hyland, The intrinsic recursion theory on the countable or continuous functionals, in J.E.Fenstad, R.O. Gandy, & G.E. Sacks (eds), Generalized Recursion Theory II, North-Holland, 1978, pp 135-145.

11. S.C. Kleene, Recursive functionals and quantifiers of finite types I, T.A.M.S. 91, 1959, pp 1-52, and II, 1963, pp 106-142.

12. S.C. Kleene, Countable functionals, in A. Heyting (ed) Constructivity in mathematics, North-Holland, 1959, pp 87-100.

13. S.C. Kleene, Recursive functionals and quantifiers of finite types revisited I, in J.E. Fenstad, R.O. Gandy & G.E. Sacks (eds), Generalized Recursion Theory II, North Holland, 1978, pp 185-222.

14. G. Kreisel, Interpretation of analysis by means of functionals of finite type, in A. Heyting (ed) Constructivity in mathematics, North-Holland, 1959, pp 101-128.

15. D. Normann, A continuous functionals with non-collapsing hierarchy, J.S.L. 43, 1978, pp 487-491.

16. D. Normann, Nonobtainable continuous functionals, to appear in the proceedings of the 6th International Conference of Logic, Methodology and Philosophy of Science, Hannover, 1979.

17. D. Normann, Recursion on the countable functionals, Springer Lecture Note, in preparation.

18. D. Normann & S. Wainer, The 1-section of a countable functional, J.S.L. 45, 1980.

19. R.A. Platek, Foundations of Recursion theory, Thesis, Stanford University, 1966.

20. J.V. Tucker, this volume.

21. S.S. Wainer, The 1-section of a non-normal type-2 object, in J.E. Fenstad, R.O. Gandy & G.E. Sacks (eds), Generalized Recursion Theory II, North Holland, 1978, pp 407-417

THREE ASPECTS OF RECURSIVE ENUMERABILITY IN HIGHER TYPES

G. E. Sacks[1]

Harvard University and

Massachusetts Institute of Technology

ABSTRACT

A set is E-closed if it is closed under the schemes of set recursion, i.e. the Kleene schemes S0-S9 (with equality) revised to allow objects of all types. Inadmissible E-closed sets admit: E-closed generic extensions via appropriate forcing notions, solutions to Post's problem via priority arguments, and sidewise E-closed extensions via type-omitting results for logic on E-closed sets. A sample theorem states: suppose $L(\kappa)$ is countable, E-closed, and satisfies "there is a greatest cardinal and its cofinality exceeds ω"; then there exist $\delta < \kappa$ and $T \subseteq \delta$ such that $L(\kappa, T)$ is the least E-closed set with T as a member.

1. INTRODUCTION

The schemes of set recursion (or E-recursion) were devised independently by Normann [1] and Moschovakis. When restricted to objects of finite type, they coincide with the Kleene schemes [2] plus equality, the so-called normal case. For each set X, let E(X) be the least set closed under application of the schemes of set recursion and containing $X \cup \{X\}$. $E(2^{\omega})$ is precisely the collection of all sets coded by relations R on 2^{ω} such

[1]The author is grateful for the support of the National Science Foundation, the suggestions of Ed Griffor and Ted Slaman, and the assistance of Vie Wiley.

that R is Kleene recursive in ${}^3E$ and some real b. The operator E yields a generalization of the Kleene theory to objects of arbitrary type. Let $T_p(\kappa)$ be the set of all sets of objects of type less than κ, and let ${}^{\kappa}E : T_p(\kappa) \longrightarrow \{0,1\}$ be 0 on x iff $x = \emptyset$. Then $E(T_p(\kappa))$ is in essence the set of all computations needed to define recursion in ${}^{\kappa}E$ in the sense intended by Kleene. There are many surprises in store for the student of E. The first, discovered by Moschovakis [3], states $E(T_p(\omega))$ is the least admissible set with $T_p(\omega)$ as a member. The same holds for $E(T_p(0))$, since the latter is the least admissible set with ω $(= T_p(0))$ as a member, and is equivalent to the hyperarithmetic hierarchy, an early result of Kleene [2] that demonstrates the adequacy (Kreisel's terminology) of his notion of relative recursiveness. In general the E-closure is smaller than the Σ_1 admissible closure. $E(T_p(n))$ is not Σ_1 admissible when $0 < n < \omega$, a theorem of Moschovakis [3] whose method of proof is central to the analysis of E recursion. A related result states: if $L(\kappa)$ is the E-closure of the greatest cardinal in the sense of $L(\kappa)$, and if that cardinal has cofinality ω in $L(\kappa)$, then $L(\kappa)$ is Σ_1 admissible.

The schemes of E-recursion are intended to capture the informal idea of passing from one stage to the next in the development of an arbitrary computation with the equality predicate treated as if it were effective. The presence of equality simply means: if the result of a computation c is a set Z, then there is a further computation whose total length is one more than that of c, and which reveals whether or not Z is empty. With equality it is possible to compare the lengths of computations and prove selection theorems. Without equality the world of computation has much less structure.

The schemes inductively define the partial function

$$\{e\}(x_0,\ldots,x_n),$$

where $e \in \omega$ and $x_0,\ldots,x_n$ is any sequence of sets. The schemes fall into two groups (details in Section 2). Closure under group I is the same as rudimentary closure, cf. Jensen [4]. The first scheme in group I is on the order of

$$\{3\}(x,y) = \{x,y\}.$$

Thus 3 is the Gödel number of an instruction that says: form the unordered pair of x and y.

The n-th member of group II is

$$(1) \quad \{7^n\}(e,x_1,\ldots,x_n) \simeq \{e\}(x_1,\ldots,x_n).$$

The most descriptive term for (1), going back to Gandy, is reflection. Another is self-reference.

The role of group I in clarifying the notion of computation needs no explication. Group II is at first translucent, if not opaque. Let a computation be visualized as a wellfounded tree, and let the nodes of the tree be called individuals. Thanks to Gödel, an individual can serve as a code for an instruction to compute. Thus the result of one instruction to compute may be another instruction to compute. In short instructions can be about instructions. This last is simply and directly expressed by group II. With II it is possible to prove fixed point theorems and carry out effective transfinite inductions. Without II there is again very little structure to the world of computation; it lacks completeness and is easily transcended by diagonal arguments.

Let V be the class of all sets. A partial function $\emptyset : V \longrightarrow V$ is said to be partial recursive if there is an e such that

$$\emptyset(x) \simeq y \longleftrightarrow \{e\}(x) \simeq y.$$

(As usual, $\emptyset(x) \simeq y$ means $<x,y>$ belongs to the graph of $\emptyset$.) The partial function $\{e\}$ is defined by a finite sequence of applications of the schemes. A set $\mathcal{E}$ is E-closed if it is closed with respect to every $\{e\}$. Thus the union of E-closed sets is E-closed, in contradistinction to the union of Σ_1 admissible sets. A class A is RE if A is the domain of a partial recursive function.

Let $\mathcal{E}$ be E-closed. Assume $A \subseteq \mathcal{E}$. A is said to be RE on $\mathcal{E}$ if A is the intersection of some RE class with A. Since $\mathcal{E}$ is E-closed, each element x of A has a computation y that belongs to $\mathcal{E}$ and puts x in A. Clearly A is Σ_1 over $\mathcal{E}$. Let $B \subseteq \mathcal{E}$ be Σ_1. B need not be RE on $\mathcal{E}$. An element x of $\mathcal{E}$ finds a place in B if and only if there exists a witness (Friedberg's terminology) that puts x in B. Kleene's notion of recursive enumerability requires the witness to be a computation which virtually proves that $x \in B$. The Σ_1 notion allows the witness to bear a tenuous relationship to x. In Section 4 it will be seen via forcing that certain naturally enumerable sets, trivially Σ_1, are not RE in the sense of Kleene.

The first aspect of recursive enumerability discussed below is inadmissible forcing. Assume $L(\kappa)$ is countable, E-closed, but not Σ_1 admissible. Section 3 deals with the problem of generically extending $L(\kappa)$ to some $L(\kappa,T)$ that is also E-closed. The usual methods associated with Σ_1 admissible $L(\kappa)$'s fail. The forcing approach of [5], suitably modified, succeeds. What are needed are some structural properties of $L(\kappa)$ lumped under the term "reflection" which go back to Moschovakis's proof [3] that co-RE (in ${}^3E$) sets of reals are Σ_1, which were further developed by Harrington [6], and still further in [7] to solve Post's problem.

The second aspect is the development of priority arguments on E-closed sets. A sample result, discussed in section 6, is a positive solution to Post's problem for every E-closed $L(\kappa)$.

The third is concerned with logic on an E-closed set E. Suppose the sentence structure of $L_{\infty,\omega}$ is restricted to E. The defining difference between logic on E and on some Σ_1 admissible set is the following requirement: a proof of a sentence F must be Kleene recursive in F; that is, there must be an e such that $\{e\}(F)$ is a proof of F. It can happen that an F in E with a proof in $L_{\infty,\omega}$ has no proof recursive in F, and none in E. Section 7 remarks on type-omitting theorems for RE collections of sentences. They involve somewhat technical hypotheses on E, but clarify results, touched on above, first proved by forcing.

2. MACHINERY

The schemes of E-recursion, as formulated by Normann [1], are as follows.

(1) $f(x_1,\ldots,x_n) = x_i$ $\qquad$ $e = \langle 1,n,i\rangle$

(2) $f(x_1,\ldots,x_n) = x_i - x_j$ $\qquad$ $e = \langle 2,n,i,j\rangle$

(3) $f(x_1,\ldots,x_n) = \{x_i,x_j\}$ $\qquad$ $e = \langle 3,n,i,j\rangle$

(4) $f(x_1,\ldots,x_n) \simeq \bigcup_{y\in x_1} h(y,x_2,\ldots,x_n)$

$e = \langle 4,n,e'\rangle$ where e' is an index for h.

(5) $f(x_1,\ldots,x_n) \simeq h(g_1(x_1,\ldots,x_n),\ldots,g_m(x_1,\ldots,x_n))$

$e = \langle 5,n,m,e',e_1,\ldots,e_m\rangle$ where e' is an index for h and $e_1,\ldots,e_m$ are indices for $g_1,\ldots,g_m$ resp.

(6) $f(e_1,x_1,\ldots,x_n,y_1,\ldots,y_m) \simeq \{e_1\}(x_1,\ldots,x_m)$

$e = \langle 7,n,m\rangle$.

(Note: scheme (4), the bounding principle, terminates only if h is total on x_1.)

The above schemes give a meaning to $\{e\}(x_1,\ldots,x_n)$ for all $e \in \omega$ and all sets $x_1,\ldots,x_n$. If e is not the Gödel number of a scheme, then $\{e\}(x_1,\ldots,x_n)$ is undefined. Otherwise e singles out a scheme to be applied to $x_1,\ldots,x_n$, and the application may lead to other schemes. The process of following out the instructions encoded by e generates a tree, each node of which is an instruction coded by a number and a finite sequence of sets to which the instruction is to be applied. The process terminates iff the tree is wellfounded. In that case $\{e\}(x_1,\ldots,x_n)$ is said to converge, written $\{e\}(x_1,\ldots,x_n)\downarrow$, and the resulting tree is called a computation. The height of the tree is an ordinal regarded as the length of the computation embodied by the tree, and is denoted by $|\{e\}(x_1,\ldots,x_n)|$. If $\{e\}(x_1,\ldots,x_n)$ diverges, written $\uparrow$, then $|\{e\}(x_1,\ldots,x_n)|$ is said to be ∞, an object greater than every ordinal. A predicate is RE if its extension is the domain of some partial recursive function, that is the class of all x such that $\{e\}(x)\downarrow$ for some e.

Theorem 2.1 (cf. Normann [1]). The predicate $|\{e_1\}(x)| < |\{e_2\}(y)|$ is RE.

The proof of 2.1 is an effective transfinite induction. It follows from 2.1 that the class of all computations ordered by length has what Moschovakis aptly calls the prewellordering property: the ordering is RE, but each proper initial segment is recursive (uniformly) in any object on top of the initial segment. Here a predicate P is recursive if there is an e such that $\{e\}(x)\downarrow$ for all x but takes the value 0 iff $P(x)$ is true. The prewellordering property suggests an analogy between recursive and "finite". The proof of 2.1 fails if the Kleene schemes do not include equality.

A set E is E-closed if it is closed under application of the schemes. Thus the union of E-closed sets is E-closed. It is readily seen that every Σ_1 admissible set is E-closed. The converse is false, since $L(\alpha_\omega)$, where α_ω is the limit of the first ω Σ_1 admissible ordinals, is not Σ_1 admissible. In short an E-closed set E is closed with respect to the process of forming computations. If $A \subseteq E$, then A is said to be RE on E if A is the intersection of some RE class with E. Thus Theorem 2.1 remains true when x, y and RE are restricted to E.

Theorem 2.2 (Gandy [8]). There exists a recursive function f such that for all e:

$$(En)[\{e\}(n)\downarrow] \longleftrightarrow [\{f(e)\}(0)\downarrow \ \& \ \{e\}(\{f(e)\}(0))\downarrow].$$

The above is termed a selection principle, since it selects by effective means an element from each nonempty RE set of numbers. It remains true if set parameters are added.

Theorem 2.3 (Grilliot Selection [9]). There exist recursive functions g and h such that for all x and e:

(i) $(Ey)_{y\in x}[\{e\}(y,2^x)\downarrow] \longleftrightarrow (Ey)_{y\in x}[\{g(e)\}(y,2^x)\downarrow]$;

(ii) $(y)_{y\in x}[\{g(e)\}(y,2^x)\downarrow$

or $\{h(e)\}(y,2^x)\downarrow$ but not both].

Theorem 2.3 provides an effective method of contracting a nonempty RE subset of X to a nonempty recursive subset of X. If "recursive" is seen to be analogous to "finite", then the power of 2.3 becomes evident. Both 2.2 and 2.3 select the "shortest" computations.

Theorem 2.4 (Moschovakis). Let $\mathcal{E}$ be the E-closure of $R(\kappa)$, the set of all sets of rank less than κ. Suppose in $\mathcal{E}$ that κ has cofinality ω. Then $\mathcal{E}$ is Σ_1 admissible. In addition the predicates RE on $\mathcal{E}$ are identical with those Σ_1 on $\mathcal{E}$.

The proof of 2.4 draws on ideas behind 2.2 and 2.3. The concept of cofinality ω plays an adversary role in the study of E-recursion. If an inadmissible E-closed set $\mathcal{E}$ has a greatest cardinal of cofinality ω, then it does not appear possible to do anything worthwhile in the way of forcing or type-omitting on $\mathcal{E}$. This last point will be discussed further in sections 3 and 4, as well as some extensions of 2.4.

Let $\mathcal{E}$ be E-closed. Certain ordinals associated with $\mathcal{E}$ are helpful in the analysis of $\mathcal{E}$. $\kappa_{\mathcal{E}}$ is the supremum of all ordinals in $\mathcal{E}$. Let $a \in \mathcal{E}$. Then $E(a) \subseteq \mathcal{E}$ and is called the E-fiber over a, or simply the <u>a-fibre</u>. Note that $E(a) \subseteq L(a)$ if a is transitive and closed under pairing. This is the first hint of the important role played by absoluteness in E-recursion. If $\{e\}(a)\downarrow$, then the associated computation is a member of the a-fiber. Let κ_0^a be the least ordinal not recursive in a. Clearly κ_0^a is at most ord(E(a)), the height of the a-fiber. A great deal of information about divergent computations can also be found in the a-fiber as in the next result.

Theorem 2.5 (Moschovakis [3]). There exists an RE predicate P(e,a,y) with the following property: assume E(a) is not Σ_1 admissible; then

$\{e\}(a)\uparrow \longleftrightarrow (Ey)[y \in E(a) \ \& \ P(e,a,y)].$

Proof (a sketch). The essential idea, originated by Moschovakis in his study of recursion on ${}^3E$ ([3]), is: $\{e\}(a)$ diverges iff the associated computation tree has an infinite descending path y. Some details are needed to see that "y is an infinite descending path in the computation of $\{e\}(a)$" is RE in e, a and y. In general the procedure for computing $\{e\}(a)$ has two parts: (1) $\{e_0\}(a)$ is computed, and if convergent, yields a set W; (2) $\{e_1\}(b)$ is computed for all $b \in W$. Thus $\{e\}(a)\downarrow$ iff $\{e_0\}(a)\downarrow$ and $\{e_1\}(b)\downarrow$ for all $b \in W_e$. Assume $\{e\}(a)\uparrow$. A witness y to divergence is defined in ω steps as follows. If $\{e_0\}(a)\uparrow$, then y_0 is $\langle e_0,a\rangle$; otherwise y_0 is $\langle e_1,b\rangle$, where b is such that $b \in W$ and $\{e_1\}(b)\uparrow$. Then y_1 is defined so as to bear the same relation to y_0 that y_0 bears to $\langle e,a\rangle$. Let y be $\lambda n \mid y_n$. The predicate $P(e,a,y)$ says: y codes $\lambda n \mid y_n$, and for each n, y_{n+1} stands in relation to y_n as above. P is RE only because statements such as "$\{e_0\}(a) = W$ and $b \in W$" are RE. □

Another ordinal associated with the a-fiber is κ_r^a. Suppose $\gamma \leqslant \mathrm{ord}(E(a))$. γ is said to be a-reflecting if for every Σ_1 sentence F with parameter a,

$$L(\gamma,a) \models F \longrightarrow L(\kappa_0^a,a) \models F.$$

Reflection properties were first applied in the study of recursion in objects of higher type in the proof of the plus-one theorem [10]. Harrington [6] found an invaluable characterization of κ_r^a as a consequence of a generalization of the Gandy basis theorem for Σ_1^1 predicates, where κ_r^a is the supremum of all a-reflecting ordinals (Corollary 2.7).

Lemma 2.6 (Harrington-Kechris Basis Theorem). Suppose A is RE on E, $b \in E$ and $b-A \neq \emptyset$. Then there exists an $x \in b-A$ such that $\kappa_0^{b,x} \leq \kappa_r^b$.

Proof. Suppose not. Then each $x \in b$ is either (1) enumerated in A via an ordinal recursive in x, or (2) combines with b to give rise to some ordinal recursive in b, x and greater than κ_r^b. The predicate "δ is an ordinal greater than κ_r^b" is recursive in b. Thus b is the union of two RE classes. By Gandy selection one can pass effectively from x $(\in b)$ to an ordinal recursive in x, b that puts x in one class or the other. The supremum of all such ordinals is some ordinal γ recursive in b by virtue of the bounding principle. But $\gamma < \kappa_r^b$, since $\kappa_0^b \leq \kappa_r^b$. Then no $x \in b$ satisfies (2), and so $b \subseteq A$, contrary to hypothesis. □

Let $P(e,a,y)$ be the RE predicate of Theorem 2.5. If $P(e,a,y)$ holds, then y is said to be a Moschovakis witness to the divergence of $\{e\}(a)$. Enumerating such a y means enumerating a computation of $P(e,a,y)$.

Theorem 2.7 (cf. Harrington [6]). κ_r^a is the least ordinal θ such that for all e, if $\{e\}(a)\uparrow$ then θ suffices to enumerate a Moschovakis witness for $\{e\}(a)$.

Proof. Let θ be as above. Then $\kappa_r^a \leq \theta$, because otherwise the complete RE (in a) set of numbers would be recursive (in a). To show that κ_r^a suffices to enumerate a witness to the divergence of $\{e\}(a)$, the argument of 2.5 is sharpened. If $\{e_0\}(a)\uparrow$, then y_0 is $\langle e_0,a\rangle$; otherwise y_0 is $\langle e_1,b\rangle$, where $b \in W$, $\{e_1\}(b)\uparrow$ and $\kappa_r^{a,b} \leq \kappa_r^a$. The last mentioned property of b can be managed with Lemma 2.6. In this manner each y_n is chosen so that $\kappa_r^{a,y_n} \leq \kappa_r^a$. Note that it may not be possible to single out a Moschovakis witness $\lambda n \mid y_n$ at level

κ_r^a, but there is no difficulty in defining a nonempty set of witnesses at level κ_r^a, and the presence of such a set at level κ_r^a is all that is claimed by the theorem. □

Corollary 2.7 (Harrington [6]). If a is recursive in b, then $\kappa_r^a \leqslant \kappa_r^b$.

Proof. If $a = \{e\}(b)$, then the Moschovakis witnesses for a can be extracted via e from those for b. □

Lemma 2.8 ([11]). Let E be E-closed. Suppose there exists an $a \in E$ such that $L(\kappa_E, a)$ is not Σ_1 admissible. Then $\kappa_r^b < \kappa_E$ for all $b \in E$.

Proof. Let $D(x,y)$ be a Δ_0 formula with parameter $p \in L(\kappa_E, a)$. Suppose $c \in L(\kappa_E, a)$, and

$$(x)_{x \in c}(Ey)D(x,y)$$

holds in $L(\kappa_E, a)$. Assume $\kappa_r^b = \kappa_E$. By 2.7, $\kappa_r^{b,a,x,p} = \kappa_E$. Hence

$$(x)_{x \in c}(Ey)[D(x,y) \ \& \ y \in L(\kappa_0^{b,a,x,p}, a)].$$

In fact

$$(x)_{x \in c}(Ey)[D(x,y) \ \& \ y \in L(\gamma(b,a,x,p), a)],$$

where γ is a recursive function of b, a, x, p. The bounding principle implies γ is less than some ordinal δ recursive in b, a, c, p as x ranges over c. Thus y is bounded below κ_E. □

Theorem 2.9 ([11]). If $L(\kappa)$ is E-closed but not Σ_1 admissible, then $\kappa_r^b < \kappa$ for every $b \in L(\kappa)$.

Proof. Immediate consequence of 2.8.

Theorem 2.9 will prove useful in priority and forcing arguments over inadmissible, E-closed $L(\kappa)$'s. It says that Moschovakis witnesses for divergent computations on elements of $L(\kappa)$ can be found inside $L(\kappa)$. The presence of internal witnesses will make it possible to settle questions of convergence by paying attention to ranked sentences only.

3. INADMISSIBLE FORCING

Let $L(\kappa)$ be countable and E-closed, but not Σ_1 admissible. Let $gc(\kappa)$ denote the greatest cardinal in the sense of $L(\kappa)$. (The inadmissibility of $L(\kappa)$ implies $gc(\kappa)$ is well defined.) Let T denote an arbitrary subset of $gc(\kappa)$. This section is a brief introduction to the technology of adjoining a T to $L(\kappa)$ via forcing so that the structure $L(\kappa,T)$ remains E-closed. Suppose $b \in L(\kappa)$. How can the ideas of forcing and generity be applied to show:

(1) if $\{e\}(b,T)\downarrow$, then $|\{e\}(b,T)| < \kappa$?

As usual a sensible approach is to try and prove much more:

(2) if $\{e\}(b,T)\downarrow$, then $|\{e\}(b,T)| < \kappa_0^b$.

There is a formlessness to (1) that is inhospitable to all notions of uniformity, but (2) appears to be open to induction for all sufficiently generic T. In addition (2) does not seem to care whether or not $L(\kappa)$ is Σ_1 admissible. The thinking behind Theorem 2.9 will be seen to lead to a proof of a weakened version of (2): there is a forcing condition p such that either p forces $|\{e\}(b,T)|$ to be an ordinal recursive in b, p or p forces the existence of an internal Moschovakis witness to the divergence of $\{e\}(b,T)$.

Not every notion of forcing is amenable to the foregoing treatment. The matter is discussed in detail in [11] and [12]. The simplest condition sufficient for proving something akin to (2) is countable closure: if $\lambda n \mid p_n$ is an ω-sequence (in $L(\kappa)$) of forcing conditions, then $\cup\{p_n \mid n < \omega\}$ is also a forcing condition. For example, suppose $gc(\kappa)$ has uncountable cofinality (in $L(\kappa)$) and by a forcing condition is meant the characteristic function of a bounded subset of $gc(\kappa)$; then the countable closure condition is met. The earliest such use of countable closure is [5], where $L(\kappa)$ is replaced by the E-closure of 2^{ω} and $gc(\kappa)$ by the cardinality of the continuum, necessarily of uncountable cofinality. Note that Levy forcing with finite conditions intended to collapse ω_1 to ω fails to satisfy countable closure. It is not difficult to verify that the adjunction of a generic Levy collapsing function to $L(\kappa)$ kills E-closure. On the other hand Cohen forcing with finite conditions on ω does not satisfy countable closure but does preserve E-closure by an argument not given below (cf. [11]).

For the remainder of this section suppose the cofinality of $gc(\kappa)$ (in $L(\kappa)$) is greater than ω. Forcing conditions are denoted by p, q, r, ...; they belong to $L(\kappa)$, and each is a function from an ordinal less than $gc(\kappa)$ into $\{0,1\}$. Let F be a sentence of the language $\mathcal{L}(\kappa,T)$, the usual ramified language associated with the structure $L(\kappa,T)$. The (strong) forcing relation $p \Vdash F$ is defined in familiar Cohen style by recursion on the ordinal rank and logical complexity of F.

Proposition 3.1. The forcing relation $p \Vdash F$, restricted to F's of rank less than δ, is recursive in δ, $gc(\kappa)$ uniformly in δ.

The proof of 3.1 is a straightforward effective transfinite induction. Less straightforward is the proof that

$$p \Vdash |\{e\}(a)| \leqslant \kappa$$

is RE. The method needed is sketched after the statement of the next lemma. Moschovakis witnesses are essential as well as the notion of weak forcing. Define $p \Vdash^{*} F$ (read p weakly forces F) by

$$(q)_{p \geqslant q}(Er)_{q \geqslant r}[r \Vdash F].$$

Lemma 3.2. Suppose $p \Vdash^{*} (E\delta)[|\{e\}(b)| = \delta]$. Then there exist q_0 and β_0, each recursive in p, b, gc(κ), such that $p \geqslant q_0$ and $q_0 \Vdash |\{e\}(b)| = \beta_0$.

The proof of 3.2 is by induction on $\succ$. Define $(p,e,b) \succ (q,e^{*},b^{*})$ by:

$$p \geqslant q,$$

$$p \Vdash^{*} (E\delta)[|\{e\}(b)| = \delta], \text{ and}$$

$$q \Vdash^{*} [\{e^{*}\}(b^{*}) \text{ is a convergent subcomputation of } \{e\}(b)].$$

The only sore point in the proof is checking the wellfoundedness of $\succ$. It is painless to verify that $\succ$ has no infinite descending sequence in L(κ). Suppose $\lambda n \mid (p_n, e_n, b)$ were one such. Then

$$p_\infty \ (= \ \{p_n \mid n \in \omega\})$$

forces the existence of an infinite descending sequence of ordinals, a supreme impossibility. Note well that p_∞ is a forcing condition because of the assumption that the cofinality of $gc(\kappa) > \omega$. It remains to be seen that if $\succ$ has an infinite descending sequence, then it has one in $L(\kappa)$.

To obtain some idea of the proof of 3.2, imagine that a method M for computing q and β from p, b and $gc(\kappa)$ is to be developed by effective transfinite recursion on $\succ$. First suppose M is well defined on the predecessors of (p,e,b) in $\succ$. Recall the conventions of the proof of Theorem 2.5. Thus

$$p \Vdash^{*} (E\delta)[|\{e_0\}(b)| = \delta]$$

$$p \Vdash^{*} (EW)[\{e_0\}(b) = W]$$

$$p \Vdash^{*} (c)(E\delta)[c \in \{e_0\}(b) \longrightarrow |\{e_1\}(c)| = \delta].$$

By supposition M yields p_0 and γ such that

$$p \geqslant p_0$$

$$p_0 \Vdash |\{e_1\}(b)| = \gamma,$$

since $(p,e,b) \succ (p,e_0,b)$. Also p_0 and γ are recursive in p, b and $gc(\kappa)$. It follows that

$$(1) \qquad (c)(q)_{p_0 \geqslant q}(Er)_{q \geqslant r}[r \Vdash c \notin \{e_0\}(b) \text{ or } r \Vdash (E\delta)(|\{e_1\}(c)| = \delta)].$$

By supposition (and Proposition 3.1), r is recursive in c, q, b, $gc(\kappa)$. Hence r, as a function of c and q, is recursive in p, b, $gc(\kappa)$. Since γ is recursive in the latter trio, it is possible to decide effectively whether or not $r \Vdash c \in \{e_0\}(b)$. Hence M yields a value for $|\{e_1\}(c)|$, when r forces such a value. Since the set of all forcing conditions is recursive in $gc(\kappa)$, the bounding principle provides a ρ such that

$$p_0 \Vdash (c)[c \in \{e_0\}(b) \longrightarrow |\{e_1\}(c)| < \rho]$$

and ρ is recursive in p, b, $gc(\kappa)$. By 3.1 the set of all pairs $\langle q,\beta\rangle$ such that $p_0 \geq q$, $\beta \leq \rho$ and

$$q \Vdash |\{e\}(b)| = \beta$$

is recursive in p, b, $gc(\kappa)$. Let $\langle q_0,\beta_0\rangle$ be the least such pair.

All went well above because it was supposed that M did not encounter any infinite descending sequences in $\succ$ below (p,e,b). Now suppose otherwise with the intent of uncovering such a sequence y_n in $L(\kappa)$. Recall the proof of Theorem 2.7. If M does not yield p_0 and γ as above, then y_0 is (p,e_0,b). Otherwise M fails for some $c \in \{e_0\}(b)$ and some $q \leq p_0$. Then y_0 is (q,e_1,c) for some such c and q with the additional property that $\kappa_r^{p,b,gc(\kappa),q,c} \leq \kappa_r^{p,b,gc(\kappa)}$. As in the proof of 2.7, $\lambda n \mid y_n$ is defined at level $\kappa_r^{p,b,gc(\kappa)}$ of $L(\kappa)$.

Another way of expressing Lemma 3.2, more in accord with the above argument is: fix p, e and b; then there exists a $q \leq p$ such that either (1) or (2) holds. (1) q forces $|\{e\}(b)|$ to be an ordinal recursive in p, b, $gc(\kappa)$. (2) q forces $\{e\}(b)$ to diverge via a Moschovakis witness constructed at level $\kappa_r^{p,b,gc(\kappa)}$.

4. LIMITS OF RECURSIVE ENUMERABILITY

In [11] a strong form of Church's thesis is attacked with the aid of the next result.

Theorem 4.1 ([11]). Suppose $L(\kappa)$ is E-closed but not Σ_1 admissible. Then (i) iff (ii).

(i) $L(\kappa) \models gc(\kappa) >$ cofinality $gc(\kappa) > \omega$.

(ii) $(Ee)(Ea)_{a \in L(\kappa)}[L(\kappa) = \{x \mid \{e\}(a,x)\downarrow]$.

(It is easy enough to devise methods of enumerating $L(\kappa)$, but difficult and sometimes impossible to find a method that stops short of κ.)

The proof of 4.1 breaks into four cases. Observe that it is safe to assume κ is countable.

Case 1: $\omega < gc(\kappa)$ is regular. Suppose $L(\kappa)$ is RE over $L(\kappa)$ via e and b. Thus for all $x \in V$,

$$x \in L(\kappa) \longleftrightarrow \{e\}(b,x)\downarrow.$$

The methods of section 3 provide a sufficiently generic T such that $L(\kappa,T)$ is E-closed and $T \notin L(\kappa)$. Hence $\{e\}(b,T)\uparrow$, and there is a p that forces $\{e\}(b,T)$ to diverge via some Moschovakis witness constructed at level $\theta < \kappa$. Since $gc(\kappa)$ is regular, there exists a T_0 in $L(\kappa)$ that satisfies p and is generic with respect to all sentences of rank at most θ. But then $\{e\}(b,x)\uparrow$ for some $x \in L(\kappa)$.

Case 2: $\omega = gc(\kappa)$. Same idea as case 1. T is now a Cohen generic real, and so the methods of section 3 require considerable alteration to succeed.

Case 3: $\omega =$ cofinality $gc(\kappa) < gc(\kappa)$. Note added in proof: Green [13] is appropriate, but there is a gap in the application (found by T. Slaman) and so the case is open.

Case 4: $\omega <$ cofinality $gc(\kappa) < gc(\kappa)$. A fine structure argument inspired by S. Friedman [14] shows that the least ordinal that constructs $x \in L(\kappa)$ is recursive in x and a fixed parameter below κ. The computation exploits the filter of closed unbounded subsets of cofinality $gc(\kappa)$. The inadmissibility of $L(\kappa)$ is needed to insure that the method given of enumerating elements of $L(\kappa)$ does not yield any nonmembers of $L(\kappa)$.

A predecessor to Theorem 4.1 can be found in [5]. There it was shown (Theorem 9) that

(A) $\{R \mid R \subseteq 2^{\omega} \ \& \ (Eb)(b \in 2^{\omega} \ \& \ R \text{ recursive in } {}^{3}E, b)\}$

is not recursively enumerable in ${}^{3}E$, c for any $c \in 2^{\omega}$, if Normann's assumption holds. The latter states there exists a wellordering $\succ$ of 2^{ω} such that $\succ$ is recursive in ${}^{3}E$, and $|\succ|$, the ordinal height of $\succ$, is regular with respect to all functions recursive in ${}^{3}E$ and any real. It is not yet clear if (A) is recursively enumerable in ${}^{3}E$ in any model of ZFC, but most likely it is. Some of the uncertainty obscuring (A) can be avoided by an absolute formulation related to the discussion of Post's problem given in [5].

Let κ_1 be the least ordinal not recursive in ${}^{3}E$, b for any $b \in 2^{\omega}$. Call a set $R \subseteq 2^{\omega}$ <u>ordinal recursive</u> in ${}^{3}E$ if R is recursive in 2^{ω}, W, where W is a prewellordering of 2^{ω} recursive in ${}^{3}E$ and some real. It can be shown ([11]) that

(B) $\{R \mid R \subseteq 2^{\omega} \ \& \ R \text{ is ordinal recursive in } {}^{3}E\}$

is not recursively enumerable in ${}^{3}E$, Q for any ordinal recursive Q under conditions substantially weaker than Normann's assumption. These conditions require a certain ordinal ρ (defined in [11,12]) to be regular with respect to functions ordinal recursive in ${}^{3}E$. There does not seem to be any escape from some assumption of cardinal regularity if the nonenumerability of (B) is desired. The presence of a singularity raises the possibility of the argument hinted at in case 4 of the proof of Theorem 4.1. At any rate the recursive enumerability of an inadmissible, E-closed $\mathcal{E}$ in some member of $\mathcal{E}$ appears to be closely tied to the E-regularity of the greatest E-cardinal. The admissible case lacks a similar resolution.

5. COUNTABLE E-CLOSED ORDINALS

Suppose $L(\kappa)$ is E-closed. Since the E-closure of an arbitrary set is better understood than an arbitrary E-closed set, it is natural to look for a $\delta < \kappa$ and a $T \subseteq \delta$ such that the E-closure of T is $L(\kappa,T)$. The main theorem of this section describes completely those countable κ's for which such a T exists. The principal tool is forcing. The machinery of section 3 runs smoothly on any countably closed set notion of forcing. One such is forcing with perfect subsets of 2^τ, where τ is an uncountable regular cardinal, developed by Baumgartner [15] as a generalization of perfect forcing on ω [16].

A perfect condition p on τ is a tree with τ nodes and levels. p is required to split "closedly often" as follows. Let λ be a limit ordinal below τ, and let Z be a path through the first λ levels of the tree. If the set of levels below λ at which a node on Z splits (i.e. has at least two immediate successors) is unbounded in λ, then the node on Z at level λ also splits. It follows that a contracting intersection of less than τ conditions is a condition. It also follows that a contracting sequence of τ conditions give rise to a so-called diagonal intersection that is almost contained in each of the given conditions. Consequently every sentence in a τ-sequence of sentences can be decided in the sense of weak forcing by a single forcing condition.

Suppose $L(\kappa)$ is countable and E-closed. If $L(\kappa)$ is Σ_1 admissible, then according to [17] there is a $T \subseteq \omega$ such that $L(\kappa,T)$ is the E-closure of T. So the next theorem fully answers the initial question of the current section. Recall that if $L(\kappa)$ is inadmissible, then inside $L(\kappa)$ there is a greatest cardinal denoted by $gc(\kappa)$.

Theorem 5.1 (Sacks-Slaman [12]). Let $L(\kappa)$ be countable and E-closed but not Σ_1 admissible. Then (i) is equivalent to (ii).

(i) $L(\kappa) \models$ cofinality $gc(\kappa) > \omega$.

(ii) $(E\delta)(ET)[T \subseteq \delta < \kappa$ & $L(\kappa,T)$ = E-closure of $T]$.

Sketch of Proof. Suppose (i). One choice for δ is the cofinality of $gc(\kappa)$ in $L(\kappa)$. The first step is to add to $L(\kappa)$ a generic collapse of $gc(\kappa)$ to $cf\ gc(\kappa)$. This is done in Levyesque fashion. A forcing condition is a one-one map from a proper initial segment of $cf\ gc(\kappa)$ into $gc(\kappa)$. Since $cf\ gc(\kappa) > \omega$, the set of forcing conditions is countably closed, and so the argument of section 3 shows that the addition of a generic collapse will not destroy the E-closedness of $L(\kappa)$.

The second step is to add a $T \subseteq cf\ gc(\kappa)$ so that $L(\kappa,T)$ is the E-closure of T. For simplicity assume $gc(\kappa) = cf\ gc(\kappa)$ so that the first step may be dropped. T is generic with respect to forcing with perfect subsets of $2^{gc(\kappa)}$. The arguments of section 3 do not quite suffice in this case, because the collection of forcing conditions is not a set in $L(\kappa)$ but rather a definable subclass of $L(\kappa)$. The proof of Lemma 3.2 is modified as follows.

Formula (1) of the proof of 3.2 still holds, but the collection of all q's such that $p_o \geq q$ is no longer a set. However the set of all c is a set of $L(\kappa)$ of $L(\kappa)$-cardinality equal to $gc(\kappa)$. With the aid of 3.2 (1), a contracting sequence of r's is developed such that for each c, some r forces either

$$c \in \{e_0\}(b) \quad \text{or} \quad (E\delta)(|\{e_1\}(c)| = \delta).$$

The diagonal intersection of the r's, call it r_∞, will then weakly force one term or the other of the above disjunction. Thus

$$r_\infty \Vdash (c)[c \in \{e_0\}(p) \longrightarrow |\{e_1\}(c)| < \rho],$$

where ρ is an ordinal computable with r_∞ from the above contracting $gc(\kappa)$-sequence of r's. The latter sequence is defined by means of an effective transfinite recursion on $gc(\kappa)$ with p_0 and b as parameters. Thus ρ is recursive in $p, b, gc(\kappa)$ and all is well.

Now suppose (ii). To obtain (i) a selection theorem is needed.

Theorem 5.2 ([12]). Suppose $L(\kappa,T)$ is the E-closure of T, where T is a subset of some $\delta < \kappa$. If in $L(\kappa,T)$ there is a greatest cardinal and its cofinality is ω, then $L(\kappa,T)$ is Σ_1 admissible.

This last is inspired by Theorem 2.4. Its proof is a combination of the ideas behind 2.2 and 2.3 and some details of the internal structure of $L(\kappa,T)$.

One last word about Theorem 5.1. If (i) holds, then δ can be taken to be the ω_1 of $L(\kappa)$. Thus recursion in ${}^3E$, i.e. the E-closure of 2^ω, seems to be the simplest universal model of Σ_1 inadmissible, Kleene recursion in an arbitrary object.

6. POST'S PROBLEM

The formulation of Post's problem for an E-closed $L(\kappa)$ requires a definition of degree for sets $A \subseteq L(\kappa)$. First a meaning is assigned to $\{e\}^A(\gamma)$. A is treated as an additional predicate. The relevant scheme, added to those of section 1, states: if a set x has been computed, then the set $A \cap x$ may be computed. Thus the universe of computation is enlarged from L to $L[A]$. $L[\kappa,A]$ is said to be E-closed over A if it is closed with respect to the computation schemes. Note that the E-closedness of $L(\kappa)$ need not imply the E-closedness of

$L[\kappa,A]$ over A. $\{e\}^A(\gamma)$ converges if instruction e applied to γ, with A as additional predicate, converges.

Suppose $A, B \subseteq L(\kappa)$. $A \leqslant B$ (read A is recursive in B) if there exists an e such that

$$\{e\}^B(\gamma) = 0 \longleftrightarrow \gamma \in A$$

$$\{e\}^B(\gamma) = 1 \longleftrightarrow \gamma \notin A$$

for all $\gamma < \kappa$. If B is RE on $L(\kappa)$ and every RE-on-$L(\kappa)$ A is recursive in B, then B is said to be complete. A is recursive if $A \leqslant \emptyset$, where $\emptyset$ is empty.

Post's problem for $L(\kappa)$, in its most primitive form, is: does there exist a nonrecursive, incomplete RE subset of $L(\kappa)$? The next theorem constitutes an affirmative answer.

Theorem 6.1 ([7]). Let $L(\kappa)$ be E-closed. Then there exist $A, B \subseteq L(\kappa)$, each RE over $L(\kappa)$ and regular, such that $A \not\leqslant B, c$ and $B \not\leqslant A, c$ for any $c \in L(\kappa)$.

A regular means $A \cap x \in L(\kappa)$ for every $x \in L(\kappa)$.

Since the A and B of 6.1 are regular, it must be that $L[\kappa,A] = L[\kappa,B] = L(\kappa)$. Since neither A nor B is complete, it follows that $L(\kappa)$ is E-closed over A, and over B. Thus any computing procedure applied to A and a parameter from $L(\kappa)$, if convergent, yields an element of $L(\kappa)$ via a computation in $L(\kappa)$. The same holds for B. Consequently the incomparability achieved by 6.1 is the most possible.

From now on assume $L(\kappa)$ is not Σ_1 admissible.

The proof of 6.1 applies some further properties of κ_r, defined in Section 2, and a bit of fine structure from the E-recursive point of view.

Lemma 6.2 ([7]). Suppose $x \in L(\kappa)$, $x^2 \subseteq x$ and

$\sup\{\kappa_0^y \mid y \in x\} < \kappa.$

Then

$$\sup\{\kappa_r^y \mid y \in x\} < \kappa.$$

Proof. Suppose otherwise. Then

$$\sup\{\kappa_r^{y,z} \mid y \in x\} = \kappa$$

for each $z \in x$. Hence each Σ_1 sentence about z true in $L(\kappa)$ is also true in

$$L(\sup\{\kappa_0^{y,z} \mid y \in x\}).$$

Since $x^2 \subseteq x$, it follows that κ z-reflects to $\sup\{\kappa_0^y \mid y \in x\}$. By supposition there is a z such that $\kappa_r^z \geqslant \kappa_0^y$ for all $y \in x$. But then $\kappa = \kappa_r^z$, an impossibility according to Lemma 2.8. □

Two RE projecta, useful in priority and forcing arguments, are now defined. They are analogous to the two Σ_1 projecta of Jensen [4].

$$\rho = \mu\gamma_{\leqslant\kappa}(Ef)[f \text{ partial rec. } \& \text{ dom } f \subseteq \gamma \ \& \text{ range } f = \kappa]$$

$$\eta = \mu\gamma_{\leqslant\kappa}(ER)[R \in RE - L(\kappa) \ \& \ R \subseteq \gamma].$$

In the definitions of η and ρ, RE and partial rec. are boldface notions, i.e. parameters from $L(\kappa)$ are allowed. It is tempting to define ρ by requiring f to map κ one-one into γ. To do so would be a false move, as has been shown by T. Slaman [18]. He has shown that a downward definition produces a larger ρ than the one desired and invalidates the next lemma. The surprise here is that the inverse of a one-one partial recursive map onto κ need not be partial recursive.

Lemma 6.3 ([7]). $\rho = \eta$.

Proof. Clearly $\eta \leqslant \rho$. Suppose $\eta < \rho$. Let $p \in L(\kappa)$ be such that $R \subseteq \eta$ is RE (on $L(\kappa)$) in p but not recursive via any parameter from $L(\kappa)$. Let Z be the E-closure of $R \cup \{p\} \cup \eta \cup \{\eta\}$. If $\overline{Z}$, the transitive collapse of Z, were bounded below κ, then R would belong to $L(\kappa)$. Hence $\overline{Z}$ is $L(\kappa)$. Note that η collapses to η. Some slight modification of the universal partial recursive function maps η onto κ, an impossibility when $\eta < \rho$. □

One more parameter is needed for the solution of Post's problem. The RE cofinality of δ, for any $\delta \leqslant \kappa$, is the least γ such that γ is the ordertype of an unbounded (in δ) RE subclass of δ, symbolically RE - cf δ.

Lemma 6.4 ([7]). RE - cf η = RE - cf κ.

Proof (sketchy). Let S be an RE subclass of ρ unbounded in ρ. With each $x \in S$, associate

$$x^r = \sup\{\kappa_r^y \mid y < x\}.$$

$x^r < \kappa$ by Lemmas 6.2 and 6.3. $\{x^r \mid x \in S\}$ has the same ordertype as S, is unbounded in κ, and is RE.

Outline of solution to Post's problem. The requirements are indexed by ordinals less than ρ. A typical requirement is

$$A(w) \neq \{f(\delta)\}^B(w)$$

where f is a partial recursive map from ρ onto κ, and $\delta < \rho$. w is a Friedberg-Muchnik witness chosen in the usual manner to minimize conflicts and overlaps between requirements. As in Shore [19], the requirements are divided into blocks. The number of blocks is RE - cf ρ. The first block consists solely of requirements of the type shown above and indexed by δ. The second block contains only the sort intended to make B fail to be recursive in A.

Within each block there is no conflict between requirements. Any conflict between blocks is resolved in favor of the block with lesser index. Work on requirement δ proceeds as follows. As B is enumerated, the computation tree associated with $\{f(\delta)\}^B(w)$ may begin to develop. Every convergent portion of the tree, as it comes into being, is preserved. Thus it may happen that the entire tree never develops, but that portions of it are preserved forever. That is all to the good, since it may be possible to meet requirement δ by developing and preserving a Moschovakis witness (defined in Section 2) to the divergence of $\{f(\delta)\}^B(w)$.

Consider the first block with the intent of showing all activity associated with that block comes to an end before the construction does. Requirement δ becomes inactive in one of two ways:

(1) $\{f(\delta)\}^B(w)$ develops a value, is preserved, and A(w) is given the opposite value;

(2) A Moschovakis witness to the divergence of $\{f(\delta)\}^B(w)$ is developed and preserved.

An event of type (1) is enumerated at some stage recursive in δ. (Assume w is recursive in δ.) An event of type (2) is enumerated at worst by stage κ_r^δ according to Theorem 2.7. Thus requirement δ settles down in either case by stage κ_r^δ. Let δ_m be the supremum of all δ in the first block. By Lemma 6.3,

$$\sup\{\kappa_0^\delta \mid \delta < \delta_m\} < \kappa.$$

Assume $\delta_m^2 \subseteq \delta_m$. By Lemma 6.2

$$\sup\{\kappa_r^\delta \mid \delta < \delta_m\} < \kappa,$$

and so all activity in the first block comes to an early end.

Now fix $\beta <$ RE - cf(η), and suppose activity in each of the first β block ends before the construction of A and B does. For each $\nu < \beta$, the stage $\sigma(\nu)$ at which activity in block ν ends can be effectively recognized; i.e. one can look down from stage $\sigma(\nu)$ and see that each requirement in block ν has been met in one of the two ways sketched for the first block. Thus $\{\sigma(\nu) \mid \nu < \beta\}$ is an RE class of ordertype less than Re - cf κ by Lemma 6.4, and consequently bounded below κ.

Note that the above solution to Post's problem can be managed without injuries. One simply works only on the first block until it is completed, then proceeds to the second block, and so on. Thus the priorities are used only to decide what to do next, and never to resolve conflicts by imposing injuries. It follows that the "incomparable" sets can be enumerated monotonically. Unavoidable injuries occur in the study of Post's Problem for E-closed sets where $\eta < \rho$ [7].

Normann [20] developed a solution for Post's problem that adapts to $L(\kappa)$ when $\rho < \kappa$ and ρ is a regular cardinal in the sense of $L(\kappa)$. His precise result is: suppose $E(2^\omega)$ has as an element a wellordering of 2^ω which is regular with respect to cofinal sequences in $E(2^\omega)$; then there exist RE subclasses of $E(2^\omega)$ incomparable with respect to every element of 2^ω. The methods of this paper make it possible to replace Normann's assumption of regularity by: $\eta = \rho$. Results of Normann and Slaman [18] show $\eta = \rho$ is weaker than regularity. Hypotheses stronger than regularity have been utilized by E. Griffor [21] to obtain minimal pair and density theorems for

RE subsets of $L(\kappa)$. His work applies the following regular sets theorem.

Theorem 6.5 ([7]). Let A be RE on $L(\kappa)$. Then there exists an RE B on $L(\kappa)$ such that B is regular, $A \leqslant B$ and $B \leqslant A$.

Only one case of the above needs proof, A complete, and that is managed by a deficiency set argument combined with an effective transfinite induction. (Note that the deficiency set of an RE set need not be RE.)

A central open problem on RE degrees is as follows. Suppose $L(\kappa)$ is E-closed and $A \subseteq \kappa$ is an RE set of order-type less than ρ. Does the supposition that A is scattered imply that A is incomplete? recursive?

7. LOGIC ON E-CLOSED SETS

Suppose $L(\kappa)$ is E-closed but not Σ_1 admissible. Let F be a sentence of $L_{\infty,\omega}$ restricted to $L(\kappa)$. An early completeness result of Barwise [22] shows that if F is valid, then F has a proof in any Σ_1 admissible set with F as a member. A later result of Stavi [23] shows that a fairly arbitrary structure A is $L_{\infty,\omega}$ complete if and only if A is the union of Σ_1 admissible structures. Thus there is in general no hope of finding proofs in $L(\kappa)$ for all valid sentences in $L(\kappa)$.

On the other hand $L(\kappa)$ does have strong closure properties. It may not be closed under the usual rules of proof for $L_{\infty,\omega}$, but it is closed under effective rules for computation which have counterparts as rules of proof. The idea is as follows. The usual rules of proof are restricted so as to allow only effective proofs. To be precise, a proof P of F is allowed only if $P = \{e\}(F)$ for some e. Of course there will be valid F's without effective proofs. It turns out that all of the usual rules have effective adaptations save one, disjunction introduction. Suppose $\bigvee_{i \in I} F_i$ is proved by first proving some F_i effectively. There is no effective method for selecting such an F_i. It can happen that no such F_i is recursive in $\bigvee_{i \in I} F_i$. If I is countable, then Gandy selection applies, but in general nothing can be done.

In contrast, if each F_i has an effective proof, then $\bigwedge_{i \in I} F_i$ has an effective proof.

Despite the limitations of effective proof procedures, it is possible to develop a type omitting theorem for $L(\kappa)$. For simplicity let $L(\kappa)$ have a greatest cardinal, let $\mathcal{L}(\kappa,T)$ be the standard ramified language for $L(\kappa,T)$ as in Section 2, and let Σ be an RE set of ranked sentences of $\mathcal{L}(\kappa,T)$. The type of theorem being sought says: if Σ has a model and some other conditions hold, then Σ has a model that omits κ. To omit κ means to find a model $L(\kappa,T)$ of Σ such that $L(\kappa,T)$ is E-closed. In the study of countable admissible structures, it was seen that type omitting could do much of the work of forcing, but that as yet is not the case for E-closed structures.

The type omitting theorem offered below imposes two conditions on Σ, strict RE-ness and ω-compactness. Σ is said to be strict RE if the set of all logical consequences of Σ is RE. The proof of the existence of strict RE sets is momentarily deferred. (For suitable $L(\kappa)$'s, the empty set is not strictly RE.) Σ is said to be ω-compact if for each ω-sequence $\{F_i \mid i \in \omega\} \in L(\kappa)$, it is the case that $\bigvee_i F_i$ is a logical consequence of Σ only if $\bigvee_{i<n} F_i$ is for some n. It is possible for Σ to be strictly RE without being ω-compact.

Theorem 7.1 ([24]). Suppose $\Sigma \subseteq L(\kappa)$ is strictly RE, ω-compact, and has a model. Then Σ has a model that omits κ.

The proof of 7.1 is not unlike that of 3.2.

The existence of strict RE Σ's is a consequence of forcing theory. Let $\Vdash$ be the forcing relation of Section 3 or any set forcing relation for $L(\kappa)$. For Σ take the RE collection of ranked sentences that defines $\Vdash$ for all ranked sentences. Then $L(\kappa,T)$ is a model of Σ if and only if T is generic with respect to all ranked sentences. A ranked sentence F holds in every model of Σ if and only if F is weakly forced by the null condition. It follows that Σ is strictly RE.

To make Σ ω-compact, choose $\Vdash$ so that the set of forcing conditions is ω-closed, as in the proof of Theorem 4.1.

Strict RE sets can be defined in purely recursion theoretic terms, but it is not clear such a definition is of any use.

A central problem consists of proving the existence of strict RE sets of sentences without recourse to forcing theory. A negative solution might sharpen the differences between, and relative strengths of, forcing and type omitting constructions on E-closed, non-Σ_1 admissible sets. On the other hand it is possible that every RE set has a strict RE extension. If that is so, then the notion of effective proof is more than a logical curiosity.

REFERENCES

1. D. Normann, Set recursion, in Generalized Recursion Theory II, North-Holland 1978, 303-320.

2. S. C. Kleene, Recursive functionals and quantifiers of finite type, Trans. Amer. Math. Soc. 91 (1959), 1-52; and 108 (1963), 106-142.

3. Y. N. Moschovakis, Hyperanalytic predicates, Trans. Amer. Math. Sco. 138 (1967), 249-282.

4. R. B. Jensen, The fine structure of the constructible hierarchy, Annals Math. Log. 4 (1972), 229-308.

5. G. E. Sacks, Post's problem, absoluteness and recursion in finite types, in 1978 Kleene Symposium volume, to appear.

6. L. Harrington, Contributions to Recursion Theory in higher types, Ph.D. Thesis, M.I.T., 1973.

7. G. E. Sacks, Post's problem for E-recursion, in preparation.

8. R. O. Gandy, General recursive functionals of finite type and hierarchies of functions, University of Clermont-Ferrand, 1962.

9. T. Grilliot, Hierarchies based on objects of finite type, Jour. Symb. Log. 34 (1969), 177-182.

10. G. E. Sacks, The k-section of a type n object, Amer. Jour. Math. 99 (1977), 901-917.

11. ______, The limits of recursive enumerability, in preparation.

12. ______ and T. Slaman, Forcing over E-closed sets, in preparation.

13. J. Green, Σ_1 compactness for next admissible sets, Jour. Symb. Log. 39 (1974), 105-116.

14. S. D. Friedman, Negative solutions to Post's problem I, in Generalized Recursion Theory II, North-Holland 1978, 127-134.

15. J. Baumgartner and R. Laver, Iterated perfect set forcing, Annals Math. Log., to appear.

16. G. E. Sacks, Forcing with perfect closed sets, in Axiomatic Set Theory, Amer. Math. Soc., Providence 1971, 331-355.

17. ______, Countable admissible ordinals and hyperdegrees, Advances in Mathematics 20 (1976), 213-262.

18. T. Slaman, Independence Results in E-Recursion Theory, Ph.D. Thesis, Harvard Univ., 1981.

19. R. Shore, Splitting an α-recursively enumerable set, Trans. Amer. Math. Soc. 204 (1975), 65-78.

20. D. Normann, Degrees of functionals, Preprint Series in Mathematics No. 22, Oslo 1975.

21. E. Griffor, E-Recursively Enumerable Degrees, Ph.D. Thesis, M.I.T., 1980.

22. J. Barwise, Admissible Sets and Structures, Springer-Verlag, Heidelberg 1975.

23. J. Stavi, A converse of the Barwise completeness theorem, Jour. Symb. Log. 38 (1973), 594-612.

24. G. E. Sacks, Logic on E-closed sets, in preparation.

COMPUTING IN ALGEBRAIC SYSTEMS

J.V. Tucker
Mathematical Centre
2e Boerhaavestraat 49
1091 AL AMSTERDAM

INTRODUCTION

Given a relational structure A, imagine an A-*register machine* which can hold in its registers a fixed, finite number of elements of A, perform the basic operations and decide the basic relations on these elements, and manage some simple manipulations and decisions such as to replace the contents of one register by those of another and to tell when two registers carry the same element. Next, picture an A-*register machine with counting* by adding a finite number of *counting registers* to an A-register machine; these carry natural numbers and the device is able to put zero into a counting register, add or subtract one from the contents of any counting register, tell if two registers contain the same number, and so on. Thirdly, there is the A-*register machine with stacking* which augments an A-register machine with a single *stack register* into which the entire contents of the ordinary algebraic registers of the basic machine can be temporarily placed at various points in the course of a calculation. Thus, the combinatorial operations of the A-register machine are extended in the first instance by permitting subcomputations on the natural numbers ω and in the second by prolonging the number and complexity of entirely algebraic subcomputations. On arranging both we have an A-*register machine with counting and stacking*.

To use one of these machines to compute a partial function on A is to write down the familiar finite program of instructions referring to whatever activities of the machine are available and containing information to stop in certain circumstances. Such a program is called a *finite algorithmic procedure*, a *finite algorithmic procedure with counting*, a *finite algorithmic procedure with stacking*, or a *finite algorithmic procedure with counting and stacking* accordingly as it includes or ignores instructions involving the counting and stack facilities. These are abbreviated in turn by *fap*, *fapC*, *fapS* and *fapCS*. The classes of all functions over A they compute, FAP(A), FAPC(A), FAPS(A) and FAPCS(A), are the subject of this paper.

The idea of the A-register machine first appeared in G.T. Herman & S.D. Isard [13] and in H. Friedman [7], there together with its counting facility. The stack mechanism belongs to

[22,23] in which J. Moldestad, V. Stoltenberg-Hansen and I investigated recursion-theoretic properties of the four families of fap-computable functions. This third article is a set of notes which introduces such models of computing into an algebraic *milieu* to settle decision problems: *The membership question for finitely generated ideals of polynomial rings over fields is fapCS-computable* (Theorem 3.2). *The membership question for single generator subgroups of the torus groups is fap-semicomputable, but not fapCS-computable* (Example 6.4). And to settle theoretical issues about computation: *For* A *an "everyday algebraic system of Mathematics"*, FAPC(A) = FAPCS(A) (Theorem 5.2). *But, for example, if* A *is the algebraic closure of some finite field* $\mathbb{Z}_p$ *then* FAPS(A) $\subsetneqq$ FAPC(A) (Theorem 5.9).

The intention is to map out *Elementary* Recursion Theory in an algebraic setting: enough to show what peculiarities arise in lifting the contents of, say, Machtey & Young's book [20], stopping short of the Turing degree theory of the various fap-semicomputable sets which seems mysterious outside the finitely generated algebras (c.f. Section 4 and Theorem 5.10). The first two sections summarise formal definitions and certain relevant properties of the functions from [22,23,24]. The next two sections establish the local algebraic character of machine computations while Section 5 is about counting and stacking. Section 6 contains a useful theorem about fapCS-computable subsets of topological algebras.

Although the emphasis is on the classes of functions computed over natural algebraic systems, and not on the structure of programs and their semantics, the material here does usefully connect with the literature of Theoretical Computer Science: it provides the sort of basic technical information needed to fit de Bakker's mathematical semantics of program control structures based on ω [2] to the ADJ Group's algebraic theory of data types [8], for example. A specific technical application to program semantics is [3] by J.A. Bergstra, J. Tiuryn and myself.

My interest in Generalised Recursion Theory I owe to my time in Oslo, to J.E. Fenstad, D. Normann, S.S. Wainer and, especially, J. Moldestad and V. Stoltenberg-Hansen; in this connection I gratefully acknowledge the support of a fellowship from the European Programme of The Royal Society, London. These notes have also profited from information and advice from P.R.J. Asveld, J.A. Bergstra, B. Birkeland, H. Rolletschek, and J.I. Zucker.

NOTATION Throughout we are concerned with relational structures of finite signature. The word function, unqualified, will mean partial function and typically these will be (n,m)-ary maps either $A^n \times \omega^m \to A$ or $A^n \times \omega^m \to \omega$, a distinction we preserve in the abbreviation $A^n \times \omega^m \to A/\omega$. For $(a,x) \in A^n \times \omega^m$, $f(a,x) \simeq g(a,x)$ means $f(a,x)$, $g(a,x)$ are both defined and equal or are both undefined. On fixing part of the argument $a \in A^n$ of a function $f:A^{n+m} \to A$ we write $f(a):A^m \to A$. On extending a unary function $f:A \to A$ to an n-fold product $f\times\ldots\times f:A^n \to A^n$, we write for $a = (a_1,\dots,a_n) \in A^n$ the value $f(a)$ for $(fa_1,\dots,fa_n)$. Relations may be identified with their characteristic functions using 0 for *true*, 1 for *false*. The complement of a set A is denoted $\neg A$.

$\mathbb{Z}$, $\mathbb{Z}_p$, $\mathbb{Q}$, $\mathbb{R}$, $\mathbb{C}$ are the integers, the integers modulo p, the rationals, reals and complex numbers respectively.

For those ideas and facts of Algebra left unexplained in the text consult the books of A.G. Kurosh [18] and Mal'cev [21] on Universal Algebra, that of Kurosh [16,17] on Group Theory and B.L. van der Waerden [25] for Field Theory.

1. FINITE ALGORITHMIC PROCEDURES

The program language for A-register machines has variables $r_0, r_1, r_2,\dots$ for *algebra registers*; its constants, function and relation symbols are those of the signature of A together with new constants T for *true*, F for *false*, H for *halt*, 1,2, ... for *instruction labels* or *markers* and new relation = for *algebraic equality*.

A *finite algorithmic procedure* is a finite ordered list of labelled instructions $I_1,\dots,I_\ell$ which are of these types. The *algebraic operational instructions* manipulate elements of A and are

$r_\mu:=\underline{a}$ meaning "*replace the contents of register* r_μ *by the element* $a \in A$ *named by* $\underline{a}$".

$r_\mu:=\sigma(r_{\lambda_1},\dots,r_{\lambda_k})$ meaning "*apply the* k-*ary operation* σ *of* A *to the contents of registers* $r_{\lambda_1},\dots,r_{\lambda_k}$ *and replace the contents of* r_μ *by this value*".

$r_\mu:=r_\lambda$ meaning "*replace the contents of register* r_μ *with that of* r_λ".

The *algebraic conditional instructions* determine the order of executing instructions and are

<u>if</u> $R(r_{\lambda_1},\dots,r_{\lambda_k})$ <u>then</u> i <u>else</u> j meaning "*if the* k-*ary relation* R *is true of the contents of* $r_{\lambda_1},\dots,r_{\lambda_k}$ *then the next instruction is* I_i *otherwise it is* I_j".

<u>if</u> $r_\mu=r_\lambda$ <u>then</u> i <u>else</u> j which takes its obvious meaning and, incidentally, gives us <u>goto</u> i.

Finally, $r_\mu:=T$, $r_\mu:=F$ marking true and false, and H meaning "*stop*" are included.

For the A-register machine with counting the language is extended by variables $c_0,c_1,c_2,\dots$ for *counting registers*, a constant 0 for *zero*, function symbols +1 for *successor*, $\dot{-}1$ for *predecessor*; and relational symbol = for *numerical equality*. The *counting instructions*, which when mixed with fap instructions make *finite algorithmic procedures with counting*, are simply the usual instructions for (natural number) register machines:

$c_\mu:=0$ $c_\mu:=c_\lambda+1$ $c_\mu:=c_\lambda\dot{-}1$ <u>if</u> $c_\mu=c_\lambda$ <u>then</u> i <u>else</u> j

and take their obvious meanings.

For the A-register machine with stacking we add to the A-register machine language a variable s for *stack register* and $\emptyset$ for *empty*. The new instructions are

<u>stack</u>$(z,r_0,\dots,r_m)$ meaning "*along with instruction*

marker z, *copy the contents of* $r_0,\ldots,r_m$ *as an* (m+1)-*tuple into the stack*".

restore($r_0,\ldots,r_{j-1},r_{j+1},\ldots,r_m$) meaning "*remove the last or topmost vector in the stack and replace the contents of the* r_i ($i \neq j$) *by the corresponding entries of the vector*".

if $s=\emptyset$ then i else marker meaning "*if the stack is empty then the next instruction is* I_i *otherwise it is the instruction labelled by the marker in the topmost element in the stack*".

A *finite algorithmic procedure with stacking* is an ordered list of fap instructions and stack instructions with the convention that all the ordinary algebraic register variables are included in every stack-instruction, and all but one of them are included in every restore-instruction; the rôle of the marker is to remember at which points in the program the basic A-register machine is cleared for an independent subcomputation. For a proper account of the stack and its operation the reader should consult [22], but the description given is adequate for what follows. The definition of a *finite algorithmic procedure with counting and stacking* is immediate though we note that the stack allows only algebra elements to be stored.

To compute $A^n \times \omega^m \to A/\omega$ with program α choose n of its algebraic variables, and m of its counting variables, as *input variables*, and name a suitable variable as *output variable*. On fixing a machine appropriate for α load input argument $(a,x) \in A^n \times \omega^m$ into the registers α reserves as inputs, and make the remaining registers empty. The instructions of α are executed on the machine in the order they are given except where a conditional instruction directs otherwise. If at some stage an instruction cannot be carried out, such as happens when one applies a fap conditional to empty registers, then the computation is said to *hang* in that state and no value is computed. If the machine halts then the output value of the computation $\alpha(a,x)$ is the element contained in the output register named by α, if this is not empty; in all other circumstances, no output value is obtained. Converging and diverging computations are distinguished by $\alpha(a,x)\downarrow$ and $\alpha(a,x)\uparrow$, respectively, and, as usual, $\alpha(a,x)$ also denotes the (defined or undefined) value of a computation.

The four types of fap-computable function, relation or set can now be formally defined in the obvious way. A set or relation $S \subset A^n \times \omega^m$ is, say, *fapCS-semicomputable* if it is the domain of a fapCS-computable function; fap/fapC/fapS-semicomputability is defined similarly, of course. The different sets of computable functions $A^n \to A$ are denoted ${}^n\text{FAP}(A)$, ${}^n\text{FAPC}(A)$, ${}^n\text{FAPS}(A)$, ${}^n\text{FAPCS}(A)$.

For those acquainted with Theoretical Computer Science it ought to be pointed out that the hierarchy of low to high-level languages characteristic of computing *praxis* is also an essential feature of *theoria*: to prove a relation on an algebra A decidable, the richer the programming language the better whereas to prove it undecidable the opposite is true. Here our needs are best met by defining the various fap-computable functions through a crude "assembler code" for A-register machines, but features like the so called *structured control statements, recursion,* and *data type creation* can be encorporated into the definitions of FAP(A), FAPS(A) and FAPCS(A) respectively. Thus, to prove the set $FO(G)$ of all elements of finite order in a group G is fap-semicomputable one may use this program wherein r_I and r_0 are input and output variables:

$r:=r_I$; while $r\neq 1$ do $r:=r.r_I$ od; $r_0:=T;H$.

In conclusion, here are some technical ideas. A *state description* in a machine computation under fapCS α is typically a list

$(k;a_1,\ldots,a_p,x_1,\ldots,x_q;(z_1;a_{11},\ldots,a_{1p}),\ldots,(z_s;a_{s1},\ldots,a_{sp}))$

where $a_1,\ldots,a_p$ are the contents of the algebra registers named by α; $x_1,\ldots,x_q$ are those of the counting registers; there are s vectors piled in the stack register, z_i is the marker of the i-th element and a_{ij} the element in the j-th register of the i-th stored vector. And, finally, k is the number of the instruction in α which is to be applied to these elements. Each state description represents a *step* in a computation. Often we need to *unfold* a computation $\alpha(a,x)$ into all its stages and then we use

$D_i(\alpha,a,x) = (m_i,a_{ij},x_{ij};\ (z_{ij},a^i_{jk}))$

to denote the i-th step in the computation $\alpha(a,x)$. The length $|\alpha,a,x|$ of the computation $\alpha(a,x)$ is by definition the ordinal number of steps in the unfolding of $\alpha(a,x)$.

All four sets of programs we assume gödel numbered in the usual way with $\Omega_0,\Omega_C,\Omega_S,\Omega_{CS}$ denoting the code sets for fap, fapC, fapS and fapCS programs respectively. We also assume the *term* or *polynomial algebra* $T[X_1,\ldots,X_n]$ of any signature to be coded uniformly in n, ${}^n\gamma_*: {}^n\Omega \to T[X_1,\ldots,X_n]$ with the abbreviation ${}^n\gamma_*(i) = [i]$. Each term $t(X_1,\ldots,X_n)$ defines a function $A^n \to A$ by substituting algebra elements for indeterminates and we define n-ary term evaluation ${}^nTE:{}^n\Omega \times A^n \to A$ by ${}^nTE(i,a) = [i](a)$. Finally, define for $t,t' \in T[X_1,\ldots,X_n]$ $t \equiv_A t'$ if, and only if, for all $a \in A^n$, $t(a) = t'(a)$.

2. THE FAP-COMPUTABLE FUNCTIONS IN THE LARGE

To sketch essential background information from [22,23,24] about the recursion-theoretic properties of fap-computable functions, we begin with an axiomatisation of the large-scale structure of the partial recursive functions on ω.

In summary, a set of functions Θ over A is a *computation theory over* A *with code set* $C \subset A$ if associated with Θ is a surjection $\varepsilon: C \to \Theta$, called a *coding* and abbreviated by $\varepsilon(e) = \{e\}$ for $e \in C$, and an ordinal valued length of computation function $|.|$ such that $|e,a|\downarrow \iff \{e\}(a)\downarrow$, for which the following properties hold.

(1) C contains a copy of ω and Θ contains copies of zero, successor, predecessor on ω.

(2) Θ contains the projection functions, the operations of A and the relations of A in the form of definition-by-cases functions.

(3) Θ is closed under composition, the permuting of arguments, and the addition of dummy arguments. And, in particular,

(4) Θ contains *universal functions* ${}^nU$ such that for $e\in C, a\in A^n$

$${}^nU(e,a) \simeq \{e\}(a).$$

(5) Θ enjoys this *iteration property*: for each n,m there is a map $S^n_m \in \Theta$ such that for $e \in C$, $a \in C^n$, $b \in A^m$

$$\{S^n_m(e,a)\}(b) \simeq \{e\}(a,b).$$

Moreover, it is required that certain uniformity hypotheses are satisfied and that the length function respect the efficiency of the functions mentioned in the definition; for full details see [23] or J.E. Fenstad's monograph [6] from which this axiomatisation is taken.

The coding of programs, mentioned in the previous section, extends to a coding of the functions they compute: choose $C=\omega$ and $\{e\}$ to be the function computed by fapC e, if $e \in \Omega_C$, or to be the everywhere undefined function otherwise. Define length of computation $|e,a,x|$ to be the number of steps in computation $\{e\}(a,x)$. To be faithful to the definition of a computation theory, the code set C is adjoined to A to make the structure A_ω whose domain is $A \dot{\cup} \omega$ and whose operations and relations are those of A together with zero, successor, predecessor, and equality on ω. It turns out that the fapC-computable functions can be identified with $FAP(A_\omega)$.

2.1 THEOREM *The fapC-computable functions over* A *constitute a computation theory over* A *with code set* $C = \omega$ *if, and only if, term evaluation* ${}^n TE$ *is fapC-computable over* A *uniformly in* n.

Now term evaluation is always fapCS-computable and so on repeating the coding constructions with Ω_{CS} it follows that

2.2 THEOREM *The fapCS-computable functions over* A *constitute a computation theory over* A *with code set* $C = \omega$.

The next step in [23] was to define a computation theory Θ over A with code set C to be *minimal* if $\bar{\Theta}$ is contained in any other computation theory Φ over A with code set C. And then to prove

2.3 THEOREM *The fapCS-computable functions over* A *constitute the minimal computation theory over* A *with code set* $C = \omega$.

2.4 THEOREM *There exists a structure* A *where the following inclusions are strict*

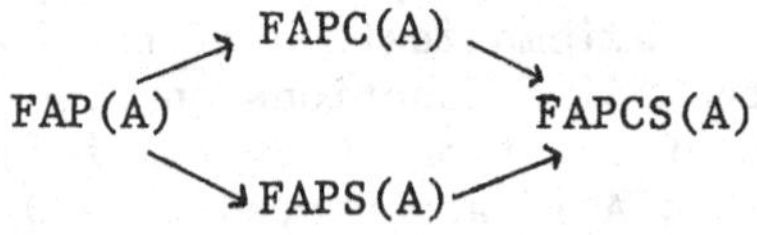

Theorem 2.3 coupled to Theorem 2.1, and Theorem 2.4 are the point of departure for Section 5. Two other basic facts are needed later on. The first comes from [23], the second, a corollary, we leave as an exercise for the reader.

2.5 THEOREM *The relation* ${}^{n,m}STEP \subset C \times A^n \times \omega^m \times \omega$ *defined by*

$${}^{n,m}STEP(e,a,x,k) \equiv |e,a,x| \leq k$$

is fapCS-computable uniformly in n,m.

2.6 THEOREM $S \subset A^n \times \omega^m$ *is fapCS-computable if, and only if,* S *and* $\neg S$ *are fapCS-semicomputable.*

In [22] it was shown that the functions *inductively definable* over A, in the sense of Platek, are precisely the fapS-computable functions. To establish a wider context for the algebraic study of computation, J. Moldestad and I have attempted to systematically classify, in terms of the fap formalism, the many disparate approaches to defining computability in an abstract setting. Thus in [24] it is shown that Normann's *set recursion* is equivalent to fapCS-computability, that a natural generalisation of Herbrand-Gödel-Kleene *equational definability* is equivalent to fapS-computability, as is computability by *flowcharts with* LISP-*like recursive procedures*. See [24] for a complete survey and a discussion of a *Generalised Church-Turing Thesis* nominating the fapCS-computable functions as *the* class of functions effectively calculable by finite, deterministic algorithms in Algebra.

3. ALGEBRAIC INFLUENCES ON FAP-COMPUTATION

3.1 LOCALITY OF COMPUTATION LEMMA *Let* $\alpha:A^n \times \omega^m \to A/\omega$ *be a fapCS and* $(a,x) \in A^n \times \omega^m$. *Then each state of the computation* $\alpha(a,x)$ *lies within* ω *and the subalgebra* $\langle a\rangle$ *of* A *generated by* $a \in A^n$. *In particular, if* $\alpha:A^n \times \omega^m \to A$ *and* $\alpha(a,x)\downarrow$ *then* $\alpha(a,x) \in \langle a\rangle$.

PROOF. Let $D_i(\alpha,a,x)$ be a state description in the unfolding of $\alpha(a,x)$. We must prove by induction on i that any algebraic element of $D_i(\alpha,a,x)$ lies in $\langle a\rangle$. This is immediate for the basis $i = 1$ by convention. The induction step from $D_i(\alpha,a,x)$ to $D_{i+1}(\alpha,a,x)$ depends upon the type of instruction numbered n_i in $D_i(\alpha,a,x)$. There are 15 cases. The conditional instructions do not alter algebraic registers, nor do the operational instructions for counting. And the algebraic operational instructions either relocate elements of $D_i(\alpha,a,x)$, assumed in $\langle a\rangle$, or apply a basic operation of A to them to create possibly new elements of $\langle a\rangle$. Q.E.D.

Let F be a field. *The function* $f(a) = \sqrt{a}$ *which picks out some square root of* a, *if it exists, is not in general fapCS-computable.* Take $F = \mathbb{R}$: if f were fapCS-computable then $f(2) = \pm\sqrt{2}$ would lie in the subfield $\langle 2\rangle = \mathbb{Q}$ which is not the case.

Let R be an Euclidean domain with degree function $\partial:R \to \omega$. *The algorithm which computes for* $a \neq 0$, $b \in R$ *elements* q,r *with* $b = qa + r$ *and either* $r = 0$ *or* $\partial(r) < \partial(a)$ *is not in general fapCS-computable over* (R,∂). Take $R = \mathbb{Z}[X]$ with ∂ the usual polynomial degree. Let $a = X^2$, $b = X^7$ so that necessarily $q = X^5$, $r = 0$. But $X^5 \notin \langle X^7,X^2\rangle$ since this subring involves only powers of the form X^{2n+7m} for $n,m \in \omega$.

In addition to treating search mechanisms with caution when formalising algorithms, one must also be prepared to sometimes dispense with pairing and unpairing functions:

A is *locally* n-*finite* if every n-generator subalgebra of A is finite. A is *locally finite* if it is locally n-finite for each n. If A is locally n-finite then the number of algebraic values for all the fapCS-computable functions at argument $a \in A^n$ is finite being bounded by the order function ${}^n\mathrm{ord}(a) = |\langle a\rangle|$. In [9], E.S.

Golod shows that for each n there is a group G which is locally n-finite but not locally (n+1)-finite. And for these groups the theory of fapCS-computable functions of $\leq$ n arguments is considerably removed from that of functions of $>$ n arguments.

The reinstatement of search and pairing (*of a local character*) is the subject of Section 4, but fapCS-computation is better thought of as "sensitive" rather than "weak":

3.2 THEOREM *Let* F *be a field. Then the membership relation for finitely generated ideals of* $F[X_1,\ldots,X_n]$, *defined*

$$^{n,k}M(q,p_1,\ldots,p_k) \equiv q \in (p_1,\ldots,p_k)$$

is fapCS-computable over F *uniformly in* k,n.

PROOF. We describe the algorithm informally for (any) fixed n,k and sketch the reasons why it stays within the realm of fapCS-computation over F, freely referencing principles which belong to the next section. The algorithm refers to the data types F, $F[X_1,\ldots,X_n]$, the ring M(s,t,F) of s $\times$ t matrices over F, ω, and a second polynomial ring $F[t_{ij}:1\leq i\leq k, 1\leq j\leq \ell]$, but in fact operates in those parts defined by the subfield L of F generated by the coefficients of the input polynomials: $L[X_1,\ldots,X_n]$, M(s,t,L) and so on. These localisations, and what the algorithm requires of them, can be fapCS-computably simulated over F, ω from the input coefficients; this claim we ask the reader to check possibly guided by the principles of Local Enumeration, Search and Pairing from Section 4.

From Satz 2 of G. Hermann [14] one can obtain this fact: *Consider equations of the form*

$$r_1p_1+\ldots+r_kp_k = q \qquad (*)$$

where $q,p_1,\ldots,p_k$ *are given and* $r_1,\ldots,r_k$ *are to be found. There exists a primitive recursive function* $f:\omega^3 \to \omega$ *such that if* (*) *has a solution in* $F[X_1,\ldots,X_n]$ *then it does so with* $\deg(r_i) \leq f(a,b,n)$ *where* $a = \deg(q)$ *and* $b = \max\{\deg(p_i): 1\leq i\leq k\}$. To decide $q \in (p_1,\ldots,p_k)$ is to decide whether or not (*) has a solution and this is done by setting up a system of linear equations over F.

Construct formal polynomials $r_1,\ldots,r_k$ of degree d = f(a,b,n) with coefficients treated as indeterminates over F,

$$r_i = \sum_{|j|\leq d} t_{ij}X^j$$

where $j = (j_1,\ldots,j_n)$, $|j| = j_1+\ldots+j_n$ and $X^j = (X_1^{j_1}\ldots X_n^{j_n})$. Substituting these into equation(*) produces a polynomial identity whose LHS has degree $\leq$ f(a,b,n) + b and whose RHS has degree = a. Comparing coefficients leads to a set of linear equations in t_{ij} over F. Thus $q \in (p_1,\ldots,p_k)$ iff this set of equations has a solution in F; the latter point is covered by this lemma:

3.3 LEMMA *The relation* $^{m,n}R \subset M(m,n,F) \times F^m$ *defined*

$$^{m,n}R(A,b) \equiv (\exists x\in F^n)(Ax=b)$$

is fapCS-computable over F *uniformly in* n,m.

PROOF. First consider the rank function $^{m,n}r:M(m,n,F) \to \omega$ defined $^{m,n}r(A)$ = rank of matrix A. This is fapCS-computable over F uniformly in m,n: calculate $r_0 \in \omega$ such that A has at least one non-singular $r_0 \times r_0$ minor and, for $s > r_0$, every $s \times s$ minor of A is singular. Then ${}^{m,n}r(A) = r_0$ and ${}^{m,n}r$ is fapCS-computable (by Local Search 4.4 and the fact that determinants are polynomials over the prime subfield of F!)

Now the lemma follows from a well known theorem of Linear Algebra: given $A \in M(m,n,F)$ and $b \in F^m$, let $[A,b]$ be A augmented by b as an (n+1)-th column. Then ${}^{m,n}R(A,b)$ iff ${}^{m,n}r(A) = {}^{m,n+1}r([A,b])$. Q.E.D.

What is striking about Hermann's bound is that it holds for *all* fields. To properly exploit the constructive implications of this uniformity one must dispense with traditional hypotheses, such as that F be computable, designed to make the problem ${}^{n,k}M$ constructively well posed, but irrelevant to its algorithmic solution which is field-theoretic in the fullest, *abstract*, sense of the term. And it is with precisely this sort of example in mind that we advocate the study of effective computability in a completely abstract algebraic setting.

3.4 INVARIANCE THEOREM *Let* A *and* B *be relational systems and* $\phi: A \to B$ *a relational homomorphism which is injective. Let* α *be a fapCS over their signature. Then*

$$\alpha(\phi a,x) \simeq \phi\alpha(a,x) \quad \text{if } \alpha: A^n \times \omega^m \to A;$$
$$\alpha(\phi a,x) \simeq \alpha(a,x) \quad \text{if } \alpha: A^n \times \omega^m \to \omega.$$

PROOF. This is best proved via a stronger, but technical, fact about state descriptions of $\alpha(a,x)$ and $\alpha(\phi a,x)$: if $D_i(\alpha,a,x) = (m_i,a_{ij},x_{ij},(z_j,a^i_{jk}))$ then $D_i(\alpha,\phi a,x) = (m_i,\phi a_{ij},x_{ij},(z_j,\phi a^i_{jk}))$. The argument is by induction on i. The basis i=1 is true by convention. Assume the identity true at step i and compare how $D_{i+1}(\alpha,a,x)$ and $D_{i+1}(\alpha,\phi a,x)$ arise from $D_i(\alpha,a,x)$ and $D_i(\alpha,\phi a,x)$, respectively; this depends on the 15 types of instruction numbered by m_i at stage i.

For example, let m_i name <u>if</u> $R(r_{\lambda_1},\ldots,r_{\lambda_k})$ <u>then</u> u <u>else</u> v. In both cases the new state descriptions differ from those at stage i only in their instruction numbers say m_{i+1} and n_{i+1}. Thanks to the induction hypothesis, the transitions are determined by these formulae

$$m_{i+1} = \begin{cases} u \text{ if } R(a_{i\lambda_1},\ldots,a_{i\lambda_k}); \\ v \text{ otherwise;} \end{cases} \qquad n_{i+1} = \begin{cases} u \text{ if } R(\phi a_{i\lambda_1},\ldots,\phi a_{i\lambda_k}); \\ v \text{ otherwise.} \end{cases}$$

Since ϕ is a relational homomorphism $m_{i+1} = n_{i+1}$.

We leave to the reader the task of checking the other cases noting only that it is the algebraic conditional with equality which requires ϕ to be a monomorphism. Q.E.D.

3.5 COROLLARY *Each fapCS-semicomputable set* $S \subset A^n$ *is invariant under the action of the automorphism group* Aut(A) *of* A.

PROOF. Let α be a fapCS such that $S = \text{dom}(\alpha)$ and let $\phi \in \text{Aut}(A)$. For $a \in A^n$,

$$\begin{aligned}\phi(a) \in S &\Leftrightarrow \alpha(\phi a)\downarrow \\ &\Leftrightarrow \phi\alpha(a)\downarrow \quad \text{by Theorem 3.4;} \\ &\Leftrightarrow \alpha(a)\downarrow \quad \text{since } \phi \text{ is total.}\end{aligned}$$

Thus $\phi(S) = S$. Q.E.D.

For example, fapCS-semicomputable subgroups must be normal. And since complex conjugation $z \mapsto \bar{z}$ is an automorphism of the field $\mathbb{C}$, the set $\{i\}$ where $i = \sqrt{-1}$ is not fapCS-semicomputable.

3.6 COROLLARY *The group of all fapCS-computable automorphisms* $\mathrm{Aut}_{fapCS}(A)$ *is a subgroup of the centre of* $\mathrm{Aut}(A)$. *Thus* $\mathrm{Aut}_{fapCS}(A) \lhd \mathrm{Aut}(A)$.

If G is a group with trivial centre then Aut(G) has trivial centre, see Kurosh [16] p. 89. Therefore, *if group* G *has trivial centre, as is the case when* G *is non-abelian and simple or when* G *is a free group of rank* > 1, *then* $\mathrm{Aut}_{fapCS}(G)$ *is trivial.* On the other hand, if G is a torsion-free divisible abelian group then Aut(G) contains the infinite family $\exp_n(g) = g^n$ for $0 \neq n \in \omega$.

4. LOCAL FAPCS-ENUMERATION, SEARCH AND PAIRING

In fapCS-computation, algebraic enumerations assume a strictly local character: *given* $a_1,\ldots,a_n \in A$ one can always fapCS-computably list the subalgebra $\langle a_1,\ldots,a_n\rangle$. And from this comes *local* fapCS-computable search and *local* fapCS-computable pairing. Here we state results which formulate these mechanisms precisely, leaving their proofs to the reader as straightforward, though instructive, exercises in manipulating the term evaluation functions by means of recursive calculations on ω. The theorems are quite fundamental, however. They strongly suggest that most results in Degree Theory and Axiomatic Complexity Theory on ω can be reproved *in a local form* for fapCS-computation, and that, in particular, Classical Recursion Theory can be reconstructed for FAPCS(A) provided A is finitely generated *by elements named as constants in its signature.*

(A study of the regularities and singularities involved in lifting the principal facts of Degree Theory and Axiomatic Complexity Theory has been undertaken by V. Stoltenberg-Hansen and myself and will appear in due course.)

4.1 LOCAL ENUMERATION THEOREM *There is a family of functions* ${}^{n}L: A^n \times \omega \to A$ *fapCS-computable uniformly in* n *such that for each* $a \in A^n$, ${}^{n}L(a): \omega \to \langle a\rangle$ *is a surjection; moreover* ${}^{n}L$ *can be so chosen as to make* ${}^{n}L(a)$ *a bijection* $\omega \to \langle a\rangle$, *if* $\langle a\rangle$ *is infinite, or a bijection* $\{0,\ldots,m-1\} \to \langle a\rangle$, if $\langle a\rangle$ *is finite of order* m.

Two corollaries of this enumeration facility which we use later on are

4.2 LEMMA *The order function* ${}^{n}\mathrm{ord}: A^n \to \omega$ *defined* ${}^{n}\mathrm{ord}(a) = |\langle a\rangle|$ *is fapCS-computable uniformly in* n.

4.3 LEMMA *The membership relation for finitely generated subalgebras of* A, *defined* ${}^{n}M(b,a) \equiv b \in \langle a\rangle$ *for* $b \in A$ *and* $a \in A^n$, *is fapCS-semicomputable uniformly in* n.

4.4 LOCAL SEARCH THEOREM *There is a family of functions* ${}^{n,m,k}v : C \times A^n \times \omega^m \to A^k$ *fapCS-computable uniformly in* n,m,k *such that if* $S \subset A^n \times \omega^m \times A^k$ *is fapCS-semicomputable by fapCS with code* $e \in C$ *and there is* $y \in \langle a\rangle^k$ *such that* S(a,x,y) *then* ${}^{n,m,k}v(e,a,x)\downarrow$ *and* $S(a,x,{}^{n,m,k}v(e,a,x))$.

4.5 LOCAL PAIRING THEOREM *There is a family of functions* ${}^{n,m}L_* : A^n \times \omega \to A^m$ *fapCS-computable uniformly in* n,m *such that for each* $a \in A^n$, ${}^{n,m}L_*(a) : \omega \to \langle a\rangle^m$ *is a surjection; moreover* ${}^{n,m}L_*$ *can be so chosen as to make a listing of* $\langle a\rangle^m$ *without repetitions.*

5. COUNTING AND STACKING: VARIETIES AND LOCAL FINITENESS

While the four kinds of fap-computable functions on algebra A are distinct (Theorem 2.4), the typical situation in Algebra is

$$FAP(A) \hookrightarrow FAPS(A) \hookrightarrow FAPC(A) = FAPCS(A)$$

Defining algebra A to be *regular* if A has uniform fapC-computable term evaluation, with Theorems 2.1 and 2.4 in mind, we substantiate this claim with a sufficient condition for A to be regular based on the proofs that groups and rings are regular. Notice regularity is an isomorphism invariant (Theorem 3.4).

First, we introduce *term width* which is used to measure the number of algebra registers required to fapC-evaluate a term on any input; c.f. Friedman [7], pp. 376-377.

An n-*ary syntactic development of width* m is a sequence $T_1,\dots,T_\ell$ such that

(i) each T_1 is a list of m terms from $T[X_1,\dots,X_n]$ $1 \le i \le \ell$;

(ii) T_1 contains only indeterminates

(iii) for $1 \le i \le \ell-1$, either T_{i+1} arises from T_i by applying some k-ary operation symbol σ to some of the terms $t_1,\dots,t_k$ in T_i and replacing one of the terms of T_i by $\sigma(t_1,\dots,t_k)$ or,

(iv) T_{i+1} arises from T_i by replacing one of the terms of T_i by an indeterminate or by another of its terms;

(v) T_{i+1} differs from T_i in at most one of its terms $1 \le i \le \ell$.

A term $t(X_1,\dots,X_n)$ is of *width at most* m if it belongs to some n-ary syntactic development of width m. Given a code for a term of width m and any $a \in A^n$ one can recursively calculate a code for some development to construct its values at each stage in m algebraic registers and output its value t(a).

Next, we adapt the notion of a *varietal normal form* as in H. Lausch & W. Nöbauer [19], p. 23.

A *recursive and complete set of term representatives of width* m(n) for an algebra A is a family $J = \{J_n : n \in \omega\}$ with $J_n \subset T[X_1,\dots,X_n]$ such that

(i) the terms of J_n are of width at most m(n);

(ii) for each $t \in T[X_1,\dots,X_n]$ one can recursively calculate $t' \in J_n$ for which $t \equiv_A t'$;

(iii) m is recursive.

5.1 THEOREM *Let* A *have a recursive and complete set of term representatives of width bounded by* m(n). *Let* M *be the largest arity of the operations of* A. *Then there is a fapC involving* (M+1). m(n) *algebraic work registers which computes term evaluation. In particular,* A *and* all its homomorphic images *are regular.*

The proof of this is routine: use Friedman's Lemma 1.6.2 and straightforward properties of homomorphisms. Theorem 5.1 is designed with those algebras which make up categories possessing free objects of all finite ranks in mind; for example, varieties and quasivarieties (among the first-order axiomatisable classes). Not only is Algebra replete with such categories but, invariably, the sets of *normal forms* constructed for the syntactic copies of their free objects are recursive and are of bounded width: from Lausch & Nöbauer [19] and P. Hall [11] we can "read off"

5.2 THEOREM *Let* A *be an algebra belonging to one of these varieties: semigroups, groups, associative rings with or without unity, Lie rings, semilattices, lattices, boolean algebras. Then* FAPC(A) = FAPCS(A).

Of course, the hypothesis of recursive sets of representatives for a variety is much weaker than that of recursive normal forms as the latter entails the free algebras of a variety have a decidable word problem. Notice, too, that fields do *not* form a variety, but recursive and complete sets of term representatives are easily made from the construction of rational functions, $\mathbb{Q}(X_1,\ldots,X_n)$ or $\mathbb{Z}_p(X_1,\ldots,X_n)$, from polynomials, $\mathbb{Z}[X_1,\ldots,X_n]$ or $\mathbb{Z}_p[X_1,\ldots,X_n]$.

Keeping within the regular algebras A, we now digress to collapse FAPCS(A) to FAP(A) by counting inside A.

An algebra A is said to have *fap counting on an element* $a \in A$ if there are functions $S,P:A \to A$ such that (i) $\{S^n(a):n\in\omega\}$ is infinite, (ii) $PS^n(a) = S^{n-1}(a)$ for each n and (iii) S,P are fap computable over (A,a).

5.3 THEOREM *Let* A *be regular. Then* A *has fap-counting on an element* $a \in A$ *if, and only if,* A *is not locally* 1*-finite and* FAP(A,a) = FAPCS(A,a).

PROOF. Assuming A to have fap-counting on $a \in A$, the Locality Lemma 3.1 shows A is not locally 1-finite. To simulate a fapC over (A,a) by a fap one rewrites α replacing each counting variable c in α by a new algebra variable u and then replacing each instruction of the left hand column by the corresponding fap instructions abbreviated in the right hand column:

c: = 0	u: = a
c: = c'+1	u: = S(u')
c: = c'$\dot{-}$1	u: = P(u')
<u>if</u> c=c' <u>then</u> * <u>else</u> **	<u>if</u> u=u' <u>then</u> * <u>else</u> **

The converse implication is an exercise in the use of the Local Enumeration Theorem 4.1. Q.E.D.

5.4 EXAMPLE *A group* G *has fap-counting on an element if, and only if,* G *is not periodic.* Since periodicity and local 1-finiteness coincide in groups one implication follows from Theorem 5.3. Conversely, if G is non-periodic with $g \in G$ of finite order then define $S(x) = gx$ and $P(x) = g^{-1}x$ which are fap-computable over (G,g).

It is not enough to assume G is not locally finite to obtain counting on an element: take any of Golod's groups [9] which are periodic but not locally finite. This is not the case for fields however:

5.5 EXAMPLE *A field* F *has fap counting on an element if, and only if,* F *is of characteristic* 0 *or is of prime characteristic but not algebraic over its prime subfield.* Let F have fap-counting on $a \in F$ and assume F has characteristic p; denote the prime subfield of F by $\mathbb{Z}_p$. The subfield $\langle a \rangle = \mathbb{Z}_p(a)$ is infinite so cannot be algebraic over $\mathbb{Z}_p$ as otherwise $\mathbb{Z}_p(a)$ would have order p^n where n is the degree of a over $\mathbb{Z}_p$. The converse is obvious in case F has characteristic 0 and if $t \in F$ is transcendental over $\mathbb{Z}_p$ then F as a multiplicative group is not periodic since t is of infinite order and we can use Example 5.4. It is easy to check F is locally finite if, and only if, it has prime characteristic and is algebraic over its prime subfield.

We now turn to the study of exclusively algebraic computation: the class FAPS(A). The reader might care to keep in mind the equivalences of fapS-computability with inductive definability, equational definability and recursive procedures. For example, the facts which follow have a bearing on the debate about inductive definability as *the* generalisation of Recursion Theory to an abstract setting, see J.E. Fenstad [5,6].

5.6 THEOREM *If* A *is locally* n*-finite then the halting problem for* ^{n}FAPS(A) *is fapCS-decidable: the relation* ${}^nH(e,a)$ *iff* $\{e\}(a)\downarrow$ *is fapCS-computable on* $\Omega_S \times A^n$.

PROOF. Let R(e), I(e), and M(e) recursively calculate the number of algebraic register variables, instructions, and markers for stack blocks appearing in the fapS coded by e. A state description in a fapS computation is a list of the form

$$(k;a_1,\ldots,a_m;\ (z_1;a_{11},\ldots,a_{1m}),\ldots,(z_s;a_{s1},\ldots,a_{sm}))$$

and if this belongs to a computation by e then $k \le I(e)$, $m = R(e)$ and each $z_i \le M(e)$.

If in unfolding a computation $\{e\}(a)$ *there are two identical ordinary algebraic state descriptions or two identical stack state descriptions then* $\{e\}(a)\uparrow$. This is clear for if either of two identical states D_i, D_j arise then the program must regenerate after D_j exactly those states intermediate between D_i and D_j to produce an infinite, but periodic, set of state descriptions. When A is locally n-finite we can bound the number of distinct state descriptions in a convergent computation $\{e\}(a)$:

There are at most $B_0(e,a) = (I(e)+1)({}^n\mathrm{ord}(a)+1)^{R(e)}$ different states for the ordinary algebraic registers and instructions. Due to the fact the stack mechanism copies ordinary algebraic state descriptions, it is easy to calculate that there can be at most

$$B_S(e,a) = \sum_{s=0}^{B_0(e,a)} \frac{(M(e).B_0(e,a))!}{(M(e).B_0(e,a)-s)!}$$

distinct stack states. Therefore, the total number of state descriptions available for $\{e\}(a)\downarrow$, and so the computation's run-time, is fapCS-computably bounded by (the total function) $B(e,a) = B_0(e,a).B_S(e,a)$.

Claim: For $e \in \Omega_S$, $a \in A^n$, ${}^nH(e,a) \Longleftrightarrow {}^nSTEP(e,a,B(e,a))$. Obviously, if ${}^nSTEP(e,a,B(e,a))$ is true then $\{e\}(a)\downarrow$. Conversely, if $\{e\}(a)\downarrow$ in ℓ steps and $\ell > B(e,a)$ then some duplication of state descriptions must have appeared and so $\{e\}(a)\uparrow$. Thus $\ell \leq B(e,a)$. Q.E.D

5.7 COROLLARY *If* A *is locally finite then the halting problem for fapS computations is fapCS-decidable.*

An algebra A is said to be *fapCS formally valued* when there is a total function $v:A \to \omega$ which is a fapCS-computable surjection.

5.8 PROPOSITION *Let* A *be* 1-*finite and fapCS formally valued by* v. *Then the set* $K_v = \{a \in A: v(a) \in \Omega_S \;\&\; \{v(a)\}(a)\downarrow\}$ *is fapCS-computable but not fapS-computable.*

PROOF. A 1-finite implies ${}^1H(e,a)$ is fapCS-decidable on $\Omega_S \times A$ by Theorem 5.6. Thus, K_v is fapCS-decidable as $a \in K_v \Longleftrightarrow v(a) \in \Omega_S \;\&\; {}^1H(v(a),a) = 0$, and v is total. Assume for a contradiction that K_v is fapS-computable. Then $\neg K_v$ is fapS-semi-computable and there exists a fapS α such that $\mathrm{dom}(\alpha) = \neg K_v$. Choose $b \in A$ so that $v(b)$ codes α.

$$\begin{aligned}\text{Then } b \in \neg K_v &\Longleftrightarrow \alpha(b)\\ &\Longleftrightarrow \{v(b)\}(b)\downarrow\\ &\Longleftrightarrow b \in K_v.\end{aligned}$$

Thus there is no such α and K_v is not fapS-computable. Q.E.D.

For illustrations we look for valuations of regular algebras A which are 1-finite where $FAPS(A) \subsetneq FAPC(A) = FAPCS(A)$.

If A contains 1-generator subalgebras of every finite order then ${}^1\mathrm{ord}:A \to \omega$ is a fapCS formal valuation *provided* A is 1-finite. For example, take the locally finite group $\mathbb{Z}_\infty$ consisting of all the complex roots of unity. More generally, if algebra A is 1-finite and $\pi(A) = \mathrm{im}({}^1\mathrm{ord})$ contains a recursive set S then choose a recursive bijection $f:S \to \omega$ and define

$$v(a) = \begin{cases} f({}^1\mathrm{ord}(a)) & \text{if } {}^1\mathrm{ord}(a) \in S,\\ {}^1\mathrm{ord}(a) & \text{otherwise.}\end{cases}$$

For an example among groups we must look for a periodic group of infinite exponent. Let p be a prime and let $\mathbb{Z}p^\infty$ be the locally finite abelian group of all complex roots of unity which are of order some power of p: for these groups $\pi(\mathbb{Z}p^\infty) = \{p^n : n \in \omega\}$, a recursive set.

Turning to fields, let F be a locally finite field of characteristic p and observe that ${}^1\mathrm{ord}(a) = p^{d(a)}$ where $d:F \to \omega$

calculates the degree of $a \in F$, that is $d(a) = \dim[\mathbb{Z}_p(a): \mathbb{Z}_p]$. If F contains elements of every degree belonging to a recursive set then d is a fapCS formal valuation. Such an F can be chosen by taking the splitting field of the polynomials $\{X^{p^n}-X \in \mathbb{Z}_p[X]:n\in\omega\}$ but the best example is the algebraic closure K of $\mathbb{Z}_p$. In summary,

5.9 THEOREM *Over the groups* $A = \mathbb{Z}_\infty$, $\mathbb{Z}p^\infty$ *and the field* $A = K$,
$$\text{FAPS}(A) \subsetneq \text{FAPC}(A) = \text{FAPCS}(A).$$

In conclusion, we mention in connection with Corollary 5.7, that fapS computability and fapS semicomputability may actually coincide.

An algebra A is said to be *uniformly locally finite, ulf* for short, if there is a function $\lambda: \omega \to \omega$ such that for any $a_1,\ldots,a_n \in A$, ${}^n\text{ord}(a_1,\ldots,a_n) < \lambda(n)$.

If A is a ulf algebra then we can replace B(e,a) in the argument of Theorem 5.6 by a function B'(e,n). One consequence of this is that for *fixed* e, ${}^nH(e,a)$ is fapS-decidable:

5.10 THEOREM *Let* A *be uniformly locally finite. Then* $S \subset A^n$ *is fapS-semicomputable if, and only if, S is fapS-computable.*

For groups, *Kostrikin's Theorem* implies any locally finite group of prime exponent is uniformly locally finite. More generally, calling a class of algebras k uniformly locally finite if there is $\lambda:\omega \to \omega$ such that for $A \in k$ and $a_1,\ldots,a_n \in A$, ${}^n\text{ord}(a_1,\ldots,a_n) < \lambda(n)$, it is a corollary of the *Ryll-Nardzewski Theorem* that if k is an ω-categorical first-order axiomatisable class then k is a uniformly locally finite class. See also *Mal'cev's Theorem* [21] p. 285 that a variety consisting of locally finite algebras is uniformly locally finite.

6. TOPOLOGICAL ALGEBRAS

A is a *topological algebra* if its domain is a non-trivial topological space on which its operations are continuous.

6.1 THEOREM *Let* A *be a Hausdorff topological algebra in the quasivariety* V. *If* A *contains a* V*-free* n*-generator subalgebra then for any fapCS-decidable relation* $S \subset A^n \times \omega^m$ *and* $x \in \omega^m$ *the sets* $\{a \in A^n: S(a,x)\}$ *and* $\{a \in A^n: \neg S(a,x)\}$ *cannot both be dense in* A^n.

This fact we obtain as a useful, palatable corollary of a more general, technical result about topological relational systems. Both theorems are suggested by remarks of Herman and Isard, in [13], concerning $\mathbb{R}$; we use them on fields, abelian groups and differential rings.

A relation $R \subset A^n$ is *continuous* at $a \in A^n$ if its characteristic function $R:A^n \to \{0,1\}$ is continuous at a between the product topology on A and the discrete topology on {0,1}.

Recalling the congruence relation $\equiv_A$ on $T[X_1,\ldots,X_n]$ from Section 1, a point $a \in A^n$ is said to be *transcendental* if for any terms $t,t' \in T[X_1,\ldots,X_n]$

$t \equiv_A t'$ if, and only if, $t(a) = t'(a)$ in A.

6.2 THEOREM *Let A be a relational structure which is a Hausdorff topological algebra, and let* $S \subset A^n \times \omega^m$ *be a fapCS-decidable relation. If A contains a transcendental point* $a \in A^n$ *on whose subalgebra* $\langle a \rangle$ *the basic relations of A are continuous then for any* $x \in \omega^m$ *there is an open subset of* A^n *containing a upon which S holds or fails accordingly as it holds or fails on a.*

To deduce Theorem 6.1, let V be a quasivariety and let $T_V[X_1,\ldots,X_n] = T[X_1,\ldots,X_n]/\equiv_V$ the V-free polynomial algebra of rank n. If $A \in V$ and $a \in A^n$ V-freely generates the V-free algebra $\langle a \rangle$ then $v_a: T_V[X_1,\ldots,X_n] \to A$, defined $v_a[t] = t(a)$, is an embedding: $t(a) = t'(a)$ in A iff $[t] = [t']$ in $T_V[X_1,\ldots,X_n]$. But $t \equiv_V t'$ implies $t \equiv_A t'$ and so a is transcendental. With this hypothesis of Theorem 6.2 satisfied, the remainder of the deduction of 6.1 is straightforward.

The Locality Lemma 3.1 prompts us to make this definition. A *syntactic state description* is a list of the form

$$(k, t_1(X), \ldots, t_p(X), x_1, \ldots, x_q; (z_1, t_{11}(X), \ldots, t_{1p}(X)), \ldots, (z_s, t_{s1}(X), \ldots, t_{sp}(X)))$$

where k and the z_i are instruction labels, $x_1, \ldots, x_q \in \omega$, and what remains are terms in $X = (X_1, \ldots, X_n)$. When we unfold a computation $\alpha(a,x)$ syntactically we obtain the i-th syntactic state description $T_i(\alpha,a,x) = (m_i, t_{ij}, x_{ij}, (z_{ij}, t^i_{jk}))$ which is a mapping A^n to state descriptions.

PROOF OF THEOREM 6.2 Let α fapCS-decide S over A. Let $a \in A^n$ be transcendental and $x \in \omega^m$ be arbitrarily chosen and fixed. Since S is fapCS-decidable iff $\neg$ S is a fapCS-decidable, we assume, without loss of generality, that S(a,x) is true and consider a computation $\alpha(a,x)$ of this fact. Let $\alpha(a,x)$ have length ℓ and syntactic state descriptions $T_i(\alpha,a,x)$ for $1 \le i \le \ell$ which we take as functions,

$$T_i(X) = (m_i, t_{ij}(X), x_{ij}, (z_{ij}, t^i_{jk}(X))),$$

of $X = (X_1, \ldots, X_n)$ only. For $b \in A^n$ we denote the i-th state of the computation $\alpha(b,x)$ by

$$D_i(\alpha,b,x) = (n_i, b_{ij}, y_{ij}, (w_{ij}, b^i_{jk})).$$

We prove there is a basic open set B(a) containing a such that for all $b \in B(a)$ and each $1 \le i \le \ell$, $D_i(\alpha,b,x) = T_i(b)$. Obviously, this entails $\alpha(b,x) = \alpha(a,x)$ for $b \in B(a)$.

Claim. For each $1 \le i \le \ell$ there is a basic open set $B_i(a)$ containing a so that for $b \in B_i(a)$, $D_i(\alpha,b,x) = T_i(b)$.

On proving the claim by induction on i, we may take

$$B(a) = \cap_{i=1}^{\ell} B_i(a).$$

The basis i=1 is true for $B_1(a) = A^n$ by convention. Assume the claim true at stage i of the computation $\alpha(a,x)$: $B_i(a)$ is constructed and for any chosen $b \in B_i(a)$, $D_i(\alpha,b,x) = T_i(b)$. Consider the passage from $D_i(\alpha,b,x)$ to $D_{i+1}(\alpha,b,x)$ which depends on the nature of the instruction n_i. We give just 3 of the 15 cases:

Let $n_i = m_i$ be $r_\mu := \sigma(r_{\lambda 1}, \ldots, r_{\lambda k})$. In applying this instruction only algebra registers are changed, the transitions of $D_i(\alpha,b,x)$ to $D_{i+1}(\alpha,b,x)$, and $T_i(X)$ to $T_{i+1}(X)$, being determined by the formulae,

$$b_{i+1,j} = \begin{cases} \sigma(b_{i\lambda_1},\ldots,b_{i\lambda_k}) & \text{if } j = \mu \\ b_{ij} & \text{if } j \neq \mu \end{cases}$$

$$t_{i+1,j}(X) = \begin{cases} \sigma(t_{1\lambda_1}(X),\ldots,t_{i\lambda_k}(X)) & \text{if } j = \mu \\ t_{ij}(X) & \text{if } j \neq \mu. \end{cases}$$

Substituting $X = b$ and applying the induction hypothesis $b_{i\lambda_j} = t_{i\lambda_j}(b)$ we get $D_{i+1}(\alpha,b,x) = T_{i+1}(b)$ for *any* $b \in B_i(a)$. So we can set $B_{i+1}(a) = B_i(a)$.

The other operational instructions also make state transitions "independently" of a, but conditional instructions do not:

Let $n_i = m_i$ be <u>if</u> $R(r_{\lambda_1},\ldots,r_{\lambda_k})$ <u>then</u> u <u>else</u> v. By the induction hypothesis $b_{ij} = t_{ij}(b)$ so the state transitions are determined by

$$n_{i+1} = \begin{cases} u \text{ if } R(t_{i\lambda_1}(b),\ldots,t_{i\lambda_k}(b)) \\ v \text{ otherwise.} \end{cases}$$

$$m_{i+1} = \begin{cases} u \text{ if } R(t_{i\lambda_1}(a),\ldots,t_{i\lambda_k}(a)) \\ v \text{ otherwise.} \end{cases}$$

By the continuity hypothesis on relations, the map $R \circ p: A^n \to \{0,1\}$ is continuous at a where $p(X) = (t_{i\lambda_1}(X),\ldots,t_{i\lambda_k}(X))$. So there is a basic open set $V_i(a)$ containing a such that for any $b \in V_i(a) \cap B_i(a)$, $R \circ p(b) = R \circ p(a)$ and $n_{i+1} = m_{i+1}$. Thus take $B_{i+1}(a) = V_i(a) \cap B_i(a)$.

Let $n_i = m_i$ be <u>if</u> $r_\mu = r_\lambda$ <u>then</u> u <u>else</u> v. Again since $b_{ij} = t_{ij}(b)$ we have

$$n_{i+1} = \begin{cases} u \text{ if } t_{i\mu}(b) = t_{i\lambda}(b) \\ v \text{ otherwise.} \end{cases}$$

$$m_{i+1} = \begin{cases} u \text{ if } t_{i\mu}(a) = t_{i\lambda}(a) \\ v \text{ otherwise.} \end{cases}$$

Consider the two cases for m_{i+1}. If $t_{i\mu}(a) = t_{i\lambda}(a)$ then, since a is transcendental, $t_{i\mu} \equiv_A t_{i\lambda}$ and $t_{i\mu}(b) = t_{i\lambda}(b)$ for any $b \in A^n$, and the passage of $T_i(X)$ to $T_{i+1}(X)$ is independent of a. Thus, set $B_{i+1}(a) = B_i(a)$ on which $n_{i+1} = m_{i+1}$.

However, if $t_{i\mu}(a) \neq t_{i\lambda}(a)$ then $\{b \in A^n: t_{i\mu}(b) \neq t_{i\lambda}(b)\}$ is an open set containing a - because A is Hausdorff - and we can choose a neighbourhood $V_i(a)$ about a within it. For $b \in V_i(a) \cap B_i(a)$, $n_{i+1} = m_{i+1}$ so here we set $B_{i+1}(a) = V_i(a) \cap B_i(a)$. Q.E.D.

FIELDS In the field of reals $\mathbb{R}$ any transcendental number is a transcendental element in our special sense. Therefore, by Theorem 6.2, sets such as the rationals $\mathbb{Q}$ and the algebraic numbers A which are dense and codense in $\mathbb{R}$ cannot be fapCS-decidable. These examples, cited for fap-decidability, are in Herman & Isard [13].

ABELIAN GROUPS Let T^n be the n dimensional torus group, the n-fold direct product of the circle group S^1.

6.3 EXAMPLE *The set $FO(T^n)$ of all elements of T^n of finite order, also known as the torsion subgroup of T^n, is fap-semicomputable but not fapCS-computable.*

PROOF It is known from Section 1 that $FO(T^n)$ is fap-semicomputable. To apply Theorem 6.1 we have to show T^n contains a 1-generator free-abelian subgroup and that $FO(T^n)$ is dense and codense in T^n. We begin by examining S^1.

Let $f:[0,1) \to S^1$ be the continuous parameterisation $f(t) = (\sin 2\pi t, \cos 2\pi t)$. It is easy to check $f(t)$ is of finite order iff t is a rational number. Thus S^1 has a free-abelian subgroup and, indeed, since the rationals are dense and codense $[0,1)$, $FO(S^1)$ is dense and codense in S^1 because the image of a dense subset of a space is dense in the image of a continuous map. Now observe that for any group G, $FO(G)^n = FO(G^n)$ and $IO(G)^n = IO(G^n)$ where $IO(G) = \neg FO(G)$, and also that in any topological space X, D dense in X entails D^n dense in X^n. Applying these facts to $FO(S^1)$ and $IO(S^1)$ we are done. Q.E.D.

6.4 EXAMPLE *The 1-generator subgroup membership relation M in T^n is fap-semicomputable but not fapCS-computable.*

PROOF. First we show $M = \{(g,t): g \in \langle t\rangle\}$ is dense in $T^n \times T^n$. Let $B(a,b)$ be a basic open set containing $(a,b) \in T^n \times T^n$ which we can take, without loss of generality, to be of the form $B_1(a) \times B_2(b)$ where $B_1(a)$, $B_2(b)$ are basic open sets about a,b respectively. Now the set of those $t \in T^n$ such that $\langle t\rangle$ is dense in T^n is itself dense in T^n, see J.F. Adams [1], p. 79. So we can choose $t \in B_2(a)$ so that $\langle t\rangle$ meets all neighbourhoods and in particular $B_1(b)$. This means there is $(g,t) \in B_1(a) \times B_2(b)$ such that $g \in \langle t\rangle$ and M is dense.

Consider $\neg M$. Let $B(a,b)$ be as before. From the argument of Example 6.3, we can choose an element $t \in B_2(b)$ of finite order and $g \in B_1(a)$ of infinite order and so a pair $(g,t) \in B(a,b)$ for which $g \notin \langle t\rangle$.

From the observations of 6.3, it is easy to show T^n contains a 2-generator free-abelian subgroup and complete the argument with Theorem 6.1. Q.E.D.

INTEGRATION Let $C^{\omega}(\mathbb{R},\mathbb{R})$ be the set of all analytic functions $\mathbb{R}\to\mathbb{R}$ which is a differential ring under pointwise addition and multiplication, and differentiation. Let E be the differential subring generated by e^x, $\sin x$, the polynomial functions $\mathbb{R}[X]$, and all their compositions, a subring of the so-called elementary functions. Let $I(f)$ be the integration relation in E, $I(f) \equiv (\exists g \in E)(Dg = f)$. For example, it is well known from a theorem of Liouville that $e^{-x^2} \notin I$, see G.H. Hardy's book [12].

6.5 EXAMPLE I *is a fapCS-semicomputable subset of E which is not* *fapCS-computable.*

PROOF. Equip $C^\omega(\mathbb{R},\mathbb{R})$ with the C^∞-topology prescribed thus: a sequence $f_n \to 0$, as $n \to 0$, iff for each k, the sequence $D^k f_n \to 0$, as $n \to 0$, in the topology of uniform convergence on compact subsets on $C^\omega(\mathbb{R},\mathbb{R})$; this means that for each k and each real $R > 0$, the sequence $\sup_{|x|<R}|D^k f_n(x)| \to 0$ in $\mathbb{R}$. With the C^∞-topology $C^\omega(\mathbb{R},\mathbb{R})$ is a topological differential ring (which is not the case with the usual topology of uniform convergence on compacts where D fails to be continuous). See M. Golubitsky & V. Guillemin [10], pp. 42-50. Consider the hypotheses of 6.2. By induction on term height it is straightforward to prove that (say) e^x is a transcendental point in E: this is omitted. That I is dense in E in the C^∞-topology follows from the fact that the polynomials $\mathbb{R}[X] < I < E$ and that the sequence of Taylor polynomials of an analytic function converge to the function in the topology of uniform convergence. That I is codense is more involved.

Let $f \in I$, we shall approximate f by the sequence of non-integrable functions $f_n(x) = f(x) + 1/n\ e^{-x^2/n}$. It is easy to see that the f_n are not integrable and that the approximation property follows from the claim that $1/n\ e^{-x^2/n} \to 0$ as $n \to \infty$ in the C^∞-topology.

On calculating the k-th derivative $D^k(e^{-x^2/n})$ we find it to be of the form $P_k(x,1/n)e^{x^2/n}$, where $P_k(x,1/n) = \Sigma_{i=1}^{n} a_i X^i/n^{f_i(k)}$ where $f_i(k) \geq [k/2]$ = largest natural number $\geq k/2$. Whence it is easy to check that $D^k(e^{-x^2/n}) \to 0$ in the topology of uniform convergence on compact sets. Q.E.D.

Two papers the reader might care to study, in the light of these notes, are A. Kreczmar [15] and E. Engeler [4], though these deal with a class of functions slightly smaller than FAP(A).

REFERENCES

1. J.F. ADAMS, *Lectures on Lie groups*, W.A. Benjamin, New York, 1969.

2. J.W. de BAKKER, *Mathematical theory of program correctness*, Prentice Hall International, London, 1980.

3. J.A. BERGSTRA, J. TIURYN & J.V. TUCKER, 'Correctness theories and program equivalence', *Mathematical Centre, Computer Science Department Research Report*, IW 119, Amsterdam, 1979.

4. E. ENGELER, 'Generalized galois theory and its application to complexity'. *ETH-Zürich, Computer Science Institute Report* 24, Zürich, 1978.

5. J.E. FENSTAD, 'On the foundation of general recursion theory: computations versus inductive definability' pp. 99-111 of J.E. FENSTAD, R.O. GANDY & G.E. SACKS (eds.) *Generalized recursion theory II*, North Holland, Amsterdam, 1978.

6. _______ , *Recursion theory: an axiomatic approach*, Springer-Verlag, Berlin, 1980.

7. H. FRIEDMAN, 'Algorithmic procedures, generalised Turing algorithms, and elementary recursion theory', pp. 316-389 of R.O. GANDY & C.E.M. YATES (eds.), *Logic Colloquim, '69*, North-Holland, Amsterdam, 1971.

8. J.A. GOGUEN, J.W. THATCHER & E.G. WAGNER, 'An initial algebra approach to the specification, correctness and implementation of abstract data types' pp. 80-149 of R.T. YEH (ed.) *Current trends in programming methodology IV. Data structuring*, Prentice-Hall, Engelwood Cliffs, New Jersey, 1978.

9. E.S. GOLOD, 'On nil-algebras and residually finite p-groups'. *American Math. Soc. Translations*, (2) 48 (1965) 103-106.

10. M. GOLUBITSKY and V. GUILLEMIN, *Stable mappings and their singularieies*, Springer-Verlag, New York, 1973.

11. P. HALL, 'Some word problems', *J. London Math. Soc.*, 33 (1958) 482-496.

12. G.H. HARDY, *The integration of functions of a single variable*, Second edition, Cambridge University Press, London, 1916.

13. G.T. HERMAN & S.D. ISARD, 'Computability over arbitrary fields', *J. London Math. Soc.* 2 (1970) 73-79.

14. G. HERMANN, 'Die Frage der endlich vielen Schritte in der Theorie der Polynomideale', *Mathematische Annalen* 95 (1926) 736-788.

15. A. KRECZMAR, 'Programmability in fields', *Fundamenta Informaticae* 1 (1977) 195-230.

16. A.G. KUROSH, *The theory of groups I*, Chelsea, New York, 1955.

17. __________, *The theory of groups II*, Chelsea, New York, 1956.

18. __________, *General algebra*, Chelsea, New York, 1963.

19. H. LAUSCH and W. NÖBAUER, *Algebra of polynomials*, North-Holland, Amsterdam, 1973.

20. M. MACHTEY & P. YOUNG, *An introduction to the general theory of algorithms*, North-Holland, New York, 1978.

21. A.I. MAL'CEV, *Algebraic systems*, Springer-Verlag, Berlin, 1973.

22. J. MOLDESTAD, V. STOLTENBERG-HANSEN & J.V. TUCKER, 'Finite algorithmic procedures and inductive definability', to appear in *Mathematica Scandinavica*.

23. ______ , 'Finite algorithmic procedures and computation theories' to appear in *Mathematica Scandinavica*.

24. J. MOLDESTAD & J.V. TUCKER, 'On the classification of computable functions in an abstract setting', in

preparation.

25. B.L. VAN DER WAERDEN, *Algebra I*, Ungar, New York, 1970.

Applications of Classical Recursion Theory to Computer Science

Carl H. Smith

Department of Computer Sciences
Purdue University
West Lafayette, Indiana 47907

Introduction

In introducing his definitive work on recursion theory, Rogers (1967) remarks

> "... our emphasis will be *extensional* , in that we shall be more concerned with objects *named* (functions) than with objects *serving as names* (algorithms)."

Computer scientists are interested in algorithms; their existence, expression, relative efficiency, comprehensibility, accuracy, and structure. Many computer scientists are more interested in algorithms themselves, rather than what is computed by them. In the sequel will emphasize and exploit the intensional aspects of recursion theory. Our intention is to substantiate the claim that the formalisms and techniques of recursive function theory can be applied to obtain results of interest in computer science. In what follows we will survey results which yield insights into the nature of programming techniques, complexity of programs, and the inductive inference of programs given examples of their intended input-output behavior. The results presented below are from the field of *intensional recursion theory* in that they make assertions concerning algorithms and their proofs use the recursion theoretic techniques of diagonalization and recursion. Furthermore, in many instances, these techniques are applied intensionally in that they are used to specify an algorithm which manipulates other algorithms syntactically without necessarily any knowledge of the function specified by the manipulated algorithms. We include proofs only to illustrate the intensionality of their techniques.

First we introduce the necessary concepts and notation of recursive function theory from the perspective of a computer scientist. We start with an enumeration of the closure under concatenation of some arbitrary, but fixed, finite alphabet. An element of the enumeration is called a *file*. Files contain text which will be interpreted as either a *program* or as *data*. The ambiguity of the representation of programs and data is a fundamental feature of nearly every modern digital computer. φ_e denotes the function computed by the program described by the eth file. If every partial recursive function is computed by at least one of the programs in the enumeration then $\varphi_0, \varphi_1, \varphi_2, \ldots$ is a *programming system* (Machtey & Young, 1978). Furthermore, a programming system is *acceptable* (Machtey & Young, 1978) if and only if (*i*) there is a universal program u such that for all i and x, $\varphi_u(i,x)=\varphi_i(x)$ and (*ii*) there is a composition program c such that for all i, j and x $\varphi_{\varphi_c(i,j)}(x)=\varphi_i(\varphi_j(x))$. Note that program u effectively interprets its datum i as a program and program c effectively manipulates programs i and j as data to produce a third program. The ability to manipulate a program as data and to interpret a datum as a program in an acceptable programming system is not merely a ramification of the ambiguity of the representations. Friedberg (1953) exhibited a programming system with programs associated one-to-one with natural numbers and without a composition program. Computers exploit the ambiguity of program and data representations during the compilation of a program written in high level language into machine code. Stored program computers seem to rely not only on the indistinguishability of program and data but also on the ability to effectively exploit that ambiguity.

In the last paragraph we saw an axiomatization of acceptable programming systems in terms of simulation and composition. It is hardly surprising that these techniques are vital to every computing system. Most users interface with a computer via an operating system which provides facilities run programs with given input (simulation). Several operating systems offer the capability to compose programs as well. For example, the UNIX operating system has a pipe operator specifically designed to facilitate program composition (Ritchie & Thompson, 1978). Together, the hardware and software of a computing system form an acceptable programming system. Other characterizations of acceptable programming systems in terms of common program writing tools is the subject of the next section.

Programming Tools

The features shared by acceptable programming systems and actual computers include the ability to effectively interpret data as program and manipulate programs as data. The universal machine axiom of acceptable programming systems is a straightforward and precise statement about interpreting data as programs. The composition axiom serves only as an example of a programming tool which is sufficiently powerful to guarantee the unrestricted ability to manipulate programs as data. Each acceptable programming system has a program to compute the s_1^1 function (Hamlet, 1974). Hence, acceptable programming systems are, more traditionally, acceptable numberings of all and only the partial recursive functions (Rogers, 1958). Since a composition program is obtainable from indices for the universal and s_1^1 functions (Rogers, 1967) the capability to effectively store arbitrary constants in programs is a sufficiently general programming technique to guarantee the unrestricted ability to manipulate programs as data. In fact, weaker constraints suffice. For any i, $store_i$ is a program such that for all e and x, $\varphi_{store_i(e)}(x)=\varphi_e(i,x)$. Royer (1981) has shown that any programming system with a universal program and $store_i$ and $store_j$ programs with $i \neq j$ is acceptable.

Recursive programming techniques have become popular. Wirth (1976) notes that "Recursion ... is an important and powerful concept in programming." There are several recursion theorems that hold in any acceptable programming system which generalize the above mentioned programming technique. Rogers (1967) maintains that these theorems constitute "a fundamental tool in the theory." Below we discuss the following two such theorems and their relationships to each other and to acceptable programming systems.

Kleene Recursion Theorem *(1938). For any program i there exists a program e, which can be found effectively from i, such that for any x,* $\varphi_e(x)=\varphi_i(e,x)$.

The intuitive content of Kleene's original recursion theorem is that we may write programs using a copy of the completed program as an implicit parameter in any effective calculation we choose. In practice, recursive programs in procedural programming languages almost always invoke themselves with arguments smaller than those of the original call. By the recursion theorem we may, in principle, write recursive programs which invoke themselves on arguments larger than those of the original call; which invoke effective distortions of themselves on a variety of arguments; which measure their own complexity; and which perform a myriad of transformations on their own description. In practice, the use of recursion, as in the recursion theorem, amounts to deciding on the name of the file which will contain the code for the recursive program to be written. Then, when writing the program, the chosen file name may be used in conjunction with simulation and edit commands to implement any use of recursion alluded to above.

Fixed Point Theorem *(Rogers, 1967). For any recursive function f there exists a program e, which can be found effectively from a program for f, such that* $\varphi_e=\varphi_{f(e)}$.

This form of recursion tells us that for any algorithmic transformation on programs, there is a program which is transformed into one exhibiting the same behavior.

The two recursion theorems stated above are interderivable in any acceptable programming system. However, the proof of the fixed point form of the recursion theorem requires the invocation of the universal function. Kleene's proof uses only the composition and s_1^1 functions: let v be a program to compute $\lambda xy[\varphi_i(s_1^1(x,x),y)]$, then the desired e is given by $s_1^1(v,v)$. In the sequel we state and interpret several results with extensional content and intensional proofs which serve to further distinguish the two stated recursion theorems.

A programming system satisfies the padding theorem if given any program it is possible to effectively enumerate infinitely many distinct programs, each of which computes the same function as the given program. Riccardi (1980) has shown that any programming system which satisfies the enumeration, padding, and Kleene recursion theorems is acceptable. His proof exploits the one-to-oneness of the algorithm which produces the appropriate self referential program. Royer (1981) shows that the proof necessarily exploits the one-to-oneness property. A programming system which satisfies the enumeration, padding, the fixed point theorems but which is not acceptable is constructed in (Machtey, Winklmann & Young, 1978). The fixed point function exhibited in their proof is also one-to-one. These results indicate the superiority of Kleene recursion over fixed fixed point recursion in a very strong sense. Here we have programming techniques serving as the building blocks of an acceptable programming system in such a way that if one form of recursion is replaced by another then the guarantee of acceptability dissolves.

There are results which indicate a sense in which fixed point recursion constitutes a more potent programming tool then Kleene recursion. However, the evidence is not as strong. Any programming system satisfying the composition theorem also satisfies the Kleene recursion theorem. The following theorem, obtained in collaboration with Machtey (private communication), is a sharpening, in the effectiveness of the exhibited composition function, of a result of (Riccardi, 1980) which generalizes, from sub recursive to acceptable programming systems, a result noticed by (Alton, 1976; Kozen, 1980; Lewis 1973). The programming system constructed below would be enhanced by the addition of fixed point recursion, but not by the addition of Kleene recursion.

Theorem. *There is a programming system satisfying the effective composition theorem but not the effective fixed point theorem. Furthermore, a program for the composition function can be effectively located in the programming system.*

Proof: Suppose p is a program in the acceptable programming system $<\varphi_i>$ for the function $\lambda x,y[4(x,y)+2]$. Let $<\psi_i>$ be the r.e. sequence of functions given by:

i) $\psi_{4i}=\varphi_i$;

ii) ψ_{4i+1} = Either $\lambda x[0]$, $\lambda x[1]$, or $\lambda x[2]$, whichever differs from both ψ_{4i} and ψ_{4i+2};

iii) $\psi_{4i+2}=\psi_j\circ\psi_k$, where i is the pairing of j and k;

iv) ψ_{4i+3} = Either $\lambda x[0]$, $\lambda x[1]$, or $\lambda x[2]$, whichever differs from both ψ_{4i+2} and ψ_{4i+4}.

By i) $<\psi_i>$ is a programming system. Furthermore, ψ_{4p} is its composition function. The constant p can easily be found by judicious choice of initial programming system and some simple pairing and coding programs. The function $\lambda x[x+1]$ has no fixed point (by ii) and iv)). Furthermore, using p, a few other simple programs, and an algorithm of (Machtey, Winklmann & Young, 1978), it is also possible to effectively locate a program for s_1^1 in the programming system $<\psi_i>$.

■

We conclude this section with a brief discussion of multiple recursion theorems. Smullyan (1961) proved a double analogue of the fixed point theorem for r.e. relations. Riccardi (1980) proved that the double analogue of the Kleene recursion theorem for partial recursive functions with the enumeration theorem implies the s-m-n theorem. Hence, acceptable programming systems can also be characterized as programming systems with enumeration and double Kleene recursion theorems. It is interesting that a form of mutual recursion can be construed as a potent program manipulating tool. Case (1974) proved an infinitary analogue of the Kleene recursion theorem which we state below and use in subsequent sections.

Theorem. *(Case, 1974). Suppose Θ is an effective operator Then one can effectively find a recursive one- to- one monotone increasing function p such that for all i and x, $\varphi_{p(i)}(x)=\Theta(p)(i,x)$.*

Intuitively, the operator recursion theorem allows one to construct a sequence of programs, p(0), p(1), ..., each of which can use its own index within p and descriptions of other programs in the range of p as implicit additional parameters. Computationally, this amounts to each program in the sequence knowing its position in the sequence and how to generate the entire sequence.

Complexity

The recursion theoretic study of the complexity of computations Began with Blum's (1967) abstraction of resource consumption by programs. In the sequel we need only an acceptable programming system and the following definition. A list of functions $<\Phi_i>$ is a *complexity measure* for the acceptable programming system $<\varphi_i>$ iff

a) for any program i, domain (Φ_i) = domain (φ_i) and
b) the set $\{(i, x, y) | \Phi_i(x) = y\}$ is recursive.

Intuitively $\Phi_i(x)$ may be thought of as the running time (or the amount of any computational resource used in the execution) of program i on input x. As an example we present a result asserting that there are arbitrarily expensive to compute any function. This result originally appeared in (Blum, 1967). The generalization below is from (McCreight, 1969).

Theorem. *For any program i and any recursive function h there is a program e which computes φ_i and for all x, $\Phi_e(x)>h(x)$.*

Proof: By implicit use of the recursion theorem there exists (effectively in i and a program for h) a program e such that

$$\varphi_e(x) = \begin{cases} \varphi_i(x), & \text{if} \Phi_e(x) > h(x); \\ undefined, & otherwise. \end{cases}$$

For any x, if $\Phi_e(x) \leq h(x)$, then $\varphi_e(x)$ is undefined, a contradiction. Hence, $\varphi_e = \varphi_i$, and Φ_e bounds h.

■

The use of recursion above is intensional. Program e uses self referencing only to compare its complexity with the given function h. If program e is equipped with a "clock" (or, more generally, a resource meter) then the clock can be interrogated at intervals and its valued compared with the given bound h. In this fashion the construction above can be carried out without using simulation. Abstract complexity theory is a well studied and well surveyed area. For more information see (Hartmanis & Hopcroft, 1971; Brainerd & Landweber, 1974; Machtey & Young, 1978).

More recently computer scientists have been less abstract in their study of the complexity of computations. Attention has been focused on algorithms whose complexity is bounded by polynomial in the input argument. Also of interest are nondeterministic algorithms which terminate in a polynomial amount of time (Garey & Johnson, 1979). A fundamental technique used contemporary complexity theory is the notion of reducibility due to Post (1944). There are also notions of completeness.

Inductive Inference

In this section we survey some results concerning the inductive inference of programs given examples of their intended input-output behavior. The recursion theoretic approach to the problem of inductive inference, as formalized by L. and M. Blum (1975). constitutes a continuation of three distinct lines of research. Many of the fundamental definitions and concepts are taken from Gold's work in grammatical inference (Gold, 1967). Inductive inference can be viewed as a problem of synthesising Turing machines. Barzdin formally investigated the synthesis of automata (Trakhtenbrot & Barzdin, 1973). Philosophers of science, most notably Carnap (1952) have investigated the process by which an empirical scientist examines some experimental data and conjectures an hypothesis intended to explain the data and to predict the outcome of future experiments. Philosophical implications of the recent recursion theoretic work on inductive inference, including a mechanistic repudiation of the principle expounded by Popper (1968) that every scientific explanation ought to be refutable, can be found in (Case & Smith, 1978), the source of of the material presented below.

An *inductive inference machine* (abbreviated: IIM) is an algorithmic device with no *a priori* bounds on how much time or memory resource it shall use, which takes as its input the graph of a function from **N** into **N** an ordered pair at a time (in any order), and which from time to time, as it's receiving its input, outputs computer programs.

We will introduce several notions of what it means for an IIM to *succeed* at eventually finding an explanation for a function. The first is essentially from (Gold, 1967), but see also (Blum & Blum, 1975). We say M EX *identifies* a recursive function f (written: f∈EX(M)) iff M, when fed the graph of f in any order, outputs over time but finitely many computer programs the last of which computes (or *explains*) f. No restriction is made that we should be able to algorithmically determine when (if ever) M on f has output its last computer program.

An IIM M is said to be *order independent* iff for any function f, the corresponding sequence of programs output by M, is independent of the order in which f is input. Clearly, any IIM M can be effectively transformed into an IIM M' which preprocesses any recursive function f and feeds it to M in the order (0,f(0)), (1,f(1)), (2,f(2)), ... An order independence result that covers the case of partial functions appears in (Blum & Blum, 1975). In what follows we shall suppose without loss of generality that all IIMs are order independent.

We define the class of sets of functions EX = { **S** |(there exists an IIM M)[**S** ⊆ EX(M)]}. EX is the collection of all sets **S** of recursive functions such that some IIM EX identifies every function in **S**. For example, (Gold, 1967) showed that {f | f is primitive recursive}∈ EX. As noted in (Blum & Blum, 1975), Gold's proof can be easily modified to show that any recursively enumerable class of recursive functions is EX identifiable.

The motivation for our first generalization of Gold's result stems from the observation that sometimes scientists will use an explanation of a phenomenon which has an anomaly in it, an explanation which fails to correctly predict the outcome of one experiment but which is correct on all other experiments. We say M EX^1 *identifies* a recursive function f (written: f∈EX^1(M)) iff M, when fed the graph of f in any order, outputs a last computer program which computes f except perhaps at one anomalous input. For recursive functions f and g, "f is a singleton variant of g" is written $f=^1g$. We analogously define the class EX^1 = { **S** |(there exists an IIM M)[**S** ⊆ EX^1(M)}.

Putnam (1963) showed that there is no general purpose robot scientist in the sense that a naturally restricted subclass of EX does not contain all the recursive functions. Gold (1967) showed that no single inductive inference machine can EX identify every recursive function. {f | $\varphi_{f(0)}=^1f$} ∈(EX^1-EX) indicating that if the goal set of mechanized scientists is relaxed to allow a possible single anomaly in explanations, then, in general, they can identify strictly larger classes of recursive functions than those that are error intolerant.

There are two possible kinds of single anomalies in an explanatory program. The first kind occurs when the program on some one input actually gives an output which is incorrect. This kind of single anomaly eventually can be found out, refuted, and patched. The second kind occurs when the program on some one input fails to give any output at all; the explanation is incomplete. This latter kind of anomaly, in general, cannot be algorithmically found out; the explanation is not, in general, (algorithmically) refutable. If we define $EX^{=1}$ *identification* just as we defined EX^1 identification but we replace "except perhaps at one anomalous input" by "except at exactly one anomalous input", we have that $EX^{=1}$ = EX. This is because exactly one anomaly (of either kind) can be patched in the limit: Patch in the correct output for input 0 until (if ever it is discovered that the output was already correct on input 0, then patch in the correct output for input 1 until ... . Eventually the patch will come to rest on the single anomaly which needed patching. It follows that the strength of EX^1 identification must come from two sources: Possibly incomplete explanations and our inability to test algorithmically for incompleteness. The proof that $EX \subset EX^1$ reflects this last observation. L. and M. Blum (1975) proved that the union of two EX identifiable sets of functions is not necessarily EX identifiable. Their result can be obtained as a simple corollary of the EX and EX^1 separation result.

The remainder of this section will cover several more general separation results. For any natural number n, define EX^n *identification* and the class EX^n analogously with EX^1 identification and the class EX^1. The notation $f =^n g$ means that the recursive function f is an n-variant recursive f the function g, e.g. $\{x | f(x) \neq g(x)\}$ has at most n members. Then, for any natural number n $\{f \mid \varphi_{f(0)} =^{n+1} f\} \in (EX^{n+1} - EX^n)$. Hence, the more tolerant a learning procedure is of errors (anomalies) in it's output the better the chances, in general, of success.

Allowing a finite but unbounded number of anomalies in a final explanation constitutes an inference criterion more general than any discussed above. We say M EX^* *identifies* a recursive function f (written: $f \in EX^*(M)$) iff M, when fed any enumeration of the graph of f, outputs but finitely many programs, the last of which computes f except perhaps on finitely many anomalous inputs. The class EX^* is defined in the usual way. For functions f and g, "f is a finite variant of g" will be written $f =^* g$. EX^* identification coincides with *almost everywhere identification* introduced in (Blum & Blum, 1975) and *subidentification* in (Minicozzi, 1967). A sharpening of a result mentioned in (Blum & Blum, 1975) is that $\{f \mid \varphi_{f(0)} =^* f\} \in (EX^* - \bigcup_{x \in N} EX^n)$.

Hence, the EX classes form a hierarchy of more and more general inference criteria. Notice that for any n the set $\{f | \varphi_{f(0)} =^n f\}$ can be EX^n identified by an IIM which, when fed the graph of f, outputs f(0) as its only conjecture, i.e. can be EX^n identified by an IIM which outputs a single conjecture and never later changes its mind. This last observation leeds to the following definitions. a, b, c and d will denote members of $N \cup \{*\}$. For any a and b we say M EX^a_b *identifies* f (written: $f \in EX^a_b(M)$) iff M EX^a identifies f after no more than b mind changes (no restriction when b=*). The class $EX^a_b = \{ S | (\text{there exists an } M)[S \in EX^a_b(M)]\}$. Observe that for any a, EX^a_* identification coincides with EX^a identification above. By convention, any natural number is <*. Then $EX^a_b \subseteq EX^c_d$ iff a≦c and b≦d. Hence, learning procedures can *not*, in general, infer more accurate solutions by making more attempts.

Next we introduce a notion of inference without convergence to a fixed explanation. We say an IIM M BC *identifies* a recursive function f (written: $f \in BC(M)$) iff M, when fed the graph of f (in any order) outputs over time an infinite sequence of computer programs all but finitely many of which compute f. The class BC={ S |(there is an IIM M)[S ⊆ BC(M)]. The class BC is defined in the usual manner. Barzdin (1974) acting on an observation of Feldman (1972) independently defined a notion, referred to as GN^{∞} in the Russian literature, which coincides with our BC. John Steel (private communication) first observed that $EX^{*} \subseteq$ BC. That the inclusion is proper is a result from (Case & Smith, 1978) obtained in collaboration with Leo Harrington (private communication). Barzdin (1974) proved that EX ⊆ BC. His proof actually shows that $EX^{*} \subset$ BC. However, Barzdin's construction makes no use of recursion theorems and is consequently longer. Also, Barzdin's proof is not as easy to modify so as to prove more general results concerning inference by finite collections of IIMs.

Theorem. *Let* S *be* $\{f \mid$ *for all but finitely many* $x,\ \varphi_{f(x)}=f\}$. *Then* S ∈*(BC -* EX^{*}*).*

Proof: Clearly S ∈ BC. Let M be given. For σ an initial segment of some function, M(σ) denotes M's last output after seeing σ as input. We suppose without loss of generality that for all σ, M(σ) is defined. It remains to exhibit a function f which∈ (S -EX^{*}(M)). By implicit use of the operator recursion theorem (Case, 1974), we obtain a repetition free r.e. sequence of programs p(0), p(1), p(2), ... such that one of these programs computes such an f. We proceed to give an informal effective construction of the $\varphi_{p(i)}$'s in successive stages s≥0. p(1) is just a program for $\varphi_{p(0)}$ which differs from p(0). $\varphi^{s}_{p(i)}$ denotes the finite initial segment of $\varphi_{p(i)}$ defined before stage s. For each i, $\varphi^{0}_{p(i)} = \phi$. $q^{s} = M(\varphi^{s}_{p(0)})$.

Begin stage s. Simultaneously execute the following three substages *until* (if ever) either suitable x and σ are found in substage (i) or a mind change is found in substage (ii).

(i) Dovetail a search for x and σ such that $\varphi^{s}_{p(0)} \subset \sigma$, *range* $(\sigma - \varphi^{s}_{p(0)}) \subset \{p(0), p(1)\}$, x is a member of *domain* $(\sigma - \varphi^{s}_{p(0)})$, and program q^{s} converges on x to a value $\neq \sigma(x)$.

(ii) See if there is a τ such that $\varphi^{s}_{p(0)} \subset \tau \subseteq$ what has been put into $\varphi_{p(s+2)}$ so far *and* $M(\tau) \neq q^{s}$. (Before stage s, $\varphi^{s}_{p(s+2)}$ is made $=\varphi^{s}_{p(0)}$.)

(iii) Make $\varphi_{p(s+2)}$ have value p(s+2) at more and more successive arguments not yet in its domain.

Condition (1). x and σ are found in substage (i) *before* a mind change is found in substage (ii). Set $\varphi^{s+1}_{p(0)}=\sigma$, do not extend $\varphi_{p(s+2)}$ any further, and set $\varphi^{s+1}_{p(s+3)}=\varphi^{s+1}_{p(0)}$.

Condition (2). A mind change is found in substage (ii) before or at the same time as suitable x and σ are found in substage (i). Set $\varphi^{s+1}_{p(0)}$= what has been put into $\varphi_{p(s+2)}$ so far, make program p(s+2) from this point on simulate program p(0) on all inputs not yet in its domain so that $\varphi_{p(s+2)}$ will be the same as $\varphi_{p(0)}$, and set $\varphi^{s+1}_{p(s+3)}=\varphi^{s+1}_{p(0)}$

End stage s.

Case (1). Some stage s never terminates. Then by substage (iii), $\varphi_{p(s+2)}$ is a (total) recursive function and for all but finitely many x, $\varphi_{p(s+2)}(x)$ = p(s+2). Set f = $\varphi_{p(s+2)}$. Clearly f ∈ S. Program q^{s} is M's last output on input f; furthermore, q^s never converges on any x not in *domain* $(\varphi^s_{p(0)})$ since, if it did, it could not converge to both p(0) and p(1) and so substage (i) *would* find suitable x and σ. It follows that φ_{q^s} is a finite function and hence not a finite variant of f. Therefore, $f \in ($ S $-EX^*(M))$.

Case (2). Not Case (1). Then $\varphi_{p(0)}$ is a (total) recursive function and *everything* in its range is a program for $\varphi_{p(0)}$: By Condition (1) p(0)'s and p(1)'s are introduced into its range and these compute $\varphi_{p(0)}$; by Condition (2) p(s+2)'s are introduced into its range, but *then* p(s+2) also computes $\varphi_{p(0)}$. Set $f=\varphi_{p(0)}$. Clearly, then, $f \in$ S. Suppose M on f outputs a last program q. Then Condition (1) holds at all but finitely many stages s. Hence, infinitely often, f is defined to differ from φ_q. Therefore, $f \in ($ S $-EX^*(M))$.

■

From the above theorem we may conclude that evolutionary learning procedures are potentially more powerful than ones which attempt to arrive at a fixed explanation. Also, if we allow our learning procedures to accumulate larger and larger size explanations, then they can, in general, infer larger classes of phenomenon. Of particular interest in the above theorem is the highly intensional use of recursion in the construction of the $\varphi_{p(i)}$'s. The range of p was used as the range of the $\varphi_{p(i)}$'s. Previously determined initial segments of $\varphi_{p(0)}$ are fed into an IIM and the resulting program is simulated. None of the programs in the range of p is simulated directly.

Summary

We presented three example applications of recursion theory to computer science. The first application concerned the comparisons of various programming techniques. Briefly mentioned was the use of recursion in abstract complexity theory. The last application was from the relatively new area of inductive inference. The above list of applications was not intended to be exhaustive, but rather illustrative of the use of the intensional aspects of recursion theory in theoretical computer science.

Acknowledgements

We gratefully acknowledge many conversations with J. Case, M. Machtey, and G. Riccardi. Financial support came from NSF grant MCS 7903912. Computer time was provided by the Department of Computer Sciences, Purdue University. The referee made some useful comments.

References

. Alton, D., "Natural complexity measures, subrecursive languages and speed-ups," Computer Science Department Technical Report 76-05, University of Iowa (1976).

. Barzdin, J., "Two theorems on the limiting synthesis of functions," pp. 82-88 in *Theory of Algorithims and Programs*, ed. Barzdin,Latvian State University, Riga, U.S.S.R. (1974). 210

. Blum, L. and Blum, M., "Toward a mathematical theory of inductive inference," *Information and Control* **28** pp. 125-155 (1975).

. Blum, M., "A machine-independent theory of the complexity of recursive functions," *JACM* **14**(1967).

. Brainerd, W. and Landweber, L., *Theory of Computation*, John Wiley & Sons, New York (1974).

. Carnap, R., *The Continuum of Inductive Methods*, The University of Chicago Press, Chicago, Illinois (1952).

. Case, J., "Periodicity in generations of automata," *Mathematical Systems Theory* **8** pp. 15-32 (1974).

. Case, J. and Smith, C., "Anomaly hierarchies of mechanized inductive inference," *Proceedings of the 10th Symposium on the Theory of Computing*, pp. 314-319 (1978).

. Feldman, J., "Some decidability results on grammatical inference of best programs," *Math Systems Theory* **10** pp. 244-262 (1972).

. Friedberg, R., "Three theorems on recursive enumeration," *Journal of Symbolic Logic* **23** pp. 309-316 (1953).

. Garey, M. and Johnson, D., *Computers and Intractability: a guide to NP- completness*, W. H. Freeman & Co., San Francisco, Calif. (1979).

. Gold, E.M., "Language identificatin in the limit," *Information and Control* **10** pp. 447-474 (1967).

. Hamlet, R., *Introduction to Computation Theory*, Intext Educational Publishers, New York (1974).

. Hartmanis, J. and Hopcroft, J., "An overview of the theory of computational complexity," *JACM* **18** pp. 444-475 (1971).

. Kleene, S., "On notation for ordinal numbers," *Journal of Symbolic Logic* **3** pp. 150-155 (1938).

. Kozen, D., "Indexing of subrecursive classes," *Journal of Theoretical Computer Science*, (1980). To appear

. Lewis, F., "Algorithmic classes of functions and fixed points," Technical report 10-73, Aiken Computation Laboratory, Harvard University, Cambridge, Mass. (1973).

. Machtey, M., Winklmann, K., and Young, P., "Simple Gödel numberings, isomorphisms and programming properties," *SIAM Journal of Computing* **7** pp. 39-59 (1978).

. Machtey, M. and Young, P., *An Introduction to the General Theory of Algorithms*, Elsevier North-Holland, Inc., New York (1978).

. McCreight, E., "Classes of computable functions defined by bounds on computation," PhD dissertation, Carnegie Mellon University, Pittsburgh, Penn. (1969).

. Minicozzi, E., "Some natural properties of strong-identification in inductive inference," *Theoretical Computer Science* **2** pp. 345-360 (1976).

. Popper, K., *The Logic of Scientific Discovery*, Harper Torch Books, N.Y. (1968). 2nd Edition

. Post, E., "Recursively enumerable sets of positive integers and their decision problems," *Bulletin of the American Mathematical Society* **50** pp. 284-316 (1944).

. Putnam, H., "Probability and confirmation," in *Mathematics, Matter and Method*, Cambridge University Press (1975). originally appeared in 1963 as a Voice of America Lecture

. Riccardi, G., "The independence of control structures in abstract programming systems," Ph.D Dissertation, State University of New York, Buffalo, New York (1980).

. Ritchie, D. and Thompson, K., "The UNIX time-sharing system," *The Bell System Technical Journal* **57** pp. 1905-1930 (1978).

. Rogers, H. Jr., "Gödel numberings of partial recursive functions," *Journal of Symbolic Logic* **23** pp. 331-341 (1958).

. Rogers, H. Jr., *Theory of Recursive Functions and Effective Computability*, McGraw Hill, New York (1967).

. Royer, J., "Effective enumerations and control structures," Ph.D Dissertation, State University of New York, Buffalo, New York (1981).

. Smullyan, R., *Theory of Formal Systems, Annals of Mathematical Studies*, Princeton University Press, Princeton, New Jersey (1961).

. Trakhtenbrot, B. and Barzdin, J. A., *Finite Automata: Behavior and Synthesis*, North Holland (1973).

. Wirth, N., *Algorithms + Data Structures = Programs*, Prentice Hall, Englewood Cliffs, New Jersey (1976).

"NATURAL" PROGRAMMING LANGUAGES AND COMPLEXITY MEASURES FOR SUBRECURSIVE PROGRAMMING LANGUAGES: AN ABSTRACT APPROACH

Donald A. Alton

University of Iowa

1 INTRODUCTION

While the complexity of a program is sometimes gauged in terms of a single number such as the number of instructions or the maximum depth of nesting of loops, much attention focuses on amounts of computational resources which may vary for different inputs to a program, for instance execution time or the amount of memory which must be created by dynamic storage allocation. The abstract theory of computational complexity uses recursion theory to model some of the principal properties of such dynamic computational resources. Emphasis is on the machine-independent features which various dynamic resources share in common; for instance, the theory is not intended to allow detailed comparisons of execution times of Turing machines and random access machines. Only general properties of such resources are modeled; for instance, the resource requirements of sorting algorithms or matrix inversion algorithms are never singled out for special consideration.

The theory is based on the notion of M. Blum (1967a) of a measure of computational complexity (complexity measure), which is discussed in Section 3 below. Although Blum's axioms appear at first glance to be quite weak, the area has been the subject of numerous papers, and surprisingly strong consequences of the axioms have been obtained. The most celebrated result is Blum's speed-up theorem, which can be paraphrased as establishing the existence of a computable function which does not possess an optimal (in the dynamic sense of fastest on infinitely many inputs, not in the static sense of shortest) program.

The theory has two principal limitations:

(1) The underlying programming language with which a complexity measure is associated must correspond to an acceptable indexing of all partial recursive functions. If we restrict our attention to "universal" programming languages—those capable of computing all partial recursive functions—this is a restriction which many logicians and virtually all computer scientists will accept

happily. However, Section 2 below argues for the potential importance of a wide variety of subrecursive programming languages—languages which are only capable of computing some of the recursive functions.

(2) If we do restrict our attention to acceptable indexings of all the partial recursive functions, <u>too many</u> things satisfy Blum's axioms. Section 3 below gives examples of complexity measures which are pathological in the sense that they violate our intuitive notions concerning the relative efficiencies with which various tasks can be performed. Thus, for instance, a theorem which asserts that all complexity measures have a given property is worth reading, but a theorem which asserts that there exists a complexity measure having a given property is of dubious value (except possibly in documenting the existence of pathological complexity measures).

Constable & Borodin (1972) focused attention on problem 1. They posed the problem of developing an abstract theory of subrecursive complexity comparable to Blum's theory, and they also emphasized the need for a definition of "acceptable" subrecursive indexings analogous to the notion of an acceptable indexing of all partial recursive functions. Hartmanis (1973) highlighted the importance of problem 2 by asking for stronger axioms to characterize the "natural" complexity measures.

This paper surveys and refines proposals of the author (1977a, 1977b,1980) concerning ways to approach these problems. If one simply slavishly adapts Blum's axioms to a subrecursive indexing, the ridiculous "complexity measure" which claims that every program requires precisely 10 units of "execution time" on every input satisfies those axioms. Section 4 discusses reasonable axioms which avoid such pathological situations by forcing a complexity measure to include large values. The proofs which are presented in Section 4 allow the derivation of subrecursive analogs of results which are proved in Section 3 for the classical case of an acceptable indexing of all partial recursive functions and a complexity measure which satisfies Blum's axioms. <u>The proofs given in Sections 3 and 4 are trivial; the point of these sections is not to illustrate sophisticated proof techniques but, rather, to illustrate appropriate ways to model intuitive notions concerning the complexity of subrecursive programming languages.</u>

Section 5 discusses axioms which view complicated programs as being built up from simpler component programs and which require that the complexity of a complicated program should be related in a reasonable fashion to the complexities of the various relevant executions of the component programs; for instance, it is reasonable to assume that the execution time of a loop will not be much larger than the product of the number of passes through the loop times the maximum execution time of any pass through the loop.

This idea provides the central contrast with Blum's axioms, which treat each program as an atomic entity, unrelated to any of the other programs. The ideas of this section can be used to gauge the "naturalness" of a complexity measure, for a programming language which can be either universal or subrecursive, by the number of ways in which new programs can be synthesized from old programs in such a fashion that the computational resource requirements of the new programs are related in a "reasonable" fashion to the computational resource requirements of the old programs.

Section 6 focuses attention on a technical lemma which embodies, in a simpler setting, the key issue which is confronted in proofs of the speed-up theorem. It is shown that a straightforward proof technique which works for acceptable indexings of all the partial recursive functions cannot possibly work for reasonable subrecursive indexings and complexity measures, and then a proof technique is developed which does apply to a rich class of subrecursive indexings and complexity measures. This proof technique, which is significantly more subtle than the proofs given in Sections 3 and 4, appears to be the key to adaptation of a number of results of classical complexity theory to subrecursive settings, and an effort is made to illustrate the manner in which the proof can be developed gradually.

Section 8 attempts to assess the extent to which our effort to develop an abstract theory of subrecursive computational complexity is done at the right "level of abstractness." It also argues, based on the viewpoint advocated in Section 5, that instead of attempting to characterize _the_ acceptable subrecursive indexings and _the_ "natural" complexity measures, it is more appropriate to talk about _how_ acceptable a given subrecursive indexing is and _how_ "natural" a complexity measure is; we envision a whole spectrum of possibilities.

This expository paper is of course concerned with an interface between logic and computer science. The arguments which are given for the reasonableness of various additional axioms are motivated by attempts to model "real" programming languages and efficiency considerations, but it is hoped that the resulting axiomatic framework will be of interest to logicians and will lead to a theory which is richer in something like the same sense that the study of noetherian rings is richer than the study of arbitrary rings.

2 SUBRECURSIVE PROGRAMMING LANGUAGES

As noted earlier, a _subrecursive_ programming language is one whose programs are only capable of computing some of the recursive functions, whereas the programs of a _universal_ programming language are capable of computing all the partial recursive functions. (Of course these aren't the only two possibilities, but they include the most interesting cases.)

The mainstay of most subrecursive programming languages is the <u>for</u> loop. A typical example is

<u>for</u> $i = 1$ <u>to</u> n <u>do</u> S

which executes statement S for n times, with the value of i taking on the integer values $1, 2, \cdots, n$ on successive passes through the loop. Here S is a statement which may refer to the values of i and n as well as to other variables. It can change the values of other variables but cannot change the value of i or n. Zero passes through the loop are executed if $n \leq 0$. (Several aspects of the above description differ from the conventions of FORTRAN and certain other popular programming languages.) Such a loop has the crucial property that the number of passes through the loop is determined by the value of n <u>at loop entrance</u>. In recursion theoretic terms, one of the crucial features of <u>for</u> loops is their capability to perform primitive recursion. For instance, consider a primitive recursion of the form

$$f(0) = 0$$

$$f(z) = h(f(z-1), z-1) \quad \text{for } z > 0.$$

A <u>for</u> loop of the form

$w \leftarrow 0$
<u>for</u> $z = 1$ <u>to</u> n <u>do</u> $w \leftarrow h(w, z-1)$

ends with $f(n)$ as the final contents of variable w. (At the start of each pass through the loop, $w = f(z-1)$.) Here $w \leftarrow h(w, z-1)$ denotes an assignment statement which uses the current values of the variables w and z to compute $h(w, z-1)$ and then assigns that value as the new value of variable w.

In contrast, universal programming languages usually contain <u>while</u> loops. In recursion theoretic terms, one of the crucial features of <u>while</u> loops is their capability to perform unbounded minimization. For instance, an unbounded minimization of the form

$$\varphi(x) = \min w[g(x,w) = 0]$$

corresponds to a <u>while</u> loop of the form

$w \leftarrow 0$
<u>while</u> $g(x,w) \neq 0$ <u>do</u> $w \leftarrow w+1$.

At the start of a pass through the loop, $g(x,w)$ is evaluated. If it evaluates to 0, execution exits from the loop and w is the desired value of $\varphi(x)$. Otherwise the value of variable w is incremented and another pass through the loop is attempted. Execution of the loop halts eventually if and only if $\varphi(x)$ is defined.

Of course when a <u>while</u> loop is entered it may be far from clear how many passes through the loop will be executed, and some <u>while</u> loops will execute forever. In contrast, the programs in a subrecursive programming language must halt on all inputs.

There is substantial evidence that subrecursive programming languages can be useful, powerful, and flexible (e.g., Meyer & Ritchie, 1967a,b; Constable, 1971a; Constable & Borodin, 1972; Pagan, 1973; Verbeek, 1973; Coy, 1976; Goetze & Nehrlich, 1978). To be sure, the comforting knowledge that all programs will halt on all inputs comes at a price; no one reasonable subrecursive programming language has enough programs to compute all of the recursive functions, since any "reasonable" programming language should have a universal function which is partial recursive. (The partial recursiveness of the universal function corresponds in the terminology of computer scientists to the existence of an "interpreter" for the language. See Hamlet (1974). If the universal function is partial recursive and the programming language is subrecursive, of course the universal function will be recursive. Of course it is well known that no indexing $f_0, f_1, \cdots$ of the recursive functions possesses a recursive universal function $u(i,x) = f_i(x)$, since if u were recursive then $f(i) = u(i,i) + 1$ would be recursive.) However, in practice this is not as ominous as it sounds: Typical subrecursive languages are rich enough to allow all computations which one really <u>wants</u> to perform in practical computing. In particular, all computations which can be performed in typical universal languages <u>using reasonable amounts of computational resource</u> such as time or memory can also be performed in typical subrecursive languages. In addition, Constable & Borodin (1972) show that those computations which can be performed <u>at all</u> in typical subrecursive languages can be performed <u>almost as efficiently</u> there as they can be performed in typical universal languages.

Roughly speaking, Blum (1967b) shows that subrecursive programming languages lack the succinctness of description available in universal programming languages. Further results along these lines appear in Constable (1971b) and Meyer (1972). While this is unfortunate, it is also true that the length of a program is a poor gauge of the "intrinsic complexity" of a program, the difficulty the programmer encountered in coding the program, and the difficulty of establishing the correctness of the code.

Besides having the virtue that their programs always halt, subrecursive programming languages frequently have the property that one can produce an upper bound on their resource requirements such as execution time in terms of syntactic and structural information about the program such as length of the program and maximum depth of nesting of loops. Moreover, Constable & Borodin (1972) have exhibited subrecursive programming languages which contain quite sophisticated features such as conditionals, <u>for</u> loops, forward-branching <u>goto</u>'s (where the number of statements you are to branch forward can be determined at execution time via the current value of a variable), and function procedures.

3 COMPLEXITY MEASURES

Let $<,>$ denote a recursive pairing function. Exercise 2.10 of Rogers (1967) or consideration of Rogers (1958) justifies the

DEFINITION 1

<u>An acceptable indexing</u> φ of the partial recursive functions of one argument is a sequence $\varphi_0, \varphi_1, \cdots$ of partial recursive functions of one argument such that each partial recursive function of one argument φ equals φ_i for some i and such that

(i) the universal function $\psi(i,x) = \varphi_i(x)$ is partial recursive and

(ii) there exists a recursive function s such that $\varphi_{s(i,x)}(y) = \varphi_i(<x,y>)$ for all $i, x,$ and y.

We will think of an index i as (the encoding of) a "program" i which computes output $\varphi_i(x)$ on input x. The notation $\varphi_i(x)\downarrow$ means that $\varphi_i(x)$ is defined, i.e., that x is in the domain of φ_i.

The recursion theoretic approach to dynamic efficiency considerations is founded on

DEFINITION 2 (Blum, 1967a)

Let φ be an acceptable indexing $\varphi_0, \varphi_1, \cdots$ of the partial recursive functions of one argument. A sequence Φ of partial recursive functions of one argument $\Phi_0, \Phi_1, \cdots$ is a <u>measure of computational complexity</u> (or <u>complexity measure</u>) for φ if

(i) for all i and x, $\varphi_i(x)\downarrow \iff \Phi_i(x)\downarrow$ and

(ii) there is a recursive function M_Φ such that for all i, x, and r,

$$M_\Phi(i,x,r) = \begin{cases} 1 & \text{if } \Phi_i(x) \le r \\ 0 & \text{if } \Phi_i(x) \not\le r. \end{cases}$$

The function M_Φ is referred to as the <u>measure function</u> for

Φ, and the function Φ_i is referred to as the run time for "program" i.

We can think of $\Phi_i(x)$ as the amount of dynamic computational resource such as time or memory required by "program" i on input x; by (i), this amount of resource is finite precisely if the computation halts, and by (ii) one can determine whether any prescribed amount r of resource suffices. (To model memory usage, slightly awkward conventions must be used, since (i) requires that the amount of memory must be undefined (or "infinite") if a computation fails to halt. This conflicts with our intuitive notion that some computations fail to halt by repeatedly reusing the same finite number of memory locations. In practice, however, the convention that all nonhalting computations use infinitely much "memory" doesn't give any trouble. For instance, there is an algorithm which determines whether or not a (deterministic one tape) Turing machine will cycle forever without visiting more than r tape squares by simulating the computation long enough to determine either that it halts after visiting r or fewer tape squares, that it visits more than r tape squares, or that it repeats a configuration of tape contents, state, and head location without visiting more than r tape squares. In the later case, the computation must cycle forever without visiting more than r tape squares. This allows us to distinguish between "uses" of more than r tape squares which actually visit more than r tape squares and "uses" of more than r tape squares associated with nonhalting computations which visit r or fewer tape squares.)

Of course the T_1 predicate of p. 281 of Kleene (1950) provides a classical prototype for the notion of a complexity measure if we define

$$\Phi_i(x) = \min m\ T_1(i,x,m) = 1.$$

Upon first glance, Blum's axioms for the definition of a complexity measure may appear too weak to be capable of proving anything of interest. In fact, they are strong enough to prove some surprising results of extraordinary beauty, the most notable being Blum's speed-up theorem.

As a first indication that the axioms are strong enough to prove at least some results which aren't entirely trivial, they can be used to prove the following four lemmas, each of which suggests in some fashion that every complexity measure has the property that some of its run times are large.

LEMMA 1 (Blum, 1967a)

Let $\varphi_0,\varphi_1,\cdots$ be an acceptable indexing of the partial recursive functions of one argument, i.e., an indexing which satisfies the conditions of Definition 1. Suppose that $\Phi_i(x) = \varphi_i(x)$ for every i and every x. Then $\Phi_0, \Phi_1, \cdots$ is _not_ a complexity measure for $\varphi_0, \varphi_1, \cdots$.

The intuitive reason why the lemma _should_ be true for a reasonable notion of a complexity measure, and the reason the lemma can be interpreted as insisting that every complexity measure has some large run times, is that there should exist some programs which take a large amount of computational resource such as execution time to choose between two small outputs, say 0 and 1.

For the proof, assume that $\Phi_i = \varphi_i$ is a complexity measure with measure function M_Φ. Since M_Φ is recursive, there is a recursive function f such that for each i, $f(i) = 0 \Leftrightarrow M_\Phi(i,i,0) = 0$. Since f is recursive, there exists i such that $f = \varphi_i$. But then $f(i) = 0 \Leftrightarrow M_\Phi(i,i,0) = 0 \Leftrightarrow \Phi_i(i) \neq 0 \Leftrightarrow \varphi_i(i) \neq 0 \Leftrightarrow f(i) \neq 0$, a contradiction.

LEMMA 2 (Blum, 1967a)

Let $\varphi_0, \varphi_1, \cdots$ be an acceptable indexing of the partial recursive functions of one argument and let $\Phi_0, \Phi_1, \cdots$ be a complexity measure for $\varphi_0, \varphi_1, \cdots$. Let b be a recursive function. Then there exists an i_0 such that it is not the case that $\Phi_{i_0}(x) \leq b(x)$ for all x.

Lemma 2 is of course a trivial consequence of the existence of partial recursive functions which are not total and of Definition 2.(i); it is dignified by being called a lemma only for purposes of comparison in Section 4 below.

Lemmas 1 and 2 imply the independence of axioms (i) and (ii) of Definition 2.

We will appeal to the recursion theorem in a form patterned after the statement on p. 352 in Kleene (1950):

THEOREM 1

Let $\varphi_0, \varphi_1, \cdots$ be an acceptable indexing of the partial recursive functions of one argument and let φ be a partial recursive function of two arguments. Then there exists i_0 such that

$$\varphi(i_0, x) = \varphi_{i_0}(x) \quad \text{for all } x.$$

Rogers (1967) has popularized an alternative statement which guarantees that if f is a recursive function of one argument then there exists i_0 such that

$\varphi_{f(i_0)}(x) = \varphi_{i_0}(x)$ for all x.

It is easy to see that the two versions are equivalent for acceptable indexings of the partial recursive functions of one argument. Our reasons for using Kleene's version will become clear in Sections 4 and 6 below.

LEMMA 3 (McCreight, 1969)

Let $\varphi_0, \varphi_1, \cdots$ be an acceptable indexing of the partial recursive functions of one argument and let $\Phi_0, \Phi_1, \cdots$ be a complexity measure for $\varphi_0, \varphi_1, \cdots$. Let b and g be recursive functions. Then there exists an i_0 such that $\varphi_{i_0} = g$ and such that $\Phi_{i_0}(x) > b(x)$ for all x.

The proof of Lemma 3 is as follows:

Let M_Φ be the recursive measure function for Φ, satisfying Definition 2.(ii). Define

$$\varphi(i,x) = \begin{cases} \varphi_i(x)+1 & \text{if } M_\Phi(i,x,b(x)) = 1 \\ g(x) & \text{if } M_\Phi(i,x,b(x)) = 0. \end{cases}$$

Since M_Φ, b, and g are recursive and the universal function for $\varphi_0, \varphi_1, \cdots$ is partial recursive, clearly φ is partial recursive. By Theorem 1, there exists i_0 such that $\varphi_{i_0}(x) = \varphi(i_0,x)$ for all x. We claim that $\Phi_{i_0}(x) \not\leq b(x)$ for all x. To see this, suppose that $\Phi_{i_0}(x) \leq b(x)$ for some x. By Definition 2.(ii), $M_\Phi(i_0,x,b(x)) = 1$. Hence $\varphi_{i_0}(x) = \varphi(i_0,x) = \varphi_{i_0}(x)+1$. Since $\Phi_{i_0}(x) \leq b(x)$, in particular $\Phi_{i_0}(x)\downarrow$. By Definition 2.(i), $\varphi_{i_0}(x)\downarrow$. But this means that $\varphi_{i_0}(x) = \varphi_{i_0}(x)+1$ is not possible.

This contradiction proves that $\Phi_{i_0}(x) \not\leq b(x)$ for all x. Thus, $M_\Phi(i_0,x,b(x)) = 0$ for all x. Thus, $\varphi_{i_0}(x) = \varphi(i_0,x) = g(x)$ for all x and $\Phi_{i_0}(x) > b(x)$ for all x.

LEMMA 4

Let $\varphi_0, \varphi_1, \cdots$ be an acceptable indexing of the partial recursive functions of one argument and let $\Phi_0, \Phi_1, \cdots$ be a complexity measure for $\varphi_0, \varphi_1, \cdots$. There exists an i_0 such that

(i) $(\forall x)[\varphi_{i_0}(x)\downarrow]$ and

(ii) $(\forall x)[x > 0 \Rightarrow \varphi_{i_0}(x) > \Phi_{i_0}(x-1)]$.

Thus, program i_0 halts on all inputs and its <u>output</u> $\varphi_{i_0}(x)$ on input x is larger than its <u>run time</u> $\Phi_{i_0}(x-1)$ on the next smallest input $x-1$ whenever $x > 0$. Clearly this is a highly self-referential property for program i_0 to have, and it's not surprising that the recursion theorem can once again be exploited to yield a simple proof.

For the proof, let M_Φ be the measure function for the complexity measure Φ and define

$$\varphi(i,x) = \begin{cases} 0 & \text{if } x = 0 \\ 1 + \min m[M_\Phi(i,x-1,m) = 1] & \text{if } x > 0. \end{cases}$$

Since M_Φ is recursive, clearly φ is partial recursive. By Theorem 1, there exists i_0 such that $\varphi_{i_0}(x) = \varphi(i_0,x)$ for all x.

First we will prove (i) by induction on x. For the base step, clearly $\varphi_{i_0}(0) = \varphi(i_0,0) = 0$. For the inductive step, suppose that $x > 0$ and that $\varphi_{i_0}(x-1)\downarrow$. Then Definition 2.(i) implies that $\Phi_{i_0}(x-1)\downarrow$. Hence by Definition 2.(ii) there exists an m such that $M_\Phi(i_0,x-1,m) = 1$. Since M_Φ is recursive, it follows that $\min m[M_\Phi(i_0,x-1,m) = 1]$ will be defined. Hence $\varphi_{i_0}(x) = \varphi(i_0,x)\downarrow$. This completes the inductive proof of (i).

If $x > 0$, clearly the definition of φ gives that $\varphi_{i_0}(x) = \varphi(i_0,x) = 1 + \min m[M_\Phi(i_0,x-1,m) = 1] = 1 + \min m[\Phi_{i_0}(x-1) \le m] = \Phi_{i_0}(x-1)+1$. Hence (ii) is satisfied.

Blum's celebrated speed-up theorem is one of the most beautiful results in the theory of computational complexity:

THEOREM 2 (Blum, 1967a)

Let $\varphi_0,\varphi_1,\cdots$ be an acceptable indexing of the partial recursive functions of one argument and let $\Phi_0,\Phi_1,\cdots$ be a complexity measure for $\varphi_0,\varphi_1,\cdots$. Let r be a recursive function of two arguments. Then there exists a recursive function f such that the range of f is a subset of $\{0,1\}$ and such that for every i

such that $\varphi_i = f$ there exists a j such that $\varphi_j = f$ and an x_0 such that

$$r(x,\Phi_j(x)) < \Phi_i(x) \quad \text{for all} \quad x \geq x_0.$$

If $r(x,m)$ is much larger than m for all x and m, then the above statement implies that "program" i fails to be an "optimal" program for computing f in a particularly dramatic fashion, since the run time of "program" j is much smaller than the run time of "program" i on all but finitely many inputs. (For instance, if $r(x,m) = 2^m$, then the run time of "program" j is less than the logarithm to the base 2 of the run time of "program" i.) Thus, no "program" which computes f can have an optimal run time.

We will not prove Theorem 2 here. Note, however, that if $r(x,m)$ is much larger than m for all x and m then each "program" which computes f must have a run time which is large on most inputs. Thus, the statement of the speed-up theorem is in the same general spirit as the statements of Lemmas 1 through 4. In fact, a detailed reading of the alternative proof of the speed-up theorem due to Young (1973) makes it clear that _Lemma 4 is a technical result whose proof embodies the key technique which is required to prove the speed-up theorem_.

The above results only hint at the richness of Blum's axioms and at the beauty of the ensuing theory. Our purpose has only been to provide enough material to provide a context for a discussion of issues of subrecursive complexity and of "naturalness" of complexity measures. For surveys of the consequences of Blum's axioms, see Hartmanis & Hopcroft (1971) and Borodin (1973).

Blum's axiomatic setting treats each "program" i as an isolated, atomic entity which bears no obvious relationships to other "programs" $j \neq i$. This is a strength of the axiomatic framework in the sense that it provides a concise and reasonably austere way of talking about some of the standard ways of gauging the efficiency of programs, but it is also a weakness in the sense that one cannot talk about useful relationships between programs (e.g., the notion that one program i is used as a subprogram by another program j) and their run times (e.g., the notion that the run time of program j isn't terribly much larger than the run time of its subprogram i).

In particular, the fact that different programs are treated as atomic, unrelated entities lets us convert one complexity measure into another complexity measure by leaving the run times of some of the programs alone and "patching in" pathological behavior for the run times of other programs. For instance, algorithms which are _structurally_ and _definitionally_ equivalent (e.g., programs which are identical except for the fact that names of variables have been changed in a uniform fashion) may require radically different amounts of computational resource. As another example, in the spirit of Baker (1973, 1978), let $\varphi_0, \varphi_1, \cdots$ be an acceptable indexing of the partial recursive functions of one argument, let $\Phi_0, \Phi_1, \cdots$ be a complexity measure for $\varphi_0, \varphi_1, \cdots$ with measure function M_Φ, let i_0 be such that φ_{i_0} is a recursive function which is not primitive recursive (and hence Φ_{i_0} is total), and let b be a recursive function which is much larger than Φ_{i_0}. Define

$$\hat{\Phi}_i(x) = \begin{cases} \Phi_{i_0}(x) & \text{if } i = i_0 \\ \min m[m \geq b(x) \ \& \ M_\Phi(i,x,m) = 1] & \text{if } i \neq i_0. \end{cases}$$

It is straightforward to verify that $\hat{\Phi}_0, \hat{\Phi}_1, \cdots$ is also a complexity measure. However, $\hat{\Phi}_0, \hat{\Phi}_1, \cdots$ has a number of properties which can be regarded as pathological or "unnatural." For instance, φ_{i_0} is a recursive function which is not primitive recursive but which can be computed "faster" (i.e., with a smaller run time) than the constant function with value 0, and φ_{i_0} can be computed vastly more rapidly than the intimately related function $g(x) = \varphi_{i_0}(x)+1$.

4 SUBRECURSIVE COMPLEXITY: A BEGINNING

A general approach to the modeling of subrecursive programming languages would consider an indexing $f_0, f_1, \cdots$ of some subset $\mathcal{J}$ of the (total) recursive functions and would develop various axioms concerning the nature of the collection of functions $\mathcal{J}$ and the nature of the indexing $f_0, f_1, \cdots$ as well as various axioms about the nature of a proposed "complexity measure" $\Phi_0, \Phi_1, \cdots$ to be associated with $f_0, f_1, \cdots$. A reasonably rich theory along these lines is possible. <u>For expository ease in this section, assume that $\mathcal{J}$ is precisely the set of primitive recursive functions of one argument</u>. This simplifies the kind of axioms which must be discussed considerably and will allow us to concentrate on central issues which should communicate the central themes and "flavors" of the subject. However, similar approaches can be made for a large variety of other classes of recursive functions $\mathcal{J}$, and in particular for <u>quite small</u> subsets of the set of primitive recursive functions of one argument. These generalizations will be discussed briefly in Section 7.

Thus, let $\mathcal{PR}$ denote the set of primitive recursive functions of one argument and let $f_0, f_1, \cdots$ be an indexing of $\mathcal{PR}$. By this we mean that the mapping which takes each nonnegative integer i into f_i is a function having domain $N = \{0,1,2,\cdots\}$ and range equal all of $\mathcal{PR}$; thus far we have made no assumptions whatsoever about any effectiveness properties for that mapping. Let Φ be a sequence $\Phi_0, \Phi_1, \cdots$ of recursive functions of one argument. What assumptions should we make in order to be able to think of index i as a "program" (which gives output $f_i(x)$ for input x) in a reasonably well-behaved subrecursive programming language and in order to be able to think of $\Phi_i(x)$ as a reasonable gauge of some dynamic computational resource requirement such as execution time or memory for "program" i on input x?

Since each function f_i is total, our insistence that each Φ_i should also be total guarantees an analog of Definition 2.(i) at the start of Section 3.

Given the sequence of functions $\Phi_0, \Phi_1, \cdots$, we may associate with it the measure function for Φ,

$$M_\Phi(i,x,r) = \begin{cases} 1 & \text{if } \Phi_i(x) \le r \\ 0 & \text{if } \Phi_i(x) \not\le r \end{cases}$$

just as in Section 3. It is certainly reasonable to require that M_Φ be recursive, and doing so represents the first effectiveness condition which we impose on the way various functions Φ_i, for various "programs" i, must be "coordinated." If our programming language and method of gauging efficiency are reasonable, then it is also reasonable to require the stronger condition that the measure function M_Φ, when suitably encoded as a function of one argument, should belong to $\mathcal{PR}$, i.e., should itself be capable of being computed in the subrecursive programming language. Of course $\mathcal{PR}$ includes the inverses to various standard pairing functions, so encoding M_Φ as a function of one argument poses no problems. In the future, we'll write expressions such as $M_\Phi \in \mathcal{PR}$ without mentioning the encoding. The assumption $M_\Phi \in \mathcal{PR}$ is a reasonable requirement, basically because of the presence of r as an argument of M_Φ. For instance, if Φ is execution time, we would expect that it should be possible to describe the effect of execution of a single instruction of a reasonable programming language (how memory values are updated, what instruction will be executed next, etc.) in a primitive recursive fashion; if this is the case, then we should be able to use primitive recursion up to value r to simulate a computation for r steps and see whether or not it halts in the course of those r steps. To rephrase things in terms of <u>for</u> loops as discussed in Section 2, we can use a <u>for</u> loop to simulate the computation of "program" i on input x for at most r steps, keeping track of whether the computation halts in the course of the simulation.

Suppose that we assume that $M_\Phi \in \mathcal{PR}$. This allows a direct translation of the statement and proof of Lemma 1 of Section 3:

LEMMA 5

Let $f_0, f_1, \cdots$ be an indexing of $\mathcal{PR}$ and let $\Phi_i = f_i$ for each i. Let M_Φ have the property that

$$M_\Phi(i,x,r) = \begin{cases} 1 & \text{if } \Phi_i(x) \le r \\ 0 & \text{if } \Phi_i(x) > r. \end{cases}$$

Then $M_\Phi \notin \mathcal{PR}$.

For the proof, assume $M_\Phi \in \mathcal{PR}$. Then there is a primitive recursive function f such that $f(i) = 0 \Longleftrightarrow M_\Phi(i,i,0) = 0$, and the proof of Lemma 1 given in Section 3 can be adapted directly.

The assumption $M_\Phi \in \mathcal{PR}$, in conjunction with the requirement that each Φ_i should be total, amounts to a reasonably slavish translation into a subrecursive setting of Blum's axioms (given in Definition 2 at the start of Section 3) for a complexity measure. As just shown, it does allow us to prove an analog of Lemma 1 of Section 3, but in general it is an abysmal failure. For instance, $\Phi_i(x) = 0$ for all i and x satisfies these axioms. It's clearly an outlandish choice which <u>shouldn't</u> satisfy a reasonable notion of a subrecursive complexity measure; it seems clear that no reasonable programming language will be capable of computing each primitive recursive function on each argument in some constant amount of "execution time." In particular, if $\Phi_i(x) = 0$ for all i and x, attempts to formulate reasonable analogs of Lemmas 2 and 3 are doomed to failure.

Thus, we need further axioms. A key step is to define a "simulation function" $S_\Phi(i,x,r)$ with the property that

$$\Phi_i(x) \le r \Rightarrow S_\Phi(i,x,r) = f_i(x)$$

and to require that $S_\Phi \in \mathcal{PR}$. The rationale given earlier for the reasonableness of requiring $M_\Phi \in \mathcal{PR}$ applies equally well to requiring $S_\Phi \in \mathcal{PR}$; for instance, if Φ corresponds to execution time, then computation of $S_\Phi(i,x,r)$ only requires that we simulate the computation for a known number of steps r; since r is known, it can be used in a primitive recursive fashion, i.e., in a <u>for</u> loop instead of in a <u>while</u> loop.

To see that the assumption that $S_\Phi \in \mathcal{PR}$ helps, note that if $S_\Phi \in \mathcal{PR}$ and $\Phi_i(x) = 0$ for all i and x, then $u(i,x) = S_\Phi(i,x,0) = f_i(x)$ is the universal function for the indexing $f_0, f_1, \cdots$. Thus, $f_0, f_1, \cdots$ would be an indexing of the primitive recursive functions whose universal function is also primitive recursive, and of course this is impossible. (If the universal function u were primitive recursive, $f(i) = u(i,i)+1 = f_i(i)+1$ would be a primitive recursive function not in the indexing.)

More generally, the same sort of argument clearly implies that no one primitive recursive function can bound all the run times, which is an analog of Lemma 2 of Section 3:

LEMMA 6

Let $f_0, f_1, \cdots$ be an indexing of $\mathcal{PR}$ and let $\Phi_0, \Phi_1, \cdots$ be a sequence of recursive functions such that there exists $S_\Phi \in \mathcal{PR}$ with the property that for all $i, x,$ and r,

$$\Phi_i(x) \le r \Rightarrow S_\Phi(i,x,r) = f_i(x).$$

Let $b \in \mathcal{PR}$. Then it is <u>not</u> the case that all i and x satisfy $\Phi_i(x) \le b(x)$.

(Of course there are reasonable programming languages which compute the primitive recursive functions and which have associated with them reasonable notions of execution time such that each program has an execution time which is primitive recursive. Thus the hypothesis $b \in \mathcal{PR}$ is essential in the statement of Lemma 6.)

A generalization of the argument also proves:

LEMMA 7

Let $f_0, f_1, \cdots$ be an indexing of $\mathcal{PR}$ and let $\Phi_0, \Phi_1, \cdots$ be a sequence of recursive functions such that there exists $S_\Phi \in \mathcal{PR}$ with the property that for all $i, x,$ and r,

$\Phi_i(x) \le r \Rightarrow S_\Phi(i,x,r) = f_i(x)$.

Then $\mathcal{PR}$ does not contain a function h with the property that $\Phi_i(x) \le h(i,x)$ for every i and x.

The proof of Lemma 7 is also straightforward; if such a function h were primitive recursive, then the closure of the primitive recursive functions under composition would make $u(i,x) = S_\Phi(i,x,h(i,x)) = f_i(x)$ a primitive recursive universal function for an indexing of $\mathcal{PR}$, and we already know this is impossible. While the proof is easy, the statement of Lemma 7 is of some interest. As noted in Section 2, many subrecursive languages have the property that one can produce an upper bound on their computational resource requirements in terms of syntactic and structural information. For instance, the LOOP language (Meyer & Ritchie, 1967a,b) which computes the primitive recursive functions has associated with it a sequence of functions g_n, each itself computable in LOOP, such that the execution time of a program i on input x is less than or equal to the iterate $g_{d(i)}^{[\ell(i)]}(x)$, where d(i) is the maximum depth of nesting of loops in i and $\ell(i)$ is the number of instructions in program i. (See the appendix to Constable & Borodin (1972).) Such bounds imply the existence of a recursive function h(i,x) which bounds the execution time of each program i on each input x. For each fixed choice d_0 for the maximum depth of nesting of loops, the function $h_{d_0}(i,x) = g_{d_0}^{[\ell(i)]}(x)$ is primitive recursive and hence is itself computable in LOOP. Unfortunately, however, h(i,x) cannot itself be computed in LOOP, i.e., h is not primitive recursive. This may be regarded as an analog, for subrecursive languages, of the undecidability of the halting problem. Lemma 7 above gives a similar kind of result in a more general abstract setting.

Another way of viewing the assumption that a "simulation function" S_Φ belongs to $\mathcal{PR}$ is to note that S_Φ bounds the "rate of growth" of computations; if a program i only uses r units of computational resource on input x, the size of the output is only $S_\Phi(i,x,r)$. This corresponds to the "speed-limit" which Constable & Borodin (1972) propose on p. 564 for ruling out, e.g., the "unnatural" complexity measure which asserts that each program requires 0 units of computational resource on each input.

Just as we cannot assume that an indexing $f_0, f_1, \cdots$ of $\mathcal{PR}$ has a primitive recursive universal function, we also cannot assume that it satisfies the "strong" version of the recursion theorem popularized by Rogers (1967) which would assert that for each $g \in \mathcal{PR}$ there exists an i_0 such that $f_{g(i_0)} = f_{i_0}$. For instance, reasonable indexings of $\mathcal{PR}$ can be expected to have the property that there exists a primitive recursive function g such that $f_{g(i)}(x) = f_i(x)+1$ for all i and x; if this is the case, then the fact that each f_i is total implies that no i_0 can satisfy $f_{g(i_0)} = f_{i_0}$.

However, reasonable indexings of $\mathcal{PR}$ do satisfy an analog of Kleene's statement of the recursion theorem, which was given as Theorem 2 in Section 3. To see this, let $<, >$ denote a primitive recursive pairing function whose inverses are also primitive recursive and note that it is reasonable to assume that a function s exists which has the property that

$$f_{s(i,x)}(y) = f_i(<x,y>)$$

for all i,x, and y and which is not only recursive but also primitive recursive. Kleene (1958) gave a version of the recursion theorem for a specific indexing of $\mathcal{PR}$, and clearly his argument applies to any indexing of $\mathcal{PR}$ for which such a primitive recursive function s exists:

LEMMA 8

Let $f_0, f_1, \cdots$ be an indexing of $\mathcal{PR}$ such that there exists $s \in \mathcal{PR}$ such that

$$f_{s(i,x)}(y) = f_i(<x,y>) \quad \text{for all } i, x, \text{ and } y.$$

Let $f \in \mathcal{PR}$. Then there exists i_0 such that

$$f(i_0,x) = f_{i_0}(x) \quad \text{for all } x.$$

Of course the assumptions $s \in \mathcal{PR}$ and $f \in \mathcal{PR}$ mean that $s' \in \mathcal{PR}$ and $f' \in \mathcal{PR}$, where $s'(z) = s(\pi_1(z),\pi_2(z))$ and $f'(z) = f(\pi_1(z),\pi_2(z))$ and π_1 and π_2 are primitive recursive inverses of a primitive recursive pairing function $<, >$.

The proof of Lemma 8, patterned after Kleene's proof, is as follows: Consider $g(i,x) = f(s(i,i),x)$. Since s and f are primitive recursive, g is primitive recursive. Let ℓ be such that $f_\ell(z) = f(s(\pi_1(z),\pi_1(z)),\pi_2(z))$ for all z, i.e., such that $f_\ell(<i,x>) = f(s(i,i),x) = f'(<s(i,i),x>)$ for all i and x. It is straightforward to verify that $i_0 = s(\ell,\ell)$ has the desired property $f(i_0,x) = f'(<i_0,x>) = f'(<s(\ell,\ell),x>) = f_\ell(<\ell,x>) = f_{s(\ell,\ell)}(x) = f_{i_0}(x)$. This completes the proof.

For acceptable indexings of all the partial recursive functions, the two versions of the recursion theorem discussed in Section 3 can easily be shown to be of equal strength. For subrecursive indexings, where the strong "Rogers" version fails, the need to satisfy the hypothesis $f \in \mathcal{PR}$ of the "Kleene" version frequently limits its applicability—and even more frequently limits the ease with which it can be applied. However, in conjunction with the axioms which we introduced earlier it does allow us to give a direct translation to subrecursive settings of the argument used to prove Lemma 3 in Section 3:

LEMMA 9

Let $f_0, f_1, \cdots$ be an indexing of $\mathcal{PR}$ such that there exists $s \in \mathcal{PR}$ such that

$$f_{s(i,x)}(y) = f_i(<x,y>) \quad \text{for all } i,x, \text{ and } y.$$

Let $\Phi_0, \Phi_1, \cdots$ be a sequence of recursive functions such that there exist $M_\Phi \in \mathcal{PR}$ and $S_\Phi \in \mathcal{PR}$ such that

$$M_\Phi(i,x,r) = \begin{cases} 1 & \text{if } \Phi_i(x) \leq r \\ 0 & \text{if } \Phi_i(x) > r \end{cases}$$

for all $i,x,$ and r and such that

$$\Phi_i(x) \leq r \Rightarrow S_\Phi(i,x,r) = f_i(x) \quad \text{for all } i,x, \text{ and } r.$$

Let $b \in \mathcal{PR}$ and $g \in \mathcal{PR}$. Then there exists an i_0 such that $f_{i_0} = g$ and such that $\Phi_{i_0}(x) > b(x)$ for all x.

(As noted just after the statement of Lemma 6, the assumption $b \in \mathcal{PR}$ is essential.)

For the proof, define

$$f(i,x) = \begin{cases} S_\Phi(i,x,b(x))+1 & \text{if } M_\Phi(i,x,b(x)) = 1 \\ g(x) & \text{if } M_\Phi(i,x,b(x)) = 0. \end{cases}$$

Since M_Φ, S_Φ, b, and g are primitive recursive, f is primitive recursive. This establishes the required hypothesis for the recursion theorem of Lemma 8. Consequently the proof of Lemma 3 given in Section 3 can be adapted in a straightforward fashion.

The statement of Lemma 9 has three principal hypotheses, $M_\Phi \in \mathcal{PR}$, $s \in \mathcal{PR}$, and $S_\Phi \in \mathcal{PR}$. The assumption $M_\Phi \in \mathcal{PR}$ is a plausible adaptation, to a subrecursive setting, of Blum's requirement in Definition 2.(ii) that a complexity measure must have a measure function which is recursive, and the requirement that each run time Φ_i must be total corresponds to his other axiom in Definition 2.(i). The requirement that $s \in \mathcal{PR}$ is a plausible adaptation, to a subrecursive setting, of the requirement that an acceptable indexing of all partial recursive functions must have a recursive function satisfying Definition 1.(ii). The requirement that $S_\Phi \in \mathcal{PR}$ relates f_i and Φ_i. In particular, although the indexing $f_0, f_1, \cdots$ cannot have a <u>primitive</u> recursive universal function, the assumption that S_Φ is (primitive) recursive (together with a similar assumption about M_Φ) implies that the indexing does have a universal function which is <u>recursive</u>, since

$$f_i(x) = S_\Phi(i,x,\min r[M_\Phi(i,x,r) = 1]).$$

This can be regarded as an adaptation, to a subrecursive setting, of the requirement in Definition 1.(i) that an acceptable indexing of all the partial recursive functions must have a partial recursive universal function.

We regard the fact that the universal function for $f_0, f_1, \cdots$ must be recursive as a <u>strength</u>, not a limitation on the generality of, the axiomatic framework being developed; any reasonable subrecursive language <u>should</u> have the property that the universal function is recursive.

5 RELATIONSHIPS BETWEEN THE COMPLEXITIES OF RELATED PROGRAMS

For ease of exposition, the previous section restricted attention to indexings $f_0, f_1, \cdots$ of the collection $\mathcal{PR}$ of primitive recursive functions of one argument. <u>The present section is concerned with attempts to characterize "natural" complexity measures, independent of whether the underlying programming language is universal or subrecursive. In this section,</u> $f_0, f_1, \cdots$ <u>will denote an indexing of some subset</u> $\mathcal{J}$ <u>of the partial recursive functions of one argument</u>. In general we do not restrict the nature of $\mathcal{J}$ further, but it is reasonable to think only about the two cases when $\mathcal{J}$ consists of all partial recursive functions of one argument or of all primitive recursive functions of one argument.

In Section 2 we noted that Blum's axioms treat distinct "programs" as unrelated atomic entities. Instead we may approach efficiency considerations by viewing complicated programs as being built up from simpler ones and by considering limits on the "overhead costs" associated with the coordination of the dynamic computational resource requirements of the simpler programs during execution of the more complicated programs.

As an example of synthesis of more complicated programs from simpler ones, consider the operation of primitive recursion without parameters and with initialization to 0. This uses a function h of two arguments to define a function f of one argument by

$$f(0) = 0$$

$$f(z+1) = h(f(z),z).$$

For many "natural" programming languages (both universal and subrecursive), it should not be very difficult to pass from a description of a "program" i which computes h to a description of a "program" p(i) which computes f; as illustrated in Section 2, the output variable of program p(i) is not a variable of program i, is initialized to 0, and is updated by a <u>for</u> loop, with each pass through the loop using program i to update it. Thus, for an indexing $f_0, f_1, \cdots$ of some collection $\mathcal{J}$ which is closed under the operation of primitive recursion, it is reasonable to require that a "very easily computable" <u>synthesizing function</u> p satisfy

$$f_{p(i)}(0) = 0 \qquad (1)$$

and

$$f_{p(i)}(z+1) = f_i(f_{p(i)}(z),z) \qquad (2)$$

for all i and z. Since we are considering indexings of functions of one argument, (2) is shorthand for

$$f_{p(i)}(z+1) = f_i(< f_{p(i)}(z), z >)$$

where <, > denotes a "very easily computable" pairing function whose inverses are also "very easily computable."

For many "natural" programming languages (both universal and subrecursive) and complexity measures, execution of the <u>for</u> loop will not be substantially harder than the individual executions of the individual passes through the loop. Thus, it is reasonable to require that a "very easily computable" <u>resource estimation function</u> P(z,w) exist such that

$$\Phi_{p(i)}(z) \le P(z, \operatorname*{Max}_{z'<z} \Phi_i(f_{p(i)}(z'),z')) \qquad (3)$$

where $\operatorname*{Max}_{z'<z}$ is interpreted as 0 if $z = 0$.

What does "very easily computable" mean? Just as Section 4 considered axioms such as $M_\Phi \in \mathcal{J}$ and $S_\Phi \in \mathcal{J}$ (for the special case $\mathcal{J} = \mathcal{PR}$), frequently the assumptions $p \in \mathcal{J}$ and $P \in \mathcal{J}$ are useful. This will be illustrated in Section 6. As usual, $P \in \mathcal{J}$

means that a suitable encoding of P as a function of one argument belongs to $\mathcal{J}$. Thus, e.g., if the programming language computes the primitive recursive functions, it is certainly reasonable to require that p and P themselves be primitive recursive. This is a very modest assumption, given that many "natural" languages (both universal and subrecursive) satisfy (3) for $P(z,w) = z \cdot (w+k_1) + k_2$ for appropriate constants k_1 and k_2.

While many "natural" languages satisfy resource estimates such as (3) for very specific functions such as the one in the previous sentence, useful generality is gained by leaving the specific nature of the functions which are used to estimate resource requirements as an unspecified "parameter." For an example where this is more apparent, consider a synthesizing function which, when given as inputs a program i and a finite table t, produces as output a program x(i,t) which behaves like t for inputs which are relevant to t and behaves like i for other inputs. A complexity measure is <u>finitely invariant</u> (Borodin, 1972) if for each i and t such a program x(i,t) exists with the additional property that the run time of program x(i,t) is less than or equal to the run time of program i for all but finitely many inputs. Such a condition holds for many "natural" languages and complexity measures but does not hold for all cases which I regard as "natural." (See also Baker (1978).) The need for checking whether the table is relevant can be expected to result in a slightly larger complexity on all inputs, even those which turn out to not be dealt with in the table. This increase can often be swept under the rug, but via "unnatural" tricks such as increasing the tape alphabet of a Turing machine without increasing the execution time of each basic operation; in real life, increased word sizes can be expected to involve increased execution times. For some "natural" subrecursive languages, the cost of checking whether a table is relevant to a given input can be substantial. Thus, results which are more machine-independent and hence more widely applicable can be obtained by introducing as a "parameter" a function which concerns a "margin of error" for how much the computational resource requirements of x(i,t) exceed those of i, without specifying the precise nature of that function.

If $f_0, f_1, \cdots$ is an acceptable indexing of all partial recursive functions of one argument and $\Phi_0, \Phi_1, \cdots$ satisfies Blum's axioms for a complexity measure (Definition 2 near the start of Section 3), it is easy to establish the existence of recursive functions r and R such that

$$f_{r(i)}(x) = \Phi_i(x) \tag{4}$$

and

$$\Phi_{r(i)}(x) \le R(i,x,\Phi_i(x)) \tag{5}$$

for all i and x. The proof, which is a slight variant of one given in Blum (1967a), uses an unbounded minimization to define

$$g(<i,x>) = \min s[M_\Phi(i,x,s) = 1] = \Phi_i(x). \qquad (6)$$

If $g = f_{j_0}$, then r is defined as $r(i) = s(j_0,i)$, where s is as in Definition 1.(ii) at the start of Section 3. This satisfies (4). To satisfy (5), R is defined as

$$R(i,x,m) = \begin{cases} 0 & \text{if } M_\Phi(i,x,m) = 0 \\ \min s[M_\Phi(r(i),x,s) = 1] & \text{if } M_\Phi(i,x,m) = 1 \end{cases}$$

where definition by cases is formulated in such a fashion that $\min s[M_\Phi(r(i),x,s) = 1]$ is only computed if $M_\Phi(i,x,m)$ really does equal 1. By Definition 2, this guarantees that R is total; if $M_\Phi(i,x,m) = 1$, then $\Phi_i(x) = f_{r(i)}(x)$ is defined and hence $\Phi_{r(i)}(x) = \min s[M_\Phi(r(i),x,s) = 1]$ is also defined.

The above proof technique is generalized in the "combining lemma" on p. 451 of Hartmanis & Hopcroft (1971). Details will not be discussed here, but the key outcome is that for any complexity measure for any acceptable indexing of the partial recursive functions of one argument, it can be used to establish the existence of various <u>recursive</u> synthesizing functions and resource estimation functions such as r,R,p,P, and x discussed above. In the present context, the existence of such resource estimation functions even for "unnatural" complexity measures may be regarded as pathological. However, such pathological cases can be excluded by placing restrictions on the sizes of the resource estimation functions allowed, i.e., by adopting hypotheses concerning "how subrecursive" such "parameters" must be.

If the above approach is used for an indexing $f_0,f_1,\cdots$ of a subrecursive set $\mathcal{J}$, the use of unbounded minimization gives no guarantee that r and R will belong to $\mathcal{J}$. Indeed, results such as Lemma 7 of Section 4 imply that frequently it will not be possible to replace the unbounded minimization (6) by a bounded minimization, i.e., it will not be possible to replace a <u>while</u> loop by a <u>for</u> loop.

While Lemma 7 blocks obvious attempts to <u>derive</u> $r \in \mathcal{J}$ and $R \in \mathcal{J}$ for subrecursive languages, it does not appear to be strong enough to establish an inconsistency if we <u>assume as additional axioms</u> that $r \in \mathcal{J}$ and $R \in \mathcal{J}$. For instance, suppose $f_0,f_1,\cdots$ is an indexing of $\mathcal{GR}$ and that r and R are primitive recursive. The obvious attempt to establish an inconsistency via Lemma 7 would be based on attempting to show that $h(i,x) = f_{r(i)}(x) = \Phi_i(x)$ is primitive recursive, but the obvious attempt to show this would need for the universal function for the indexing to itself be primitive recursive, which we know is impossible.

Indeed, Lemma 7 cannot be used to establish a contradiction if we assume $r \in \mathcal{J}$ and $R \in \mathcal{J}$; many reasonable subrecursive languages and complexity measures have the property that they can compute functions r and R which satisfy (4) and (5). To see the reasonableness of such axioms, note that typically, "tracing" facilities can be added to program i, initializing a new variable to 0 and updating it prior to the execution of each instruction of i in order to account for the computational resources which will be used to execute that instruction. In particular, many "natural" subrecursive languages and complexity measures are closed under resources in the sense that $\Phi_i \in \mathcal{J}$ for each i. In addition, typically the introduction of tracing facilities does not increase the program complexity terribly radically, so that (5) will be satisfied for a "very subrecursive" R. For instance, for execution time typically we would expect that $R(i,x,s) \leq 2s + c$ for some constant c. If $R(i,x,s) = s$, Φ is proper (Landweber & Robertson, 1972). We do not require this. Allowing R as a "parameter" allows for a "margin of error" which acknowledges that the addition of tracing facilities may increase the complexity of the computation somewhat. The need for such a "margin of error" is supported by Baker (1978).

If we substitute (4) into (5), we see that

$$\Phi_{r(i)}(x) \leq R(i,x,f_{r(i)}(x)). \tag{7}$$

This attests to the R-honesty (McCreight, 1969; McCreight & Meyer, 1969) of run times; the complexity of computing program r(i) is reflected in the size of the output eventually produced, to within an amount related to R. Of course we do not expect all programs to be R-honest; as noted in connection with Lemma 1 in Section 3, we expect some programs to take a great deal of execution time to decide whether to output a 0 or a 1. But (7) requires that all programs which "obviously" compute run times (where "obviously" means they are in the range of r) must be R-honest.

We may view a programming language as a collection of basic programs together with a collection of synthesizing functions which allow us to build more sophisticated programs. Typically, the basic programs might correspond to assignment statements, tests of the values of boolean-valued variables, etc. We may use analogs of Blum's axioms to associate complexities with the basic programs, and then resource estimation functions allow us to determine upper bounds on the complexities of the synthesized programs. It can be argued that this allows all the pathologies allowed by Blum's axioms, since one can make all programs basic. (For instance, basic programs which are structurally similar, differing only in the names of the variables, could still be assigned dramatically different execution times.) But this is clearly the wrong viewpoint. Our expectation is that someone who wants to use the axioms to evaluate a particular programming language will *explicitly* examine the *basic* programs and their associated complexities, hence will *explicitly* confront any pathology inherent in the basic programs. The important thing is to be able to be confident that the synthesized programs will not exhibit any pathological properties which were not already explicitly discernible in the basic programs.

Typically, a programming language will *explicitly* offer a relatively small number of synthesizing functions to the programmer for building more complicated programs from simpler ones. However, these synthesizing functions can be combined by the programmer to form still other synthesizing functions. For instance, a programming language which contains conditional *go to* statements can be used to write *for* loops even if the programming language's syntax doesn't explicitly include *for* loops. We feel that the evaluation of a programming language should pay attention to the efficiency of such synthesizing functions as well. As such, we are focusing on the potential efficiency of the programs which can be written in the language and are ignoring the ease with which the programmer can compose those programs. (The later issue is of course also vitally important to "software engineering." A "high level" programming language which contains syntactic entities corresponding to many sophisticated synthesizing functions may be a tremendous help. In addition, some programming languages contain "macro expansion" facilities which allow the programmer to formally define new synthesizing functions in terms of the primitive constructs of the language; roughly speaking, this allows the programmer to describe operators and functionals.)

We feel that *the "naturalness" or "acceptability" of a programming language should be gauged in part in terms of the various synthesizing functions which belong to various subrecursive classes*. Thus, for instance, we might gauge how "natural" or how "acceptable" an indexing $f_0, f_1, \cdots$ is partly in terms of whether or not there exists a primitive recursive function c such that

$$f_{c(i,j)}(x) = f_i(f_j(x))$$

for all $i,j,$ and x; another gauge of how "natural" or how "acceptable" it is would be partly in terms of whether or not such a function c exists which belongs to the class $\mathcal{E}^2$ of Grzegorczyk (1953).

Similarly, we feel that *the "naturalness" of a complexity measure should be gauged in part in terms of the various synthesizing functions which have resource estimation functions which belong to* various subrecursive classes. Thus, for the synthesizing function c referred to in the previous paragraph, we might gauge how "natural" a complexity measure $\Phi_0, \Phi_1, \cdots$ is partly in terms of whether or not there exists a primitive recursive function C such that

$$\Phi_{c(i,j)}(x) \le C(i,j,\Phi_j(x),\Phi_i(f_j(x)))$$

for all i,j, and x; another gauge of how "natural" it is would be partly in terms of whether or not such a function C exists which belongs to the class $\mathcal{E}^2$.

Such an approach to efforts to characterize "natural" complexity measures is of course implicit in the work of numerous authors. For instance, Bečvář (1972) discussed a proposal (which I seem to recall he attributed at least in part to Chytil) that if $\varphi_0, \varphi_1, \cdots$ is an acceptable indexing of the partial recursive functions of one argument then attention should be focused on complexity measures $\Phi_0, \Phi_1, \cdots$ such that there exist recursive functions c and h such that $\varphi_{c(i,j)}(x) = \varphi_i(\varphi_j(x))$ and

$$\max\{\Phi_j(x), \Phi_i(\varphi_j(x))\} \le \Phi_{c(i,j)}(x)$$
$$\le \Phi_j(x) + \Phi_i(\varphi_j(x)) + h(x, \varphi_j(x), \varphi_i(\varphi_j(x))).$$

Much of the work of Wechsung (1975) and his student and colleague Lischke (1975a,1975b,1976a,1976b,1977) concerns similar notions and generalizations. Other manifestations of such themes and/or alternative attempts to characterize "natural" complexity measures are implicit or explicit in works including, but surely not limited to, Symes (1971), Lynch (1972), Constable (1973), Ausiello (1974), and Biskup (1977). In particular, the last chapter of Symes (1971) proposes a framework which associates "structure" with an indexing by specifying a basic set of subroutine operators and defining a program as any operator formed by composition of basic operators. He proposes axioms which relate the "cost" of a computation with the "costs" of its subcomputations. Alton (1977a) makes a more general proposal (in a dreadfully oppressive notation, unfortunately) concerning the form of axioms about general synthesizing functions. It is general enough to allow axiomatic distinctions between "consumable" computational resources such as execution time and "reusable" computational resources such as memory, and it also introduces the possibility of dealing with synthesizing functions which do not preserve equivalence of programs. The most obvious example of the latter is (4); equivalent programs don't in general have equivalent run times. For a first reading of that paper, every effort should be made to ignore that extra generality. (In the notation of that paper, this corresponds roughly to assuming all programs are Φ-independent, to assuming all subcomputation descriptions are complete, entire, and currently relevant, to assuming all subcomputations and answers are preserved, and to ignoring the number of activations!)

None of the above references other than Alton (1977a) acknowledges the potential applicability of such an approach to a subrecursive setting and none of them analyzes or adapts results of complexity theory such as the speed-up theorem, a task which can be approached with the technique of Section 6 below.

6 A PROOF TECHNIQUE FOR SUBRECURSIVE COMPLEXITY

Section 5 was concerned with efforts to describe how "natural" a programming language and complexity measure are, independent of whether the language is universal or subrecursive. In this section we will return to the special problems which arise in dealing with subrecursive programming languages and complexity measures. Once again, *for ease of exposition, in this section we will restrict our attention to programming languages which are capable of computing precisely the class* $\mathcal{PR}$ *of primitive recursive functions of one argument*.

Section 4 illustrated the usefulness of several possible axioms which are of a reasonably general sort. In contrast, most of the axioms which were discussed in Section 5 are more specialized, concerning various specific ways of synthesizing new programs from old programs. However, the functions r and R of (4) and (5) specifically relate the indexing and the complexity measure, and for this reason they are included in our principal definitional proposal:

DEFINITION 3

Let f be an indexing $f_0, f_1, \cdots$ of $\mathcal{PR}$ and let Φ be a sequence $\Phi_0, \Phi_1, \cdots$ of some of the functions in $\mathcal{PR}$. Then f and Φ are *a self-simulating indexing and complexity measure for* $\mathcal{PR}$ if there are *primitive recursive* functions s, M_Φ, S_Φ, r, and R such that

(i) $f_{s(i,x)}(y) = f_i(\langle x,y \rangle)$

(ii) $M_\Phi(i,x,m) = \begin{cases} 1 & \text{if } \Phi_i(x) \le m \\ 0 & \text{if } \Phi_i(x) > m \end{cases}$

(iii) $\Phi_i(x) \le m \Rightarrow S_\Phi(i,x,m) = f_i(x)$

(iv) $f_{r(i)}(x) = \Phi_i(x)$

(v) $\Phi_{r(i)}(x) \le R(i,x,\Phi_i(x))$

for all i,x,m, and y.

In particular, Lemmas 5 through 9 concern consequences of this definition.

Suppose that we now attempt to establish an analog of Lemma 4 of Section 3 for such a subrecursive setting. Thus, we wish to establish the existence of a "program" i_0 such that the <u>output</u> $f_{i_0}(x)$ on input x is greater than the <u>run time</u> $\Phi_{i_0}(x-1)$ on input $x-1$ for all $x > 0$. Prompted by the proof given in Section 3 for the case of an acceptable indexing of all partial recursive functions of one argument and a complexity measure, the obvious attempt is to define

$$f(i,x) = \begin{cases} 0 & \text{if } x = 0 \\ 1 + f_{r(i)}(x-1) & \text{if } x > 0 \end{cases} \tag{8}$$

and to attempt to apply the version of the recursion theorem given in Lemma 8 of Section 4. Unfortunately, however, Lemma 8 requires that f be primitive recursive. It is not obvious that f is primitive recursive; its definition refers to $r(i)$ <u>as a subscript</u>, so the <u>obvious</u> way of computing f would employ the universal function for the indexing $f_0, f_1, \cdots$, and that universal function cannot be primitive recursive. Indeed, there's a very good reason for the failure of this attempt to show that f is primitive recursive; f cannot possibly be primitive recursive! For if f were primitive recursive, then the function $h(i,x) = f(i,x+1)-1$ would be a primitive recursive function with the property that $h(i,x) = f_{r(i)}(x) = \Phi_i(x)$ for all i and x, and this would contradict Lemma 7 of Section 4.

Thus, the obvious attempt to adapt the proof of Lemma 4 by applying the recursion theorem of Lemma 8 to the function f of (8) fails.

The most prominent features of most subrecursive programming languages are <u>for</u> loops and primitive recursion. Is there any way that we can use primitive recursion to establish an analog of Lemma 4?

Suppose that we wish to define f_{i_0} by primitive recursion from some primitive recursive function h, say by

$$f_{i_0}(0) = 0$$

$$f_{i_0}(z+1) = h(f_{i_0}(z),z).$$

The desired property is that

$$h(f_{i_0}(z),z) > \Phi_{i_0}(z).$$

Thus, our desired program i_0 should be <u>h-honest</u>, i.e., its <u>run time</u> $\Phi_{i_0}(z)$ can only be large if its <u>output</u> $f_{i_0}(z)$ on the same input is also reasonably large.

What does the required h-honesty of the desired program i_0 tell us about the nature of the function h which is going to be used to define i_0 by primitive recursion? To obtain an answer to this (obviously self-referential) question, let's begin by phrasing things in the notation of (1), (2), and (3) of Section 5, using $f_{i_1} = h$ and $i_0 = p(i_1)$. We seek information about $h = f_{i_1}$ such that

$$f_{p(i_1)}(0) = 0 \tag{9}$$

and

$$f_{p(i_1)}(z+1) = f_{i_1}(f_{p(i_1)}(z),z) > \Phi_{p(i_1)}(z). \tag{10}$$

As argued in Section 5, it is also reasonable to assume that

$$\Phi_{p(i_1)}(z+1) \leq P(z+1, \max_{z' \leq z} \Phi_{i_1}(f_{p(i_1)}(z'),z')). \tag{11}$$

The desired honesty of program $p(i_1)$ requires that $\Phi_{p(i_1)}(z+1)$ not be too much larger than the output $f_{p(i_1)}(z+1)$ on the same input. The only obvious way to limit the size of $\Phi_{p(i_1)}(z+1)$ is to appeal to inequality (11) and limit the size of $\max_{z' \leq z} \Phi_{i_1}(f_{p(i_1)}(z'),z')$. In particular, this requires that we limit the size of $\Phi_{i_1}(f_{p(i_1)}(z),z)$. Thus, if we want to keep $\Phi_{p(i_1)}(z+1)$ from being much larger than $f_{p(i_1)}(z+1)$, this approach suggests that we should try to keep $\Phi_{i_1}(f_{p(i_1)}(z),z)$ from being much larger than $f_{p(i_1)}(z+1)$. But by (10) this means that we want to keep the <u>run time</u> $\Phi_{i_1}(f_{p(i_1)}(z),z)$ of program i_1 from being much larger than the <u>output</u> $f_{i_1}(f_{p(i_1)}(z),z)$ of the same program i_1 on the same inputs. Thus, <u>in order to make program i_0 honest, the program i_1 that we use in defining i_0 by primitive recursion should itself be honest</u>! Fortunately, we have access to a large number of honest programs, thanks to the functions r and R of Definition 3 at the start of this section; as noted earlier in (7),

$$\Phi_{r(i)}(x) \le R(i,x,\Phi_i(x)) = R(i,x,f_{r(i)}(x)),$$

so that the <u>run time</u> $\Phi_{r(i)}(x)$ can't be much larger than the <u>output</u> $f_{r(i)}(x)$. This suggests that we <u>try to choose</u> $i_1 = r(i_2)$ <u>for some program</u> i_2.

Having decided to try to use a run time in our primitive recursion, something extremely useful happens. By Definition 3.(iv), condition (10) becomes the condition

$$\Phi_{i_2}(f_{p(r(i_2))}(z),z) > \Phi_{p(r(i_2))}(z) \tag{12}$$

which requires that we choose i_2 so that its <u>run time</u> is large! This is reminiscent of Lemma 3, hence is a hopeful sign. At the same time, things are more complicated than in Lemma 3 because the lower bound on the run time of i_2 is itself a run time—and the run time of a program which is related to i_2.

If we already knew a value of i_2 which worked, we could attempt to <u>prove</u> that it worked by induction on z. (This is a reasonable proof strategy to try, since primitive recursion defines the value of $f_{p(r(i_2))}(z+1)$ in terms of the value of $f_{p(r(i_2))}(z)$.) Let's look at the form which an inductive proof might take and see what it tells us about how we should choose i_2.

For the base step of an induction, $\Phi_{p(r(i_2))}(0) \le P(0,0)$ and $f_{p(r(i_2))}(0) = 0$, so to satisfy (12) it suffices to choose i_2 so that

$$\Phi_{i_2}(0,0) > P(0,0). \tag{13}$$

For the inductive step, we could assume that we knew the analog of (12),

$$\Phi_{i_2}(f_{p(r(i_2))}(z'),z') > \Phi_{p(r(i_2))}(z'), \tag{14}$$

for all $z' \le z$. We would then wish to prove

$$\Phi_{i_2}(f_{p(r(i_2))}(z+1),z+1) > \Phi_{p(r(i_2))}(z+1). \tag{15}$$

The right hand side of (15) corresponds to (11). Motivated by the form of (11), note that by Definition 3.(v),

$$\max_{z'\le z} \Phi_{r(i_2)}(f_{p(r(i_2))}(z'),z') \le \max_{z'\le z} R(i_2,f_{p(r(i_2))}(z'),z',\Phi_{i_2}(f_{p(r(i_2))}(z'),z')). \tag{16}$$

(Recall that we have been using (10) without explicit reference to use of a pairing function to encode the arguments to f_{i_1}. Really the middle two arguments of the right hand side of (16) should be written as $\langle f_{p(r(i_2))}(z'),z'\rangle$, but once again we'll suppress this.) If our inductive assumption (14) is true, then Definition 3.(iii) allows us to eliminate explicit reference to $f_{p(r(i_2))}(z')$ from the right hand side of (16) by substituting

$$f_{p(r(i_2))}(z') = S_{\Phi}(p(r(i_2)),z',\Phi_{i_2}(f_{p(r(i_2))}(z'),z')).$$

The resulting expression which is equivalent to the right hand side of (16) can be simplified in two different places by using the equality

$$\Phi_{i_2}(f_{p(r(i_2))}(z'),z') = f_{p(r(i_2))}(z'+1).$$

This equality allows us to exploit our decision to use a <u>run time</u> in our definition by primitive recursion in a crucial way, by getting rid of explicit mention of a run time! Performing the indicated substitutions, the right hand side of (16) is equivalent to

$$\max_{z'\le z} R(i_2,\ S_{\Phi}(p(r(i_2)),z',f_{p(r(i_2))}(z'+1)),\ z',\ f_{p(r(i_2))}(z'+1)).$$

If in addition we are able to guarantee that

$$f_{p(r(i_2))}(z'+1) \le f_{p(r(i_2))}(z+1) \tag{17}$$

for all $z' \le z$ and if R is monotone nondecreasing in its last argument, then the right hand side of (16) is less than or equal to

$$\max_{z'\le z} R(i_2,\ S_{\Phi}(p(r(i_2)),z',f_{p(r(i_2))}(z+1)),\ z',\ f_{p(r(i_2))}(z+1)). \tag{18}$$

So under the assumptions we have made the left hand side of (16) is less than or equal to (18). If we substitute this inequality into (11) and assume that P is monotone nondecreasing in its second argument, we see that one way to make the inductive proof of (15) work is to choose i_2 so that

$$\Phi_{i_2}(f_{p(r(i_2))}(z+1),z+1) > P(z+1, \max_{z'\le z} R(i_2, S_\Phi(p(r(i_2)),z',f_{p(r(i_2))}(z+1)), z', f_{p(r(i_2))}(z+1))).$$

And of course one way to guarantee this is to guarantee that

$$\Phi_{i_2}(w,z+1) > P(z+1, \max_{z'\le z} R(i_2, S_\Phi(p(r(i_2)),z',w), z', w)) \tag{19}$$

for all w and z. This is a vastly simpler task than (12) appeared to be, since the right hand side of (12) contained a run time (and of a program related to i_2) and the right hand side of (19) does not.

In addition to satisfying (13) and (19), i_2 must also be chosen so that our assumption (17) is true. To satisfy (17), of course it suffices to make $f_{p(r(i_2))}(z) \le f_{p(r(i_2))}(z+1)$ for all z. Since $f_{p(r(i_2))}(z+1) = \Phi_{i_2}(f_{p(r(i_2))}(z),z)$ by (2) and Definition 3.(iv), it suffices to guarantee that i_2 has the additional property that

$$\Phi_{i_2}(w,z) \ge w \tag{20}$$

for all w and z.

If the original primitive recursive functions R and P which satisfy Definition 3.(v) and (3) are not monotone nondecreasing in their last argument, they can of course be replaced by other primitive recursive functions which satisfy both the original conditions and the monotonicity conditions. Combining conditions (13), (19), and (20), it suffices to choose i_2 so that

$$\Phi_{i_2}(w,z) > \max\{P(0,0),q(i_2,w,z),w\} \tag{21}$$

where

$$q(i,w,z) = P(z, \max_{z'<z} R(i, S_{\Phi}(p(r(i)),z',w), z', w)).$$

Since we are assuming that $P,R,S_{\Phi},p,$ and r are primitive recursive, q is also primitive recursive. Thus, an i_2 which satisfies (21) can be shown to exist by a straightforward adaptation of the proof of Lemma 9. (The presence of i_2 on the right hand side of (21) now causes no special problems, since it is an argument to a primitive recursive function, not a subscript to a run time.)

We have spent most of this section motivating an approach to establishing a subrecursive analog of Lemma 4 of Section 3. In particular, we have motivated the choice of the hypotheses for the

THEOREM 3

Let f and Φ be a self-simulating indexing $f_0, f_1, \cdots$ and complexity measure $\Phi_0, \Phi_1, \cdots$ for $\mathcal{PR}$ such that in addition there exist primitive recursive functions p and P which satisfy (1), (2), and (3). Then there exists i_0 such that $f_{i_0}(z) \geq \Phi_{i_0}(z-1)$ for all $z > 0$.

Moreover, the proof to Theorem 3 is easy to write down on the basis of the above motivation, roughly speaking by reading that motivation backwards.

As noted in Section 3, the above result embodies the principal technique which is required to prove a subrecursive analog of Blum's speed-up theorem. Details will be contained in Alton (1980).

7 OTHER SUBRECURSIVE COLLECTIONS OF FUNCTIONS $\mathcal{J}$

It is reasonably easy to generalize the lemmas of Section 4 so that they apply to indexings of various collections of functions $\mathcal{J}$ other than $\mathcal{PR}$. The pertinent axioms are straightforward, concerning the membership of functions such as those of Definition 3 in $\mathcal{J}$ rather than in $\mathcal{PR}$, closure of $\mathcal{J}$ under operations such as composition, and membership of certain specific functions, such as inverses to a pairing function, in $\mathcal{J}$. (Of course one could also treat functions of several arguments directly, without encoding them as functions of one argument. This brings with it a different set of notational problems, however.)

When one attempts to generalize the proof of Theorem 3 of Section 6, it appears at first glance that $\mathcal{J}$ must include $\mathcal{PR}$, since the operation of primitive recursion plays a crucial role in that proof. This would be a shame, as one can envision subrecursive languages which are capable of computing quite small subsets of $\mathcal{PR}$.

Fortunately, we can avoid restrictions to the case $\mathcal{PR} \subseteq \mathcal{J}$ in analogs of Theorem 3 in many interesting cases. For instance, suppose that the programs in a subrecursive programming language are only allowed to be built up by primitive recursions to a maximum depth d_0 and that each program i has associated with it the depth of nesting $d(i) \leq d_0$ of the primitive recursions used to define program i. We can now consider a modified synthesis function p' for primitive recursions which only satisfies (2) for those i such that $d(i) < d_0$. A modified resource bound function P' satisfying a condition analogous to (3) can also be formulated. If p',P', and the functions mentioned in Definition 3 not only are capable of being computed in the programming language but also are capable of being computed <u>by programs having depth of nesting of primitive recursions</u> $< d_0$, then an analog of Theorem 3 can be stated and proven.

8 CONCLUSIONS AND QUESTIONS

For an indexing to model a programming language which is at all reasonable, certain basic conditions must be satisfied. Certainly Definition 3.(i) of Section 6 should be satisfied by a "very subrecursive" function s, and certainly the universal function of the indexing should be partial recursive. If the indexing is of a subrecursive collection of functions $\mathcal{J}$, we could also require that the universal function be "reasonably close to" belonging to $\mathcal{J}$; for instance, we might require that the universal function be doubly recursive for an indexing of the primitive recursive functions.

For indexings which satisfy such conditions and which index the same collection of functions, it is reasonable to compare how "acceptable" or "natural" they are by comparing the set of operators which have "very subrecursive" synthesizing functions with respect to each indexing. (To phrase things in classical terminology, synthesizing functions correspond to effective operators.) This suggests that instead of asking for a definition of <u>the</u> "acceptable" indexings of a subrecursive set, as Constable & Borodin (1972) asked, we should instead ask <u>how</u> "acceptable" various indexings are and should compare their relative acceptability with respect to the partial order of set inclusion on their sets of operators which have "very subrecursive" synthesizing functions.

For indexings which satisfy the kinds of conditions suggested in the first paragraph of this section, we can gauge how "natural" a complexity measure is in terms of how many "very subrecursive" synthesizing functions have "very subrecursive" resource estimation functions. In Section 5 we only gave examples of specific operators and specific synthesizing functions. A general approach, based on the effective continuity of recursive operators, is implicit in Symes (1971) and, with more generality at the expense of more notation, Alton (1977a). Instead of attempting to characterize the "natural" complexity measures, a problem posed in Hartmanis (1973), this suggests we should ask how "natural" a complexity measure is.

Elaborate definitional frameworks are suspect unless they suggest theorems. Fortunately the ideas of Sections 4 and 5 prompted the axioms of Theorem 3 of Section 6. This theorem embodies the basic technique required to establish a subrecursive analog (Alton, 1977b,1980) of the speed-up theorem of Blum (1967a). As such, this appears to begin to answer the problem of finding an abstract theory of subrecursive complexity, as requested in Constable & Borodin (1972). The axioms embodied in Definition 3 at the start of Section 6 appear to be at a "level of abstractness" roughly equal to that of Blum's original axioms. The additional hypotheses of Theorem 3, concerning the existence of subrecursive synthesizing and resource estimation functions for the operation of primitive recursion, are much less abstract. We suspect that this occasional use of a more "specific" axiom is a necessary characteristic of successful efforts to adapt Blum's theory to a subrecursive setting.

Frequently we have shown that "obvious" attempts to establish something on the basis of certain axioms fail and then we have gone on to formulate additional axioms which are strong enough to establish the desired conclusion. A careful study of the independence of various axioms would be of interest and might shed light on the suspicion expressed at the end of the previous paragraph. One possible approach might concern efforts to establish subrecursive analogs of results of Riccardi (1980). For indexings of all the partial recursive functions, it is essential that Riccardi's study is not limited to acceptable indexings. For subrecursive indexings, can interesting results be obtained when attention is restricted to indexings having the kinds of properties suggested in the first paragraph of this section?

Our proofs have used the recursion theorem without apology, but computer scientists have sometimes viewed appeals to it as a suspect form of magic. For acceptable indexings of the partial recursive functions and complexity measures, an alternative pattern of proof is available. This is based in part on the result of Rogers

(1958,1967) that any two acceptable indexings of the partial recursive functions can be related by a recursive isomorphism which takes each index in one indexing to an index which computes the same partial function in the other indexing. The alternative approach also depends on a result of Blum (1967a) known as the recursive relatedness theorem, the proof of which uses unbounded minimization. Reasonable subrecursive analogs of such results cannot be expected to hold, so it appears that a reasonably abstract approach to subrecursive complexity will need to rely upon the recursion theorem.

9 ACKNOWLEDGMENTS

This research was partially supported by N.S.F. grants MCS76-15648 and MCS78-20047. Much of the exposition is the result of talks given at Novosibirsk State University and the Algebra and Logic Seminar of the Mathematical Institute of the Siberian Academy of Sciences during a Fulbright-Hays lectureship in May, 1979. The author is grateful to Professors Ju. L. Eršov and B. A. Trakhtenbrot for their generous hospitality and stimulating conversations during his visit there. He also wishes to thank Professor Juris Hartmanis for his continuing encouragement and for stimulating conversations.

REFERENCES

Alton, D.A. (1977a). "Natural" complexity measures and time versus memory: Some definitional proposals. In Automata, Languages, and Programming, ed. A. Salomaa & M. Steinby, pp. 16-29. Springer-Verlag, New York.

Alton, D.A. (1977b). "Natural" complexity measures and a subrecursive speed-up theorem. In Information Processing 77, ed. B. Gilchrist, pp. 835-838. North-Holland, New York.

Alton, D.A. (1980). A subrecursive speed-up theorem. In preparation.

Ausiello, G. (1974). Relations between semantics and complexity of recursive programs. In Proceedings of the Second Colloquium on Automata, Languages, and Programming, ed. J. Loeckx, pp. 129-140. Springer-Verlag, New York.

Baker, T. (1973). Computational complexity and nondeterminism in flowchart programs. Technical report TR 73-185. Department of Computer Science, Cornell University, Ithaca, N.Y.

Baker, T. (1978). "Natural" properties of flowchart step-counting measures. J. Comput. Syst. Sci., 16, pp. 1-22.

Běcvář, J. (1972). Programs and complexity measures. Talk at Symposium on the Theory of Algorithms and their Complexity. Oberwolfach, West Germany.

Biskup, J. (1977). Über Projektionsmengen von Komplexitätsmaßen. Elektronische Informationsverarbeitung und Kybernetik, 13, pp. 359-368.

Blum, M. (1967a). A machine-independent theory of the complexity of recursive functions. J. ACM, 14, pp. 322-336.

Blum, M. (1967b). On the size of machines. Information and Control, 11, pp. 257-265.

Borodin, A.B. (1972). Computational complexity and the existence of complexity gaps. J. ACM, 19, pp. 158-174.

Borodin, A.B. (1973). Computational complexity: Theory and practice. In Currents in the Theory of Computing, ed. A.V. Aho, pp. 35-89. Prentice-Hall, Englewood Cliffs, N.J.

Constable, R.L. (1971a). Subrecursive programming languages III, the multiple-recursive functions $\mathcal{R}^n$. In Proceedings of the Symposium on Computers and Automata, ed. Jerome Fox, pp. 393-410. Vol. XXI, Microwave Research Institute Symposia Series. Polytechnic Press of the Polytechnic Institute of Brooklyn, N.Y.

Constable, R.L. (1971b). Subrecursive programming languages II: On program size. J. Comput. Syst. Sci., 5, pp. 315-334.

Constable, R.L. (1973). Type two computational complexity. In Proceedings of Fifth Annual ACM Symposium on Theory of Computing, pp, 108-121. Association for Computing Machinery, New York.

Constable, R.L. & Borodin, A.B. (1972). Subrecursive programming languages I: Efficiency and program structure. J. ACM., 19, pp. 526-568.

Coy, W. (1976). The logical meaning of programs of a subrecursive language. Information Processing Letters, 4, pp. 121-126.

Goetze, B. & Nehrlich, W. (1978). Loop programs and classes of primitive recursive functions. In Mathematical Foundations of Computer Science 1978, ed. J. Winkowski, pp. 232-238. Springer-Verlag, New York.

Grzegorczyk, A. (1953). Some classes of recursive functions. Rozprawy Mat., 4.

Hamlet, R. (1974). Introduction to Computation Theory. Intext, New York.

Hartmanis, J. (1973). On the problem of finding natural complexity measures. In Proceedings of Symposium on Mathematical Foundations of Computer Science, pp. 95-103. Mathematical Institute of the Slovak Academy of Sciences and Computing Research Centre, United Nations D.P. Bratislava. Also, Technical Report TR 73-175. Department of Computer Science, Cornell University, Ithaca, N.Y.

Hartmanis, J. & Hopcroft, J. (1971). An overview of the theory of computational complexity. J. ACM, 18, pp. 444-475.

Kleene, S.C. (1950). Introduction to Metamathematics. Van Nostrand, Princeton, N.J.

Kleene, S.C. (1958). Extension of an effectively generated class of functions by enumeration. Colloquium Mathematicum, 6, pp. 67-78.

Landweber, L. & Robertson, E. (1972). Recursive properties of abstract complexity classes. J. ACM, 19, pp. 296-308.

Lischke, G. (1975a). Über die Erfüllung gewisser Erhaltungssätze durch Kompliziertheitsmasse. Zeitschr. f. math. Logik und Grundlagen d. Math., 21, pp. 159-166.

Lischke, G. (1975b). Flußbildmaße—Ein Versuch zur Definition näturlicher Kompliziertheitsmaße. Elektronische Informationsverarbeitung und Kybernetik, 11, pp. 423-436.

Lischke, G. (1976a). Näturliche Kompliziertheitsmaße und Erhaltungssätze I. Zeitschr. f. math. Logik und Grundlagen d. Math., 22, pp. 413-418.

Lischke, G. (1976b). Erhaltungssätze in der Theorie der Blumschen Kompliziertheitsmaße. Dissertation, Jena. (Date of 1976 is not known for certain.)

Lischke, G. (1977). Natürliche Kompliziertheitsmaße und Erhaltungssätze II. Zeitschr. f. math. Logik und Grundlagen d. Math., 23, pp. 193-200.

Lynch, N.A. (1972). Relativization of the theory of computational complexity. Technical report MAC TR-99. Project MAC, Massachusetts Institute of Technology, Cambridge, Mass.

McCreight, E.M. (1969). Classes of computable functions defined by bounds on computation. Ph.D. thesis. Carnegie-Mellon University, Pittsburgh, Pa.

McCreight, E.M. & Meyer, A.R. (1969). Classes of computable functions defined by bounds on computation. In _Proceedings of ACM Symposium on Theory of Computing_, pp. 79-88. Association for Computing Machinery, New York.

Meyer, A.R. (1972). Program size in restricted programming languages. _Information and Control_, 21, pp. 382-394.

Meyer, A.R. & Ritchie, D.M. (1967a). The complexity of Loop programs. In _Proceedings of the 22nd National ACM Conference_, pp. 465-469. Thompson Book Co., Washington, D.C.

Meyer, A.R. & Ritchie, D.M. (1967b). Computational complexity and program structure. Technical Report RC-1817. IBM Thomas J. Watson Labs, Yorktown Heights, N.Y.

Pagan, F.G. (1973). On the teaching of disciplined programming. _ACM SIGPLAN Notices_, 8, pp. 44-48.

Riccardi, G.A. (1980). The independence of control structures in abstract programming systems. Ph.D. dissertation. State University of New York at Buffalo.

Rogers, H. (1958). Gödel numberings of partial recursive functions. _J. Symbolic Logic_, 23, pp. 331-341.

Rogers, H. (1967). _Theory of Recursive Functions and Effective Computability_. McGraw-Hill, New York.

Symes, D. (1971). The extension of machine-independent computational complexity theory to oracle machine computation and to the computation of finite functions. Research report CSRR 2057. Department of Applied Analysis and Computer Science, University of Waterloo, Ontario, Canada.

Verbeek, R. (1973). Erweiterungen subrekursiver Programmiersprachen. In _Gesellschaft für Informatik Fachtagung über Automatentheorie und Formale Sprachen_, pp. 311-318. Springer-Verlag, New York.

Wechsung, G. (1975). Zulässige Untermaße von Kompliziertheitsmaßen. In _Kompliziertheit von Lern- und Erkennungsprozessen_, Vol. II, pp. 239-244. Sektion Mathematik, Friedrich-Schiller Universität, Jena.

Young, P.R. (1973). Easy constructions in complexity theory: Gap and speed-up theorems. _Proc. AMS_, 37, pp. 555-563.

COMPLEXITY THEORY WITH EMPHASIS ON THE COMPLEXITY OF LOGICAL THEORIES

Richard E. Ladner
Department of Computer Science
University of Washington
Seattle, Wa. 98195 U.S.A.

Table of Contents

LECTURE 1. BASIC COMPLEXITY THEORY

1.1 Motivation.

Complexity theory, at least as understood by computer scientists, is the study of the intrinsic difficulty of solving problems on a computer. In this series of lectures we attempt to give an advanced course in complexity theory with emphasis on some mathematical abstractions of computation such as nondeterminism, alternation and pushdown stores and their application to the understanding of the computational complexity of some logical theories. This course is not intended to be a survey of all that is known about the complexity of logical theories, but is instead intended to provide a student of the subject with some of the key ideas and methods so that he or she can pursue the subject further.

To clarify our definition of complexity theory we need to be more precise about what we mean by "problem", "solving a problem", "intrinsic difficulty" and "computer".

Let Σ be a finite character set with at least two characters in it. The set Σ^* represents the set of finite length strings in the alphabet Σ. If $x \in \Sigma^*$ then $|x|$ is the length of x. Members of Σ^* represent potential inputs or outputs of a computer program. A problem is a function from Σ^* into Σ^*. A 0-1 problem is a function from $\Sigma^* \to \{0,1\}$. A 0-1 problem is commonly defined by a set, namely the set of strings whose image is 1 under the mapping. To solve a problem means to construct a computer program which has the same input/output behavior as the function defining the problem. The <u>intrinsic difficulty of a problem</u> is the time or storage required by computer programs which solve the problem, measured as a function of the input length.

There are two aspects to the complexity of a problem. First, an <u>upper bound</u> for a problem is a function f from natural numbers to natural numbers with the property that there is a program which solves the problem and for all inputs x the program runs in time (space) $\le f(|x|)$. The number $f(|x|)$ represents an upper bound on the number of steps the program can run on input x. Second, a <u>lower bound</u> for a problem is a function f with the property that for <u>any</u> program solving the problem there are infinitely many inputs x such that the program runs in time (space) $> f(|x|)$ on input x. An upper bound for a problem is usually demonstrated by exhibiting an algorithm. A lower bound for a problem requires a demonstration that every program has a certain property. The methods for showing lower bounds is a principal topic of this course.

Although the complexity of arbitrary problems is of interest to computer scientists, their main interest is in the complexity of "natural" problems. A problem is natural to a computer scientist if it is one he or she would want to solve on a computer. We give some examples of natural problems and give their currently best known upper and lower bounds. For clarification the symbol "n" always indicates input size, $O(f(n))$ indicates any function g such that there is a $c > 0$ such that $g(n) \le cf(n)$ for all but finitely many n, and $\Omega(f(n))$ indicates any function g such that $f(n) = O(g(n))$.

<u>Examples</u>:

1. Matrix Multiplication

Input: $n \times n$ matrices A and B.
Output: a matrix C such that $C = A \times B$.

Upper bound: $O(n^{2.61})$ time [Sc 79], [Pa 79],

Lower bound: $\Omega(n^2)$ time.

2. CLIQUE

Input: Graph G, integer k.

$$\text{Output: } \begin{cases} 1 & \text{if G has a clique of size} \ge k, \\ 0 & \text{otherwise.} \end{cases}$$

(A clique of size k is a set of k nodes each of which is adjacent to each of the others.)

Upper bound: $O(c^n)$ time for some $c > 1$,

Lower bound: $\Omega(n)$ time.

3. SAT (Satisfiability)

Input: Boolean formula F in conjunctive normal form.

$$\text{Output: } \begin{cases} 1 \text{ if F is satisfiable,} \\ 0 \text{ otherwise.} \end{cases}$$

Upper bound: $O(c^n)$ time for some $c > 1$,

Lower bound: $\Omega(n)$ time.

4. RA (real addition)

Input: First-order formula F over +, <, =.

$$\text{Output: } \begin{cases} 1 \text{ if F true in } (\mathbb{R}, +, <), \\ 0 \text{ otherwise.} \end{cases}$$

Upper bound: $O(c^n)$ space for some $c > 1$ [FeR 75],

Lower bound: $\Omega(d^n)$ time for some $d > 1$ [FiR 74].

Notice that the gap between upper and lower bounds for matrix multiplication is quite small compared with that for CLIQUE and SAT. In this and the next lecture we will demonstrate that CLIQUE and SAT are equivalent problems in a certain sense. If we could find better upper or lower bounds for one of the problems then we could for the other. The situation for RA is much better. In fact in the third lecture we will use the concept of alternation to even better characterize the complexity of RA.

1.2 The Computer.

In order to investigate lower bounds we need an accurate mathematical model of a computer and computer program. Not by coincidence such a model, the Turing machine, was defined by Turing in 1936 [Tu 36] almost a decade before the advent of high-speed electronic computers. A Turing machine models both the computer and the computer program. The computer consists of an infinite tape partitioned into cells. Each cell holds one of a finite number of characters. There is a read head which is scanning one of the cells. The computer program consists of a finite set of instructions for reading a tape cell, writing in a tape cell and moving the read head left or right.

The program can be specified by a partial function

$$\delta : Q \times \Gamma \to Q \times \Gamma \times \{1, -1\}$$

where Q is a finite set of <u>states</u> and Γ is the <u>character set</u> of the machine. If the machine is in state q reading the symbol a then it goes to state p, writes symbol b and moves d cells from its current position, where $\delta(q,a) = (p,b,d)$. The <u>initial configuration</u> of the machine is one with the input of length n in the first n tape cells, the read head on the first tape cell and the machine in a specified <u>start</u> state. The tape cells in

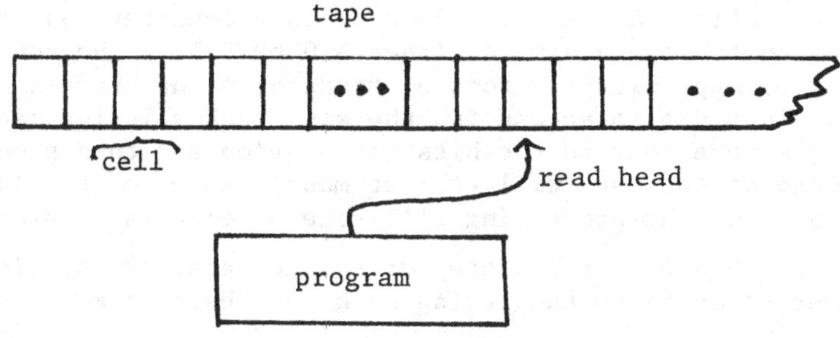

TURING MACHINE

positions n + 1, n + 2, . . . are filled with a special blank symbol. The machine runs according to the partial function δ until it can no more because δ is undefined. The output of the machine can be specified in several different ways. The non-blank portion of the tape can indicate the output or there may be an accepting state which if entered indicates that the input has been <u>accepted</u>. The latter specification is used for solving 0-1 problems.

Let x be an input. A Turing machine T <u>runs in time</u> t on input x if started in the initial configuration for x it makes $\leq$ t steps before halting. The machine T <u>runs in space</u> s on input x if started in the initial configuration for x it scans $\leq$ s distinct tape cells.

It is sometimes difficult for students to believe that the Turing machine is anything like a real computer but there is plenty of evidence that it is. Unfortunately it does take a leap of faith. The dogma of this faith can be summarized in what we call

The Practical Church - Turing Thesis

The Turing machine accurately reflects the time and space of computer computations to within a polynomial.

To be more precise, given any computer there are constants c and k such that any problem solved on the computer in time t(n) and space s(n) can be solved by a Turing machine in time $c(t(u))^k$ and space cs(u). If we allow our Turing machines to have multiple tapes or heads there is strong evidence that the constant k is actually 2. Let me give a brief argument supporting the contention that k = 2. In t steps a computer can produce no more than t bits (a bit is either a 0 or a 1). The computer can access perhaps directly some of these bits and process them in one step. In order to accomplish the same task a Turing machine must scan its tape to find the bits in question and then process them. The scan of the tape will take at most t moves of the Turing machine, and the processing will take at most a constant amount of time. Hence, the tth step of the computer can be simulated in at most ct moves of the Turing machine. Hence t moves of the computer can be simulated in at most ct^2 moves of the Turing machine.

Once we have accepted the premise that Turing machine time and space is worthy of study we can proceed. For simplicity we will now restrict our attention to 0-1 problems or equivalently sets of strings. A problem A is solved by a Turing machine T if for all $x \in \Sigma^*$, $x \in A$ if and only if T accepts x, that is, T starting in the initial configuration for x halts in its accepting state.

Define

$$\mathrm{TIME}(t(n)) = \{A \subseteq \Sigma^* : \text{there is a Turing machine T that solves A and for all } x \in \Sigma^*, \text{T runs in time} \le t(|x|)\}$$

$$\mathrm{SPACE}(s(n)) = \{A \subseteq \Sigma^* : \text{there is a Turing machine T that solves A and for all } x \in \Sigma^*, \text{T runs in space} \le s(|x|)\}$$

$$\mathcal{P} = \bigcup_{c,k} \mathrm{TIME}(cn^k).$$

The class $\mathcal{P}$ is the set of problems that can be solved in polynomial time. There is general agreement in the computing field that if $A \notin \mathcal{P}$ then A cannot be solved in practice except on small inputs and special cases of longer inputs.

1.3 Hierarchy Theorems.

One of the first questions one might ask about time or space complexity is, "Are there problems solvable in time, say, $O(n^3)$ but not solvable in time $\Omega(n^2)$?" The answer is "yes" although the only such problems we know of do not qualify as natural problems. The following "Hierarchy theorems" are basic in understanding time and space complexity on multitape Turing machines.

Time Hierarchy Theorem (Hartmanis & Stearns, 1965). If $T_2(n)$ is fully time-constructable and $\liminf_n \frac{T_1(n)\log(T_1(n))}{T_2(n)} = 0$ then $TIME(T_2(n)) - TIME(T_1(n)) \neq \emptyset$. [HS 65].

Space Hierarchy Theorem (Hartmanis, P. Lewis, & Stearns, 1965). If $S_2(n)$ is fully space-constructable and $\liminf_n S_1(n)/S_2(n) = 0$ then $SPACE(S_2(n)) - SPACE(S_1(n)) \neq \emptyset$. [HLS 65].

The hypotheses of fully time- or space-constructable are technical hypotheses which hold for any reasonable function like n^2, 2^k, n log n and so on. The import of these hypotheses will be given in our outline of the proof of the latter theorem.

Proof of Space Hierarchy Theorem.

Let $\liminf_n S_1(n)/S_2(n) = 0$. We describe a Turing machine M which accepts a set in $S_2(n)$ storage. Let us assume for convenience that $S_2(n) \geq n$ although the theorem holds for $S_2(n) < n$ if we only count the storage used by tapes other than the input tape. Let x be an input of length n. M begins by laying out $S_2(n)$ storage (this is where the hypothesis of fully space-constructable is used). The input x is now treated both as an input and as the encoding of a Turing machine. The machine x is then simulated on input x until one of the following happens.

(i) $S_2(n)$ storage is exceeded,

(ii) $2^{S_2(n)}$ steps of x have been simulated,

(iii) x is rejected by x in the allotted storage,

(iv) x is accepted by x in the allotted storage.

The second condition is enforced by setting up a binary counter in the allotted storage. In the first three cases M accepts x and in the fourth case M rejects x. Let K be the set accepted by M. By the construction $K \in SPACE(S_2(n))$.

Suppose $K \in SPACE(S_1(n))$ then there is our encoding x_K of a Turing machine which runs in space $S_1(n)$. Now, x_K runs in time $\leq c^{S_1(n)}$ for some constant $c > 1$, because there are $\leq c^{S_1(n)}$ distinct configurations the machine can be in while using $\leq S_1(n)$

storage. Choose, $n \geq |x_K|$ such that $2^{S_2(n)} > c^{S_1(n)}$ using the hypothesis that $\liminf_n S_1(n)/S_2(n) = 0$. The encoding x_K can be "padded" into an encoding x_K' where $|x_K'| = n$. On input x_K' the machine M which presumably accepts K does not do what it is supposed to. Hence we have a contradiction. Notice that the counter which counts to $2^{S_2(n)}$ is basically used to force M to halt on all inputs. Thus M is able to detect if the machine x is "looping" in its allotted storage.

The Time Hierarchy Theorem has the added log factor because multitapes must be simulated in a similar diagonalization. Unfortunately, we only know that a multitape Turing machine that runs in time T(n) can be simulated by a two-tape Turing machine that runs in time T(n)log(T(n)) [HeS 66]. A multitape Turing machine that runs in space S(n) can be simulated by a one-tape Turing machine that also runs in space S(n).

The set K does not qualify as a natural problem because it is not one that people are interested in solving on a computer. Surprisingly, however, knowing that sets like K exist enables us in some cases to establish lower bounds on the complexity of natural problems. This is done by reducing efficiently K to our natural problem of interest.

1.4 Efficient Reducibilities.

A reduction procedure for our purposes is an oracle Turing machine. Such a procedure is time t(n), space s(n), and size z(n) bounded if for all inputs of length n and for all oracles the procedure runs in time t(n), space s(n), and its queries are bounded in length by z(n).

Efficient reduction procedures have played an important role in the classification of the complexity of problems. There are two primary uses of efficient reducibility.

First, efficient reducibilities are used to relate the complexity of problems even though we may not know their actual complexity. The class of NP-complete problems, which we discuss in the next lecture, includes a large number of natural problems all of which are polynomial time reducible to each other. Hence if just one of these problems is solvable in polynomial time then all of them are.

As an example of this use of efficient reducibilities we show

Theorem (Cook, 1971). SAT is reducible to CLIQUE in polynomial time [Co 71a].

Proof. Let $\bigwedge_{i=1}^{p} \bigvee_{j=1}^{\ell(i)} \sigma_{ij}$ be an instance of the satisfiability problem where $\sigma_{ij} \in \{x_1, \overline{x}_1, \ldots, x_m, \overline{x}_m\} = L$. For convenience if $\sigma \in L$ then $\overline{\sigma}$ denotes the complement of σ. We map this instance of SAT to the following instance of CLIQUE.

Graph:

Nodes: $\{(\sigma_{ij}, i) : 1 \le i \le p,\ 1 \le j \le \ell(i)\}$,

Edges: $\{\{(\sigma,i)(\gamma,j)\} : \sigma \ne \overline{\gamma}$ and $i \ne j\}$,

Clique size: p.

This mapping amounts to a polynomial time bounded many-one reduction of SAT to CLIQUE.

The second use of efficient reducibilities is to establish an upper or lower bound on the complexity of a problem by relating it to the known complexity of another problem. For example in the third lecture we will show that every problem in $\bigcup_c \mathrm{TIME}(2^{cn})$ is reducible to RA (real addition) in polynomial time and linear size. Let $K \in \mathrm{TIME}(2^{2n}) - \mathrm{TIME}(2^n)$ using the Time Hierarchy Theorem. Since $K \in \mathrm{TIME}(2^{2n})$ then there is a polynomial $p(n)$ and a constant c such that K is reducible to RA in time $p(n)$ and size cn.

Choose $d > 1$ in such a way that $(2^{cdn} + 1)p(n) \le 2^n$. If $\mathrm{RA} \in \mathrm{TIME}(2^{dn})$ then using the reduction procedure of K to RA we could decide K in time $(2^{cdn} + 1)p(n)$ by applying the 2^{dn} time decision procedure for RA whenever the oracle is queried. But this contradicts the fact that $K \notin \mathrm{TIME}(2^n)$. Hence $\mathrm{RA} \notin \mathrm{TIME}(2^{dn})$.

The general principle is that if A is difficult to decide and A is efficiently reducible to B then B is also difficult to decide.

LECTURE 2. CONCEPTS BEYOND THE COMPUTER

2.1 Nondeterminism.

The concept of a nondeterministic computation would never have had such broad interest except for the fact that many natural problems can be solved quickly nondeterministically but do not seem amenable to efficient deterministic solution. For example, the SAT problem can be solved nondeterministically by "guessing" a Boolean assignment to the variables, then checking to see if it satisfies the formula. Enumerable problems have this character: that a proposed solution to the problem is easily checked while finding a solution seems to involve an exhaustive search.

A nondeterministic Turing machine is one with the ability to make more than one possible move from a given configuration. We formalize the notion by defining the program of the nondeterministic Turing machine T to be a function

$$\delta : Q \times \Gamma \to 2^{Q\times\Gamma\times\{1,-1\}}$$

where 2^S represents the power set of S. We use the notation

$$C \underset{T}{\to} D$$

to indicate that the configuration D follows from the configuration C in one move of T. We let $I_T(x)$ denote the initial configuration of T on input x and ACC_T denote a unique accepting configuration. We drop the subscript T when no confusion could arise. A <u>computation</u> is a sequence

$$C_0 \to C_1 \to \dots \to C_m,$$

where each C_i is a configuration. The Turing machine T <u>accepts</u> an input x in time t (space s) if there is a computation sequence $I(x) = C_0 \to C_1 \to \dots \to C_m = ACC$ such that $m \le t$ (the read head visits no more than s distinct tape cells in this computation sequence). We say that the nondeterministic Turing machine T solves a problem $A \subseteq \Sigma^*$ in time t(n) (space s(n)) if for all $x \in \Sigma^*$, $x \in A$ if and only if x is accepted by T in time $t(|x|)$ (space $s(|x|)$).

$$\text{NTIME}(t(n)) = \{A \subseteq \Sigma^* : \text{there is a nondeterministic Turing machine which solves A in time } t(n)\}$$

$$\text{NSPACE}(s(n)) = \{A \subseteq \Sigma^* : \text{there is a nondeterministic Turing machine which solves A in space } s(n)\}$$

$$NP = \bigcup_{c,k} \text{NTIME}(cn^k).$$

The class *NP* represents those problems whose solutions can be "checked" in polynomial time. Probably the most important open problem in theoretical computer science is: does

$$P = NP?$$

In 1971, Cook discovered that there are hardest problems in NP [Co 71a]. Later Karp named these hardest problems in NP, NP-complete problems and greatly expanded Cook's original list of natural NP-complete problems [Ka 72].

A problem L is <u>NP-complete</u> if

(i) $L \in NP$,

(ii) every problem in NP is reducible to L in polynomial time.

It is not hard to argue that $P = NP$ if and only if some NP-complete problem is in P. Hence the open question "$P = NP$?" can be solved by showing one of them is in P or one of them is not in P.

<u>Cook's Theorem</u>. SAT is NP-complete.

<u>Proof</u>. Let L be a member of NP and suppose that L is solved by T in time p(n) where p is a polynomial. Let x be an input of length n. We may assume T is a one-tape Turing machine whose configurations can be represented by sequences of the form

$$a_1 a_2 \ldots a_{i-1} (q, a_i) a_{i+1} \ldots a_m$$

where $m = p(n)$, q is a state and a_i is a tape symbol. A configuration symbol is either a tape symbol or a state-symbol pair. Given x and T we construct a formula with propositional variables

$P_{it\sigma}$ where $0 \leq i, t \leq m$ and σ is a configuration symbol.

Informally $P_{it\sigma}$ will be assigned "true" if the i-th configuration symbol of the t-th configuration is σ. The formula asserts the conjunction of the following:

(i) There is exactly one configuration symbol per position and time,

(ii) There is exactly one state-symbol pair per time,

(iii) The initial configuration is right,

(iv) The last configuration is accepting,

(v) Consecutive configurations follow the rules of the Turing machine T.

For example (ii) can be expressed as

$$\bigwedge_{t} \bigvee_{i,q,a} P_{it(q,a)} \wedge \bigwedge_{\substack{t,i,j\\p,q,a,b\\(i,p,a)\neq(j,q,b)}} \sim P_{it(p,a)} \vee \sim P_{jt(q,b)}$$

The clause (v) is the most interesting to try to construct. We define a relation CONSIST of six-tuples $(\sigma_1, \sigma_2, \sigma_3, \tau_1, \tau_2, \tau_3)$ of configuration symbols with the property that if $\sigma_1\sigma_2\sigma_3$ are symbols in three consecutive positions of a configuration then $\tau_1\tau_2\tau_3$ are consistent according to the rules of T for the same consecutive positions in the next configuration. Now (v) can be expressed

$$\bigwedge_{i,t} \bigvee_{\substack{(\sigma_1,\sigma_2,\sigma_3,\\ \tau_1,\tau_2,\tau_3)\\ \in \text{CONSIST}}} (P_{i,t,\sigma_1} \wedge P_{i+1,t,\sigma_2} \wedge P_{i+2,t,\sigma_3} \wedge P_{i,t+1,\tau_1} \wedge P_{i+1,t+1,\tau_2} \wedge P_{i+2,t+1,\tau_3})$$

Technically this formula is not in conjunctive normal form but it can be transformed into that form with only a constant factor increase in size.

Corollary CLIQUE is NP-complete.

Proof. It is easy to see that CLIQUE is in NP by guessing a clique of the desired size. By our reduction of SAT to CLIQUE mentioned in Lecture 1 CLIQUE must also be NP-complete.

The proof technique of the corollary is basic in demonstrating the NP-completeness of new problems. To show a new problem is NP-complete first show it is in NP then show some already known NP-complete problem is reducible to it in polynomial time. The excellent book Computer and Intractability, A Guide to the Theory of NP-Completeness, by Garey and Johnson provides a thorough study of the aspects of computational complexity we have covered so far [GJ 79].

2.2. Alternation.

Nondeterministic Turing machines have the ability to "guess" a next move, that is, just make one of several possible next moves. Suppose we add the capability that in a certain state, reading a certain symbol, the machine must make all of its next possible moves. Such machines, called alternating Turing machines, were originally defined and studied by Chandra and Stockmeyer [CS 76] and Kozen [Ko 76] independently. Later the three authors collaborated to produce a single paper titled simply "Alternation" [CKS 79]. At first glance the notion of an alternating Turing machine may seem contrived and they certainly would have languished in obscurity had they not been useful in understanding the complexity of natural problems. Alternation has allowed us to have a more precise measure on the complexity of certain problems like RA (real addition) which we will discuss in the next lecture. More importantly, the intimate relationship between alternation and normal time and space complexity on deterministic machines has given us a needed tool in proving lower bounds for certain problems, like the decision problem for the $\exists^*\forall\exists^*$ predicate calculus, which we discuss in the last lecture. Other applications of alternation can be found in the study of the complexity of games, systems of communicating processes, programming logics, and context free language recognition.

To formally define an alternating Turing machine, T, we partition the states Q into disjoint sets E and U. The states in E are called existential states and the states in U are called universal states. An existential (universal) configuration is one whose state is existential (universal). As usual there is an initial configuration I(x) for the input x and a unique accepting configuration ACC. Instead of the notion of a computation sequence as we defined for nondeterministic machines we define the notion of a computation tree. A computation tree is a tree with its nodes labeled with configurations and with the properties

(i) Each internal node (non-leaf) labeled with a universal configuration C has a child labeled D for each configuration D which follows from C in one move of T,

(ii) Each internal node labeled with an existential configuration C has exactly one child labeled D, where D is a configuration which follows from C in one move of T.

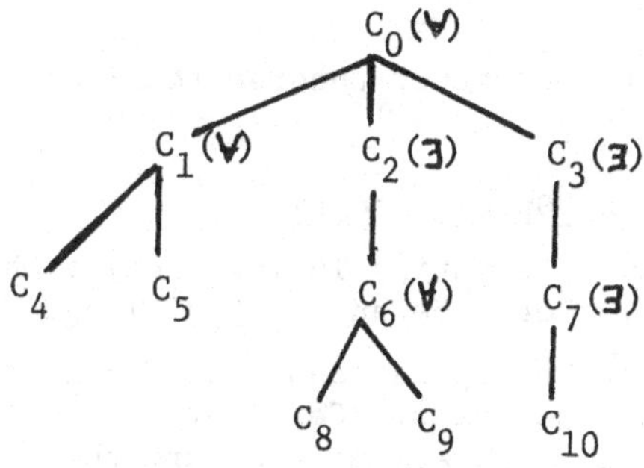

A COMPUTATION TREE

An input x is accepted by T in time t (space s) if there is a computation tree with root labeled I(x), leaves all labeled ACC, and height $\leq$ t (the read head visits no more than s distinct tape cells in the computation tree). The time of an alternating machine represents the parallel time to do an alternating computation. The space represents the local storage required by each of the processors active in the parallel execution of an alternating machine. These interpretations of time and space have led researchers to study the alternating machine as a model of parallel computation.

We say that an alternating Turing machine T solves a problem $A \subseteq \Sigma^*$ in time t(n) (space s(n)) if for all $x \in \Sigma^*$, $x \in A$ if and only if x is accepted by T in time $t(|x|)$ (space $s(|x|)$).

$$\text{ATIME}(t(n)) = \{A \subseteq \Sigma^*: \text{ there is an alternating Turing machine which solves A in time } t(n)\}$$

$$\text{ASPACE}(t(n)) = \{A \subseteq \Sigma^*: \text{ there is an alternating Turing machine which solves A in space } s(n)\}$$

The intimate relationship between alternating time and space, and deterministic time and space can be summarized in the following two theorems.

Alternating Time Theorem (Chandra, Kozen & Stockmeyer, 1976). If $t(n) \geq n$ and $s(n) \geq n$, then

(i) $\text{ATIME}(t(n)) \subseteq \text{SPACE}(t(n))$

(ii) $\text{NSPACE}(s(n)) \subseteq \bigcup_c \text{ATIME}(c(s(n))^2)$.

Alternating Space Theorem (Chandra, Kozen & Stockmeyer, 1976). If $s(n) \geq \log n$ then

$$\text{ASPACE}(s(n)) = \bigcup_c \text{TIME}(2^{cs(n)}).$$

We give a proof of the latter theorem to illustrate the ideas involved

Proof of Alternating Space Theorem.

($\subseteq$) Let T be an alternating Turing machine which solves A in space $s(n) \geq \log n$. Let x be an input of length n. There are at most $2^{cs(n)}$ distinct configurations that T can attain in an accepting computation tree for x. Consider the following "proof system" for verifying configurations that are the root of computation trees all of which have leaves all labeled ACC. We assume that T always has a move unless it is in the configuration ACC.

Proof System for Acceptance

(i) $\dfrac{}{\text{ACC}}$,

(ii) $\dfrac{D}{C}$ if C is existential and D follows from C in one move of T,

(iii) $\dfrac{D_1, D_2, \ldots, D_k}{C}$ if C universal and $D_1, \ldots, D_k$ are the the configurations that follow from C in one move of T.

We write $\vdash C$ if C is provable in this proof system. Further we write $\models C$ if C is the root of a computation tree whose leaves are all ACC.

Completeness Lemma. $\vdash C$ if and only if $\models C$.

This is proved in one direction by induction on proof length and in the other by induction on the size of the computation trees realizing $\models C$.

Complexity Lemma. $\{C: \vdash C\}$ can be constructed in time 2^{dn} for some $d > 1$.

Start with $S = \emptyset$. Now try to find a new "theorem" using the proof rules. If a new theorem is found enlarge S to include it and continue trying to find another new theorem. If no new theorem can be found then S must be $\{C: \vdash C\}$ There are at most 2^{cn} new theorems to be found. Each one requires no more than $b(2^{cn})^k$ time to find where b is a constant and k is an upper bound on the number of next possible configurations of a configuration. Hence the whole process takes no more than $b(2^{cn})^k\, 2^{cn} \leq 2^{dn}$ time.

Now x is accepted by T in space s(n) if and only if $\models I(x)$. By the two Lemmas this can be discovered in time 2^{dn}.

($\supseteq$) Let T be a deterministic Turing machine that solves A in time $2^{cs(n)}$. Let x be an input of length n. A configuration can be represented by a string $a_1 a_2 \ldots a_{i-1} (q, a_i)\, a_{i+1} \ldots a_m$ where $m = 2^{cs(n)}$. Define γ_{it} to be the configuration symbol in the i-th position of the t-th configuration. We design an alternating Turing machine M which tests if $\gamma_{i,t} = z$ where x,i,t,z are inputs and i,t are written in binary. Notice that γ_{it} depends only on $\gamma_{i-1,t-1}$, $\gamma_{i,t-1}$, and $\gamma_{i+1,t-1}$.

Test for $\gamma_{i,t} = z$.

If $t = 0$ then examine x if $1 \le i \le n$ otherwise z is blank.

If $t > 0$ then "guess" existentially z_1, z_2, z_3 then "check" universally all of

(i) z is correct given z_1, z_2, z_3,

(ii) $\gamma_{i-1,t-1} = z_1$,

(iii) $\gamma_{i,t-1} = z_2$,

(iv) $\gamma_{i+1,t-1} = z_3$.

The checks (ii) - (iv) are done in exactly the same way that the initial test for $\gamma_{i,t} = z$ is done.

Now, x is accepted by T if and only if there exists $i,t \le 2^{cs(n)}$ such that $\gamma_{i,t} = a$, where a is the accepting state. Because i,t are written in binary then they require no more than s(n) storage to write.

LECTURE 3. COMPLEXITY OF REAL ADDITION

3.1 Complexity of Logical Theories.

The complexity of decision problems for logical theories has played a special role in the discovery of lower bounds on the complexity of natural problems. Cook's Theorem explains in part why deciding the propositional calculus may be difficult, but until more is known about the $P = NP$ question we cannot say definitively that the propositional calculus is intrinsically difficult to decide. Meyer in 1972 discovered that the weak monadic second order theory of one successor (WSIS) required non-elementary time and space to decide. To be more precise he showed that WSIS $\notin$ TIME(f(n)) where $f(n) = F(n,cn)$ for some $c > 0$. ($F(n,0) = n$, $F(n,m+1) = 2^{F(n,m)}$). Meyer published his result in 1975 [Me 75]. Decision problems of logical theories provided some of the first examples of natural problems that are intrinsically difficult to decide. Some of the methods used in showing lower bounds for logical theories can also be used for showing lower bounds on natural problems in other areas.

In this lecture we will concentrate on the complexity of RA (real addition). M. Fischer and Rabin proved that RA $\notin$ NTIME(d^n)

for some $d > 1$ [FiR74] and Ferrante and Rackoff proved that $RA \in SPACE(c^n)$ for some c [FeR74]. The nondeterministic time lower bound and space upper bound do not match nicely. Using the concept of alternation Berman discovered that RA can be characterized "exactly". Define

$$\text{TIME-ALT}(t(n), a(n)) = \{A \subseteq \Sigma^* : \text{ there is an alternating Turing machine which solves } A \text{ in time } t(n) \text{ using } a(n) \text{ alternations}\}$$

For clarification, we say that an input x is accepted by T in time t using a alternations if there is a computation tree with root labeled I(x), leaves all labeled ACC, of height $\leq t$ and along each root to leaf path in the tree the state of the configurations change from universal to existential or vice versa at most a times. Further, T solves A in time t(n) using a(n) alternations if for all $x \in \Sigma^*$, $x \in A$ if and only if x is accepted by T in time $t(|x|)$ using $a(|x|)$ alternations.

Theorem (M. Fischer and Rabin, 1974, Ferrante and Rackoff, 1975, Berman, 1977).

RA is complete in the class $\bigcup_c$ TIME-ALT $(2^{cn}, n)$ is the sense that both

(i) $RA \in \bigcup_c \text{TIME-ALT}(2^{cn}, n)$,

(ii) Every problem in $\bigcup_c \text{TIME-ALT}(2^{cn}, n)$ is reducible to RA in polynomial time and linear size.

The credit for the theorem is dispensed among all five authors of the three papers [FiR 74] [FeR 75] [Be 77] because all are needed to properly understand the proof.

3.2 Upper Bound on RA.

Let $Q_1x_1Q_2x_2 \ldots Q_mx_m \Psi(x_1, \ldots, x_m)$ be a prenex formula in the language $\{+, <, =\}$. Any formula of RA can be put into prenex form by increasing its length only slightly. Define $S(w) = \{a/b : |a|, |b| \leq w \text{ and } b \neq 0\}$. Here $|a|$ refers to the absolute value of a. We will argue in a moment that there is a constant $c > 0$ such that $Q_1x_1 \ldots Q_mx_m\Psi(x_1,\ldots,x_m)$ is true in RA if and only if $(Q_1x_1 \in S(w_1))\ldots.(Q_mx_m \in S(w_m))\Psi(x_1,\ldots,x_m)$ is true in RA, where

$$w_k = \left(2^{2^{cm}}\right)^{2^{k-1}}.$$

An alternating machine that solves RA can be defined easily now. It has m + 1 stages with the i-th stage universal if $i \leq m$ and

$Q_i = \forall$ and the i-th stage existential if $i \leq m$ and $Q_i = \exists$. Stage i consists of writing down a fraction a_i/b_i with $|a_i|, |b_i| \leq w_i$. The final stage consists of evaluating $\Psi(a_1/b_1,\ldots,a_m/b_m)$ and accepting if it is true. Of course the fractions are written in binary, so that each fraction can be written in about 2^{dm} time for some d. If n is the length of the original formula then the machine runs in time 2^{cn} for some $c > 0$.

The reason that the quantifiers can be made to be bounded can be explained in part by the following quantifier elimination technique. Extend the language of RA to include terms of the form $\Sigma(a_i/b_i)\,x_i$, and restrict atomic formulas to be of the form $s < t$ where s and t are terms. To eliminate $\exists x_k$ from the formula $\exists x_k\ \theta(x_1,\ldots,x_k)$ where θ is quantifier free there are two steps.

1. Solve for x_k : Write each atomic formula containing x_k in one of the two forms $x_k < t$ or $t < x_k$ and let $T = \{t : x_k < t$ or $t < x_k$ is atomic in $\theta\}$.
2. $\exists x_k\ \theta(x_1,\ldots,x_k)$ is equivalent to $\theta(x_1,\ldots,x_{k-1},\infty)$ $\vee\ \theta(x_1,\ldots,x_{k-1}, -\infty) \vee \bigvee_{s,t\in T} \theta(x_1,\ldots,x_{k-1} \frac{s+t}{2})$.

There is no reason to write out explicitly the formula with the eliminated quantifier if we could simply bound the quantifier instead. The bound on x_k depends on the bounds on $x_1,\ldots,x_{k-1}$ as well as bounds on the coefficients of the terms already in θ. Suppose by induction that the coefficients of the terms in the quantifier eliminated form of $Q_{k+1}\ x_{k+1} \ldots Q_m x_m\ \Psi(x_1,\ldots,x_m)$ are in $S(2^{2^{d(m-k)}})$. This holds for $k = m$ because the coefficients of the terms in $\Psi(x_1,\ldots,x_m)$ are in $S(2)$. If $x_1 \varepsilon S(w_1),\ldots,$ $x_{k-1} \varepsilon S(w_{k-1})$ then we can argue that x can be written in the form

$$\frac{1}{2}\sum_{i=1}^{k-1} \left(\frac{a_i}{b_i}\cdot\frac{e}{f} + \frac{c_i}{d_i}\cdot\frac{g}{h}\right) x_i$$

$$\text{where } -2^{2^{d(m-k)}} \leq a_i,b_i,c_i,d_i,e,f,g,h \leq 2^{2^{d(m-k)}}.$$

The form arises from solving for x then combining terms as presented in the quantifier eliminated form. To continue the induction notice that the coefficients can be written in the form

a/b where $|a|,|b| \le (2^{2^{d(m-k)}})^5$ (which is not more than $2^{2^{d(m-(k-1))}}$ if d is suitably chosen). Further,

$x_k \in S\left((2^{2^{d(m-(k-1))}})^{k-1} \cdot \prod_{i=1}^{k-1} w_i\right)$ $(\subseteq S\,(2^{2^{dm}} \cdot \prod_{i=1}^{k-1} w_i)$ if d is suitably chosen). The choice of $w_k = (2^{2^{dm}})^{2^{k-1}}$ satisfies the recurrence $w_k \ge 2^{2^{dm}} \cdot \prod_{i=1}^{k-1} w_i$. Hence, the quantifiers can be bounded.

There is a slicker but less informative argument that uses Ehrenfeucht games. The method of Ehrenfeucht games for bounding quantifier is used heavily by Ferrante and Rackoff in their book The Computational Complexity of Logical Theories [FeR 79]. This book is a rich source of ideas and methods for determining upper and lower bounds on the complexity of logical theories.

3.3. Lower Bound on RA.

In this section we show that any problem in $\bigcup_c$ NTIME(2^{cn}) can be reduced to RA in polynomial time and linear size. Using the method described in Lecture 1, and a hierarchy theorem for non-deterministic machines it can be shown that RA $\notin$ NTIME(2^{cn}) for some $c > 0$. In the next section we will show how to extend these techniques to show that any problem $\bigcup_c$ TIME-ALT(2^{cn}, n) is reducible to RA in polynomial time and linear size. Let T be a non-deterministic Turing machine that solves a problem A in time 2^{cn}. Let w be an input of length n. We show how to construct a formula ψ_w of linear size with the property that w is accepted by T in 2^{cn} time if and only if ψ_w is true in RA.

Think of configurations as strings of the form $a_1a_2\ldots a_{i-1}(q,a_i)a_{i+1}\ldots a_m$ where $m = 2^{cn}$ and computation sequences as strings of the form $c_1\#c_2\#\ldots.c_m\#$ where the c_i's are configurations, c_{i+1} follows from c_i by a move of T and $m \le 2^{cn}$. Such strings are just integers written in d-ary notation for some d. Configurations and computations are actually integers $\le d^{2^{cn}}$ (which are $\le 2^{2^{en}}$ for some fixed e).

We would like to define integers I(w) and ACC whose d-ary expansion are the initial configuration and the accepting configuration. We would also like formulas of linear size

CONF(x) ---- x is a configuration,

COMP(x) ---- x is a computation,

FIRST(x,y) - x is a computation and y is the first configuration in x,

LAST(x,y) -- x is a computation and y is the last configuration in x.

Now, ψ_w can be expressed as

$$\exists x\ (\mathrm{FIRST}(x,I(w)) \wedge \mathrm{LAST}(x, \mathrm{ACC}))$$

The secret to writing such formulas is to have linear size formulas which can talk about the "dits" of integers $< 2^{2^{en}}$. The i-th dit of a d-ary number x is the integer a with the properties $md^i + ad^{i-1} \le x < md^i + (a+1)d^{i-1}$ for some integer m and $0 \le a < d$. That is, in the d-ary expansion of x the i-th low order dit is a. Now, suppose we have a formula D(i,x,a) which states that the i-th dit of x is a and $x < 2^{2^{en}}$ and suppose also we can express certain constants like $m = 2^{cn}$. To express for instance CONF(x) (x is a configuration) we write

$$\forall i > 2^{cn}\ D(i,x,0) \qquad \text{(x is the right length)}$$

$$\wedge \quad \exists i \le 2^{cn} \bigvee_{a \in S} D(i,x,a) \qquad \text{(S is the set of state-symbol pairs)}$$

$$\wedge \sim \exists i,j \le 2^{cn} \bigvee_{a,b \in S} (i \ne j \wedge D(i,x,a) \wedge D(j,x,b)) \qquad \text{(there is at most one state-symbol pair).}$$

Once we have the ability to name constants and talk about the dits of integers $< 2^{2^{en}}$ we have the ability to express the formulas concerning configurations and computations.

In order to talk about integers $< 2^{2^{en}}$ and the dits within them we need a form of bounded arithmetic. Let us assume we have the constant 1 even though 1 is not definable in ($\mathbb{R}$, +, <). If 1 is not available then a "base" constant different from zero can be used.

Lemma 1. There is a linear size formula $\mathrm{MULT}_n(x,y,z)$ which is equivalent to "x is an integer, $0 \le x < 2^{2^n}$, and $xy = z$" in RA.

Proof. Notice that $x < 2^{2^{n+1}}$ if and only if there exists x_1,x_2,x_3,x_4 such that $0 \leq x_1,x_2,x_3,x_4 < 2^{2^n}$ and $x = x_1 \cdot x_2 + x_3 + x_4$. Thus $x \cdot y$ can be expressed as $x_1 \cdot (x_2 \cdot y) + x_3 \cdot y + x_4 \cdot y$. This is the basis for a recursive definition of $MULT_n\ (x,y,z)$

$MULT_0\ (x,y,z) = (x = 0 \wedge z = 0) \quad \vee \quad (x = 1 \wedge y = z)$
$MULT_{n+1}\ (x,y,z) = \exists x_1,x_2,x_3,x_4,u_1,u_2,u_3,u_4,u_5$
$(MULT_n(x_2,y,u_1) \wedge MULT_n(x_1,u_1,u_2) \wedge MULT_n(x_3,y,u_3)$

$\wedge\ MULT_n(x_4,y,u_4) \wedge MULT_n(x_1,x_2,u_5) \wedge x = x_3 + x_4 + u_5$

$\wedge\ z = u_2 + u_3 + u_4)$

Unfortunately as it stands the size of $MULT_n(x,y,z)$ is exponential. If there were only one instance of $MULT_n$ instead of five in the recursive definition then we would have the recurrence $SIZE(MULT_{n+1}) \leq SIZE(MULT_n) + c$. Combining this with the fact that $SIZE(MULT_0) \leq c$ we obtain $SIZE(MULT_n) \leq c(n+1)$.

To reduce the number of occurrences of $MULT_n$ we employ a collapsing trick.

The Collapsing Trick Example

$$P(x,y) \wedge P(u,v) \equiv \forall w,z\ ((\ (w = x \wedge z = y) \vee (w = u \wedge z = v)) \supset P(w,z)\ )$$

Lemma 2. There is a linear size formula $POW_n(x.y,z)$ which whenever a,b,c are integers with $0 \leq a,\ b^a,\ c < 2^{2^n}$, $POW_n(a,b,c)$ is equivalent to $b^a = c$ in RA.

Proof. The proof is similar to the proof of Lemma 1. In this case we define formulas $E_k(x,y,z,u,v,w)$ with the property that if $0 \leq a < 2^{2^k}, 0 \leq b^a,\ c < 2^{2^n}$ then $E_k(a,b,c,A,B,C)$ if and only if A is an integer $0 \leq A < 2^{2^n}$, $b^a = c$ and $AB = C$. Now, $POW_n\ (x,y,z) \equiv E_n(x,y,z,0,0,0)$. The reason that M_n is "built into" E_k is so we can define E_{k+1} strictly in terms of E_k with only a constant amount of additional length. Using the fact that if $0 \le x < 2^{2^{k+1}}$ can be represented as $x_1 \cdot x_2 + x_3 + x_4$ where $0 \le x_1, x_2, x_3, x_4 < 2^{2^k}$ and $y^x = (y^{x_1})^{x_2} \cdot y^{x_3} \cdot y^{x_4}$, E_{k+1} can be defined from E_k.

The definitions of MULT_n and POW_n give us enough power to talk about constants like 2^n and talk about the i-th dit of integers $< 2^{2^{en}}$ in linear size formulas.

3.4. Extension to Linear Alternations.

In this section we show that every problem in $\bigcup_c$ TIME-ALT (2^{cn}, n) is reducible to RA in polynomial time and linear size. Let T be an alternating machine that solves A in time 2^{cn} using n alternations. Let w be an input of length n. Using the methods of the previous section we can define a linear size formula $C_n(x,y)$ which holds if and only of x and y are configurations and there is a computation sequence with first configuration x, last configuration y, all configurations in the sequence except y are of the same type (all universal or all existential), and either y is of a different type than x or y = ACC.

Also we can define a linear size formula $S_n(x,y)$ which holds in RA if and only if $y \in \{0,1\}$, x is a configuration, and $y = 0$ if and only if x is existential.

Lemma 3. There is a linear size formula $A_n(u,v,w,x,y)$ with the property that $A_n(0,v,w,x,y)$ is equivalent to $C_n(x,y)$ and $A_n(1,0,w,x,y)$ is equivalent to "$S_n(w,0)$ and w is the root of an accepting computation tree with $\le$ n alternations" and $A_n(1,1,w,x,y)$ is equivalent to "$S_n(w,1)$ and w is the root of an accepting computation tree with $\le$ n alternations."

Proof. We define $A_k(u,v,w,x,y)$ inductively on k

$$A_0(u,v,w,x,y) \equiv (u = 0 \wedge C_n(x,y))$$
$$\vee\ (u = 1 \wedge v = 0 \wedge S_n(w,0) \wedge C_n(w, \text{ACC}))$$
$$\vee\ (u = 1 \wedge v = 1 \wedge S_n(w,1) \wedge \forall z(C_n(w,z) \supset z = \text{ACC}))$$

$$A_{k+1}(u,v,w,x,y) \equiv (u = 0 \wedge A_k(0,0,0,x,y))$$
$$\vee\ (u = 1 \wedge v = 0 \wedge \exists z(A_k(0,0,0,w,z) \wedge A_k(1,1,z,0,0)))$$
$$\vee\ (u = 1 \wedge v = 1 \wedge \forall z(A_k(0,0,0,w,z) \supset A_k(1,0,z,0,0)))$$

To finish the job put A_{k+1} into prenex form then employ a collapsing trick similar to the one used for $M_n(x,y,z)$. The formula $A_n(1,0, I(w), 0,0)$ expresses the fact that w is accepted by T in time 2^{cn} using n alternations, assuming the initial state is existential.

LECTURE 4. ALTERNATING PUSHDOWN AUTOMATA WITH AN APPLICATION.

4.1. Pushdown Automata and Ackermann's Theory.

A pushdown automaton is a two tape Turing machine with some limitations. First, the input tape is read only and has end markers ¢ and $ for convenience. Second, the storage tape acts as a pushdown store, that is, only the "top" character on the tape can be read and a new character can be "pushed" onto the top to become the new top or the top character can be "popped" to reveal the next-to-top character as the new top.

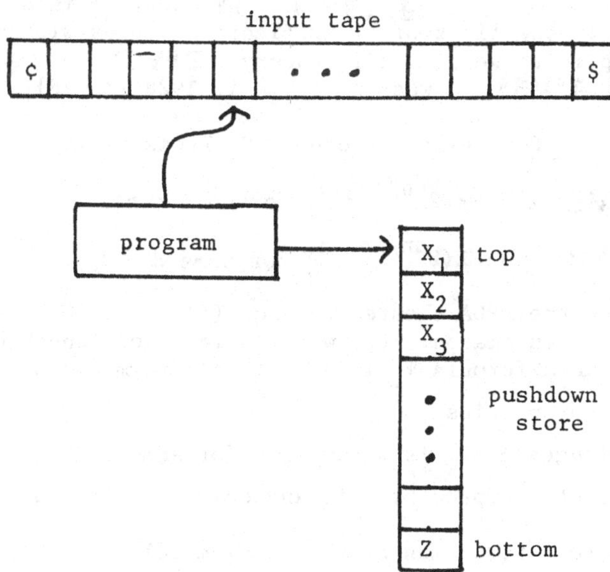

PUSHDOWN AUTOMATON

Pushdown automata with a one-way input have played an important role in computer science. It was understood by Chomsky [Ch 62] and Evey [Ev 63] that nondeterministic one-way pushdown automata were equivalent to context-free grammars. Context-free grammars have heavily influenced the design of modern programming languages to the extent that pushdown automata and context-free grammars are an essential part of a computer science education.

It is a natural mathematical question to wonder what the power of alternating pushdown automata (APDA) are. Chandra and Stockmeyer [CS 76] observed that APDA's (even with a one-way input) have the power of deterministic exponential time Turing machines, and Ladner, Lipton and Stockmeyer [LLS 78] showed that deterministic exponential time Turing machines can simulate APDA's. Hence we have the

APDA Theorem (Chandra & Stockmeyer, 1976, Ladner, Lipton & Stockmeyer, 1978).

$$\text{APDA} = \bigcup_c \text{TIME}(2^{cn}).$$

Is this theorem a mathematical curiosity or is it useful in some way? It is useful, for H. Lewis used it to show that Ackermann's $\exists^*\forall\exists^*$ predicate calculus requires time $2^{cn/\log n}$ to decide for some $c > 1$ [Le 79]. Ackermann's $\exists^*\forall\exists^*$ predicate calculus is the set of satisfiable formulas in the prenex form $\exists\ldots\exists\forall\exists\ldots\exists\,\Psi$ where Ψ is a quantifier free formula in the first-order predicate calculus without equality. For simplicity we call this theory $\exists^*\forall\exists^*$ Ackermann showed that $\exists^*\forall\exists^*$ was decidable in 1928 [Ac 28].

$\exists^*\forall\exists^*$ Complexity Theorem (H. Lewis, 1979).

(i) $\exists^*\forall\exists^* \in \text{TIME}(2^{cn/\log n})$ for some $c > 1$,

(ii) $\exists^*\forall\exists^* \notin \text{TIME}(2^{dn/\log n})$ for some $d > 1$.

Lewis used the APDA theorem to show (ii). Let M be an alternating pushdown automaton. Let w be an input of length n. In polynomial time a formula Ψ_w in Ackermann's form can be written down with the properties

(a) The length of Ψ_w is $\le cn \log n$ for some c,

(b) w is not accepted by M if and only if Ψ_w is satisfiable.

This amounts to a reduction of the problem solved by M to $\exists^*\forall\exists^*$ in polynomial time and $O(n \log n)$ size. By the APDA Theorem every problem in $\bigcup_c \text{TIME}(2^{cn})$ can be reduced to $\exists^*\forall\exists^*$ in polynomial time and $O(n \log n)$ size. By the technique mentioned at the end of the first lecture we can argue that $\exists^*\forall\exists^* \notin \mathrm{TIME}(2^{cn/\log n})$ for some c.

The APDA Theorem provides an alternative characterization of the problems solvable in exponential time. This alternative characterization may be easier to encode in a logical theory or some other problem whose complexity is of interest than a direct encoding of exponential time. In the next section we provide a detailed account of the reduction of APDA's to $\exists^*\forall\exists^*$ and in the last section we prove the APDA Theorem.

4.2. Reduction of APDA Problems to $\exists^*\forall\exists^*$

Let M be an alternating pushdown automata with states Q. We can assume its states are partitioned in to PUSH, POP, E, U where PUSH and POP are sets of deterministic states which manipulate the pushdown store, E is the set of existential branching states and U is the set of universal branching states. Let Γ be the pushdown store alphabet and Σ be the input alphabet. The transition function is also partitioned into $\delta_1, \delta_2, \delta_3, \delta_4$. In the definition $a \in \Sigma \cup \{¢, \$\}$, $X, Y \in \Gamma$ and $d = \pm 1$.

PUSH MOVES: $\delta_1(q,a,X) = (p,d,Y)$ if $q \in$ PUSH, $Y \neq Z$, (push Y)

POP MOVES: $\delta_2(q,a,X) = (p,d)$ if $q \in$ POP, $X \neq Z$, (pop X)

EXISTENTIAL MOVES: $\delta_3(q,a,X) = \{(p_1,d_1), \ldots, (p_k,d_k)\}$ if $q \in E$,

UNIVERSAL MOVES: $\delta_4(q,a,X) = \{(p_1,d_1), \ldots, (p_k,d_k)\}$ if $q \in U$.

There is a special bottom symbol $Z \in \Gamma$ which is never pushed or popped. There is a start state s and an accepting state f. We assume the machine always has a next move unless the machine is in state f reading the \$ on the input and Z on the pushdown store. This configuration is the unique accepting configuration.

Let w be an input to M. We construct a quantifier free formula θ_w in the language containing

constant symbol: B,

unary function symbols: g_X for $X \in \Gamma$,

unary predicate symbols: $C^{p,i,X}$ for $p \in Q, 0 \leq i \leq n+1, X \in \Gamma$.

The formula θ_w will have exactly one free variable v. It will turn out that θ_w is just the "Skolemized" form of a formula $\exists B \forall v \exists X_1 \ldots \exists X_m \psi$ where $\{X_1, \ldots, X_m\} = \Gamma$. Thus we may think of θ_w as being in $\exists^*\forall\exists^*$ form, with the constraints we have given the all atomic formulas having the form

$$C^{p,i,X}(q_\gamma(B))$$

where γ is a string of pushdown alphabet symbols. The notation $g_\gamma(B)$ is a shorthand for $g_{X_1} g_{X_2} \ldots g_{X_m}(B)$ if $\gamma = X_1 \ldots X_m$ and if γ is the empty string then $g_\gamma(B)$ is defined to be B. This atomic formula corresponds to the configuration of M where M is in state p, reading the i-th symbol of the input, and with pushdown store $X\gamma$ where X is at the top.

Assume $a_0 a_1 \ldots a_{n+1} = ¢w\$$ then θ_w is the conjunction of the following formulas.

(i) $\sim C^{s,0,Z}(B)$,

(ii) $C^{f,n+1,Z}(B)$,

(iii) $C^{p,i+d,Y}(g_X(v)) \supset C^{q,i,X}(v)$ if $\delta_1(q,a_i,X) = (p,d,Y)$,

(iv) $C^{p,i+d,Y}(v) \supset C^{p,i,X}(g_Y(v))$ if $\delta_2(q,a_i,X) = (p,d)$,

(v) $(C^{p_1,i+d_1,X}(v) \vee \ldots \vee C^{p_k,i+d_k,X}(v)) \supset C^{q,i,X}(v)$

if $\delta_3(q,a_i,X) = \{(p_1,d_1),\ldots,(p_k,d_k)\}$,

(vi) $(C^{p_1,i+d_1,X}(v) \wedge \ldots \wedge C^{p_k,i+d_k,X}(v)) \supset C^{q,i,X}(v)$

if $\delta_4(q,a_i,X) = \{(p_1,d_1),\ldots,(p_k,d_k)\}$.

Since we are dealing with predicate calculus without equality we can assume the so-called "Herbrand Universe". That is, the elements of any model of θ_w are simply the terms $g_\gamma(B)$.

Claim 1. Consider the model M_0 with the property that $C^{q,i,X}(g_\gamma(B))$ is satisfied if and only if the configuration $(q,i,X\gamma)$ is the root of an accepting computation tree of M on input w. In M_0 all the formulas (ii) -- (vi) are satisfied for any assignment to v.

Claim 2. In any model that satisfies (ii) -- (vi) (for all assignments to v) the atomic formula $C^{q,i,X}(g_\gamma(B))$ is satisfied whenever $(g,i,X\gamma)$ is the root of an accepting computation tree of M on input w.

Claim 1 holds because the formulas (ii) -- (vi) accurately reflect the behavior of M. Claim 2 is proven by induction on the size of accepting computation trees.

We can now argue that θ_w is satisfiable if and only if w is not accepted by M. Suppose θ_w is satisfiable. If w were accepted by M then $C^{s,0,Z}(B)$ is satisfied by Claim 2. But this directly contradicts (i) of θ_w. Suppose w is not accepted by M. The model M_0 satisfies (ii) -- (vi). It also satisfies (i) otherwise w is accepted by M by the definition of M_0. Hence M_0 satisfies θ_w.

It should be clear that θ_w can be constructed in polynomial time. Its length is O(n log n) because it requires log n bits to distinguish the O(n) predicate symbols from each other.

4.3. Proof of the APDA Theorem.

In this section we show that APDA = $\bigcup_c$ TIME(2^{cn}). We already know from Lecture 2 that ASPACE(n) = $\bigcup_c$ TIME(2^{cn}). We show ASPACE(n) $\subseteq$ APDA in order to show the inclusion $\bigcup_c$ TIME(2^{cn}) $\subseteq$ APDA.

Let M be an alternating Turing machine that solves the problem A in space n. Let w be an input of length n. Assume that from each configuration of M there are at most two possible moves. The alternating pushdown automaton that simulates M will use the pushdown store to store a path in a computation tree of M. In general the pushdown store will contain a string

$$c_0 m_0 c_1 m_1 \dots c_t m_t$$

where c_i's are configurations written as a string of length n and

$$m_i = \begin{cases} 0 \text{ if } c_{i+1} \text{ is left son of } c_i \text{ in the computation tree,} \\ 1 \text{ if } c_{i+1} \text{ is right son of } c_i \text{ in the computation tree.} \end{cases}$$

The alternating pushdown automaton proceeds now by calling NEW.

NEW : Guess a new configuration c_{t+1}; call TOP.

TOP : Call CONTINUE $\wedge$ call MATCH $\wedge$ call LENGTH.

CONTINUE : If c_{t+1} is accepting then accept,
if c_{t+1} is universal then (push 0 $\wedge$ push 1); call NEW,
if c_{t+1} is existential then (push 0 $\vee$ push 1); call NEW.

MATCH : Check that c_{t+1} follows from c_t by the branch m_t by universally checking each position in c_{t+1} against the corresponding three relevant positions in c_t. The input head is used to measure the correct distances. The pushdown store is popped during MATCH.

LENGTH : Check that the length of c_{t+1} is n using the input head to measure.

The $\wedge$, $\vee$ symbols indicate universal and existential moves respectively. Initially a configuration is written down. Universally it is checked to see if it is the initial one for w and CONTINUE is called. This shows that M can be simulated by an alternating pushdown automaton.

We now show that APDA $\subseteq \underset{c}{\cup}$ TIME(2^{cn}). Let M be an alternating pushdown automaton with a description like that given in Section 4.2. Let w be an input of length n. The method of proof we are about to describe is a generalization of a method of Cook who used it to show that nondeterministic pushdown automata with an additional s(n) - bounded storage tape can be simulated by a deterministic Turing machine in time $2^{cs(n)}$ for some c [Co 71b].

Define a surface configuration to be a triple (q,i,X) where $q \in Q$, $0 \le i \le n+1$, $X \in \Gamma$. A <u>surface configuration</u> represents all of a configuration except what is below the top symbol on the pushdown store. A <u>surface computation tree</u> is a tree whose nodes are labeled with surface configurations and with the following properties.

(i) An internal node labeled with an existential configuration C has exactly one child labeled with a surface configuration D where D follows from C in one move of M.

(ii) An internal node labeled with a universal surface configuration C has a child labeled D for each surface configuration which follows from C in one move of M.

(iii) An internal node labeled with a push surface configuration C has exactly one child labeled D where D follows from C in move of M. Notice that the top symbol of D may differ from that of C.

(iv) An internal node labeled with a pop surface configuration C has exactly one child labeled D where D follows from C in one move of M. The top symbol of D is calculated by following the path from C to the root to find the corresponding push move of M. Thus the sequence of push's and pop's along any root to leaf path are matched as left and right parentheses.

Let D be a surface configuration and $\mathcal{E}$ a set of surface configurations. We write $\models D \to \mathcal{E}$ if there is a surface computation tree with root labeled D and leaves whose labels are in $\mathcal{E}$. Note that w is accepted by M if and only if $\models (q,0,Z) \to \{(f,n+1,Z)\}$.

We now define a proof system for terms of the form $D \to \mathcal{E}$.

<u>Proof System</u>

(1) $\dfrac{}{C \to \{C\}}$,

(2) $\dfrac{C \to \mathcal{D}, \quad \mathcal{D} \subseteq \mathcal{E},}{C \to \mathcal{E}}$

(3) $\dfrac{C \to \mathcal{D} \cup \{D\}, \; D \to \mathcal{E}}{C \to \mathcal{D} \cup \mathcal{E}}$,

(4) $\dfrac{}{C \to \{D\}}$ if C existential and D follows from C in one move of M,

(5) $\dfrac{}{C \to \{D_1,\ldots,D_k\}}$ if C universal and $\{D_1,\ldots,D_k\}$ is set of surface configurations that follow from C in one move of M,

(6) $\dfrac{C \to \{D_1,\ldots,D_m\}}{B \to \{E_1,\ldots,E_m\}}$ if B is a push surface configuration, C follows from B in one move, $D_1,\ldots D_m$ are pop surface configurations and E_i follows from D_i in one move where the top symbol of E_i is the same as the top symbol of B.

<u>Completeness Lemma</u> $\vdash D \to \mathcal{E}$ if and only if $\models D \to \mathcal{E}$.

This is proved in one direction by induction on proof length and in the other direction by induction on the size of surface computation trees.

<u>Complexity Lemma</u> $\{D \to \mathcal{E} : \; \vdash D \to \mathcal{E}\}$ can be constructed in time 2^{dn} for some d.

There are at most cn distinct surface configurations for some c. So there are at most $cn2^{cn}$ distinct terms of the form $D \to \mathcal{E}$. The process of trying to find new theorems derivable in the proof system can be completed in time 2^{dn} for some d. For further explanation we refer to the proof of a similar complexity lemma in the second lecture.

This completes the series of lectures. We have tried to provide a glimpse into complexity theory and its application to the complexity of logical theories.

Addendum

In the lectures we attempted to give some of the key ideas and methods in computational complexity with an emphasis on the complexity of logical theories. Because of the limited scope of the lectures we were not able to reference certain books and papers that students would find useful.

Complexity Theory: We already mentioned Garey and Johnson's book, Computers and Intractability, A Guide to the Theory of NP-Completeness [GJ 77]. It is particularly useful in motivating the study of complexity theory because it brings together both theory and practice. The book contains a list of more than 300 natural NP-complete or otherwise complex problems. There is also an extensive bibliography of more than 600 references.

Hopcroft and Ullman's book Introduction to Automata Theory, Languages and Computation [HU 79] provides an introduction to automata based complexity theory and many other topics in theoretical computer science. Their bibliographic notes at the end of each chapter provide an interesting historical account of the subject.

Machtey and Young's book An Introduction to the General Theory of Algorithms [MY 78] provides an introduction to complexity theory from a more abstract or axiomatic point of view. They cover the complexity of logical theories, in particular, their approach to proving lower bounds involves a complexity version of the recursion theorem rather than the diagonalization of the hierarchy theorems that we used in the lectures.

Brainerd and Landweber's Theory of Computation has a chapter on axiomatic complexity theory [BL 74].

Savage's book The Complexity of Computing covers complexity theory with an emphasis on circuit complexity [Sa 76].

Decidability of Logical Theories: Rabin's contribution "Decidable Theories" to the Handbook of Mathematical Logic [Ra 77] is a thorough introduction to the methods for deciding logical theories. He also discusses the complexity of such decision procedures. Erschov, Lavrov, Talmanov, and Taitslin's survey "Elementary Theories" [ELTT 65] contains a very large table of decidability and undecidability results as well as a survey of methods for showing decidability and undecidability.

Alternation: Chandra, Kozen, and Stockmeyer's paper "Alternation" [CKS 79] provides a thorough study of the basic properties of alternating Turing machines while Ladner, Lipton, and Stockmeyer's paper "Alternating Pushdown Automata" [LLS 78] provides a thorough study of alternating auxiliary pushdown and stack automata. Paul, Praub, and Reischuk's paper "On Alternation" [PPR 78] has some interesting theorems concerning 1-tape alternating Turing machines. Ruzzo's paper "Treesize Bounded Alternation" [Ru 79] gives a new measure on the complexity of an alternating device and uses it to classify the complexity of context free language recognition. Peterson and Reif's paper "Multiple-Person Alternation" [PR 79] gives interesting generalizations of alternating Turing machines based on the analogy that alternation is a game. The main feature of their generalization is that it incorporates the notion of games with imperfect information.

Complexity of Logical Theories: Ferrante and Rackoff's book The Computational Complexity of Logical Theories [FeR 79] is perhaps the most complete reference on the subject. It contains the material in Ferrante's thesis "Some Upper and Lower Bounds on Decision Procedures in Logic" [Fe 74] and Rackoff's thesis "Complexity of Some Logical Theories" [Ra 75]. Stockmeyer's thesis "The Complexity of Decision Problems in Automata Theory and Logic" [St 74] contains many complexity results concerning regular expressions and also linear orders.

More recently, Kozen in "Complexity of Boolean Algebras" [Ko 79] and H. Lewis in "Complexity Results for Classes of Quantificational Formulas" [Le 79] use alternating machines as a basis for studying the complexity of logical theories.

Other Applications of Alternation: Stockmeyer and Chandra show that for certain games the problem of deciding which player has a winning strategy requires exponential time in their paper "Provably Difficult Combinatorial Games" [SC 79]. M. Fischer and Ladner show that the validity problem for propositional dynamic logic requires exponential time in "Propositional Dynamic Logic of Regular Programs" [FL 79]. Ladner in "The Complexity of Problems in Systems of Communicating Sequential Processes" [La 79] and Peterson and Reif in "Multiple-Person Alternation" [PR 79] argue that certain problems in communicating processes like the lockout problem are intrinsically complex by applying results about alternating machines.

References

[Ac 28] Ackermann, W. (1928). Über die Erfüllbarkeit gewisser Zählausdrücke. Mathematische Annalen 100, pp. 638-649.

[Be 77] Berman, L. (1977). Precise bounds on Presburger arithmetic and the reals with addition: preliminary report. Proceedings of 18th Annual Symposium on Foundations of Computer Science, IEEE Computer Society, pp. 95-99.

[BL 74] Brainerd, W.S. & Landweber, L.H. (1974). Theory of Computation. John Wiley & Sons, New York.

[CKS 78] Chandra, A.K., Kozen, D.C., & Stockmeyer, L.J. (1978). Alternation. IBM Thomas J. Watson Research Center Technical Report, RC 7489.

[CS 76] Chandra, A.K. & Stockmeyer, L.J. (1976). Alternation. Proceedings of 17th Annual Symposium on Foundations of Computer Science, IEEE Computer Society, pp. 98-108.

[Ch 62] Chomsky, N. (1962). Context-free grammars and pushdown storage. Quarterly Progress Report No. 65. pp. 187-194, MIT Res. Lab. Elect., Cambridge, Mass.

[Co 71a] Cook, S.A. (1971). The complexity of theorem proving procedures. Proceedings of 3rd Annual ACM Symposium on Theory of Computing, pp. 151-158.

[Co 71b] Cook, S.A. (1971). Characterizations of pushdown machines in terms of time bounded computers. Journal of the ACM 18, pp. 4-18.

[ELTT 65] Ershov, Y.L., Lavrov, I.A., Taimanov, A.D., & Taitslin, M.A. (1965). Elementary theories. Russian Math. Surveys, 20, pp. 35-105.

[Ev 63] Evey, J. (1963). Application of pushdown store machines. Proceedings of 1963 Fall Joint Computer Conference, pp. 215-227, AFIPS Press, Montvale, N.J.

[Fe 74] Ferrante, J. (1974). Some upper and lower bounds on decision procedures in logic. Doctoral Dissertation, Dept. of Mathematics, MIT, Cambridge, Mass., Project MAC Technical Report TR-139.

[FeR 75] Ferrante, J. & Rackoff, C. (1975). A decision procedure for the first order theory of real addition with order.

[FeR 79] Ferrante, J. & Rackoff, C.W. (1979). The Computation Complexity of Logical Theories. Springer-Verlag Lecture Notes in Mathematics. New York.

[FL 79] Fischer, M.J. & Ladner, R.E. (1979). Propositional dynamic logic of regular programs. Journal of Computer and System Sciences 18, pp. 194-211.

[FiR 74] Fischer, M.J. & Rabin, M.O. (1974). Super-exponential complexity of Presburger arithmetic. Complexity of Computation, ed. R.M.Karp, pp. 27-41. Proceedings of SIAM-AMS Symposium in Applied Mathematics.

[GJ 79] Garey, M.R. & Johnson, D.S. (1979). Computers and Intractability: A Guide to the Theory of NP-Completeness, H. Freeman, San Francisco.

[HS 65] Hartmanis, J. & Stearns, R.E. (1965). On the computational complexity of algorithms. Transactions of the AMS 117, pp. 285-306.

[HLS 65] Hartmanis, J., Lewis II, P.M., & Stearns, R.E. (1965). Hierarchies of memory limited computations. Proceedings of 6th Annual IEEE Symposium on Switching Circuit Theory and Logical Design, pp. 179-190.

[HeS 66] Hennie, F.C. & Stearns, R.E. (1966). Two-tape simulation of multitape Turing machines. Journal of the ACM 13, pp. 533-546.

[HU 79] Hopcroft, J.E. & Ullman, J.D. (1979). Introduction to Automata Theory, Languages and Computation. Addison-Wesley, Reading, Mass.

[Ka 72] Karp, R.M. (1972). Reducibility among combinatorial problems. Complexity of Computer Computations, pp.85-104, eds. R.E. Miller & J.W. Thatcher, Plenom Press, New York.

[Ko 76] Kozen, D. (1976). On parallelism in Turing machines. Proceedings of 17th Annual Symposium on Foundations of Computer Science, IEEE Computer Society, pp. 89-97.

[Ko 79] Kozen, D.C. (1979). Complexity of Boolean Algebras. To appear in Theoretical Computer Science.

[La 79] Ladner, R.E. (1979). The complexity of problems in systems of communicating sequential processes. Proceedings of the 11th Annual ACM Symposium on Theory of Computing, pp. 214-223.

[LLS 78] Ladner, R.E., Lipton, R.J., & Stockmeyer, L.J. (1978). Alternating Pushdown Automata. Proceedings of 19th Annual Symposium on Foundations of Computer Science, IEEE Computer Society, pp. 92-106.

[Le 79] Lewis, H.R. (1979). Complexity results for classes of quantificational formulas. Technical Report, Aiken Computation Laboratory, Harvard University.

[MY 78] Machtey, M. & Young, P. (1978). An Introduction to the General Theory of Algorithms. North-Holland, New York.

[Me 75] Meyer, A.R. (1975). Weak monadic second order theory of successor is not elementary-recursive. Proceedings of Boston University Logic Colloquium, Springer-Verlag, pp. 132-154.

[Pa 79] Pan, V.Y. (1979). Field extension and trilinear aggregating, uniting and caneling for acceleration of matrix multiplications. Proceedings of 20th Annual Symposium on Foundation of Computer Science, IEEE Computer Society, pp. 28-38.

[PPR 78] Paul, W.J., Praub, E.J., & Reischuk, R. (1978). On alternation. Proceedings of 19th Annual Symposium on Foundations of Computer Science, IEEE Computer Society, pp. 113-122.

[PR 79] Peterson, G.L. & Reif, J.H. (1979). Multiple-person alternation. Proceedings of 20th Annual Symposium on Foundations of Computer Science, IEEE Computer Society, pp. 348-363.

[Ra 77] Rabin, M.O. (1977). Decidable Theories. Handbook of Mathematical Logic, ed. J. Barwise, pp. 595-629. North-Holland, Amsterdam.

[Ra 75] Rackoff, C.W. (1975). Complexity of some logical theories. Doctoral Dissertation, Dept. of Electrical Engineering, MIT, Cambridge, Mass., Project MAC Technical Report TR-144.

[Ru 79] Ruzzo, W.L. (1979). "Tree-size bounded alternation." Proceedings of the 11th Annual ACM Symposium on Theory of Computing, pp. 352-359.

[Sa 76] Savage, J.E. (1976). The Complexity of Computing. John Wiley & Sons, New York.

[Sc 79] Schönhage, A. (1979). Total and partial matrix multiplication. Technical Report, Mathematisches Institute, Universität Tübingen.

[St 74] Stockmeyer, L.J. (1974). The complexity of decision problems in automata theory and logic. Doctoral Dissertation, MIT, Cambridge, Mass., Project MAC Technical Report TR-133.

[SC 79] Stockmeyer, L.J. & Chandra, A.K. (1979). Provably difficult combinatorial games. SIAM Journal on Computing 8, pp. 151-174.

[Tu 36] Turing, A.M. (1936). On computable numbers with an application to the Entscheidungs-problem. Proceedings of the London Math. Soc. 2, pp. 230-265.

Preparation of this paper was supported by the National Science Foundation of the United States under Grant No. MCS77-02474.